The Social Face of Complexity Science

A Festschrift for
Professor Peter M. Allen

The Social Face of Complexity Science

A Festschrift for
Professor Peter M. Allen

Edited by
Mark Strathern & James McGlade

3810 N 188th Ave
Litchfield Park, AZ 85340

Cover graphic is of Jackson Pollock's "Mural" painted in 1943.

The Social Face of Complexity Science:
A Festschrift for Professor Peter M. Allen

Edited by: Mark Strathern & James McGlade

Library of Congress Control Number: 2014932496

ISBN: 978-1-938158-13-1

Copyright © 2014 3810 N 188th Ave, Litchfield Park, AZ 85340, USA

All rights reserved. No part of this publication may be reproduced, stored on a retrieval system, or transmitted, in any form or by any means, electronic, mechanical, photocopying, microfilming, recording or otherwise, without written permission from the publisher.

Printed in the United States of America

Contents

Encomium xv
    Sander Van de Leeuw: Homage to Peter Allen xxi
    James McGlade: Peter brings the gospel to the natives in Nova Scotia xxiii
    Kurt Antony Richardson: DERA and beyond xxiv
    On Peter Allen by Mike Batty xxv

A. Knowledge, complexity and understanding

Paul Cilliers

Introduction 1
The traditional trap 2
Complexity and understanding 3
Knowledge and the subject 4
Implications 5
References 7

B. Complexity science: The integrator

Liz Varga

Introduction 11
Socioeconomic structures: How systems emerge from the interactions of individuals 12
Philosophical perspectives 15
Integrated methods 17
Environments 18
Learning and adaptation 19
Conclusion 21
References 21

Γ. Levels of abstraction and cross-cutting skills: Making sense of context in pursuit of more sustainable futures

Mark Lemon, Paul Jeffrey & Richard Snape

Introduction: A systematic approach to complex systems...................27
Conceptual frame: Desperately seeking consensus.............................30
Case study: A story of retrofit..34
      Boundaries, scale and connectivity..37
      Agency, knowledge and multiple perspectives.........................40
Conclusions: Models, meaning and narratives 43
References 44

Δ. Scaling in city systems

Michael Batty

Regularities in city systems...49
Power laws explained...54
The dynamics of power laws: Macro-regularities, micro-volatilities 59
The concept of the rank-clock..61
Related distributions:
   High buildings, transport hubs, firm sizes......................................67
Next steps: The rank clock visualizer...73
References 76

E. Robustness in complex systems: The role of information 'barriers' in network functionality

Kurt A. Richardson

Introduction...79
Boolean networks: Their structure and their dynamics.......................81
      Dynamical robustness..85
More on structure and dynamics: Walls of constancy, dynamics cores and modularization...87
      Nodes/loop types and structural attractors..............................87
      The emergence of a dynamic core..90

The role of non-conserving (structural) information loops 93
    Modularization in Boolean networks 93
    Dynamic robustness of complex networks 95
State space compression and robustness 97
    Balancing response 'strategies' and system robustness 99
Discussion 103
References 107

Z. Noise and information: Terrorism, communication and evolution

Robert Artigiani

Introduction 111
Modern societies 115
Modern individuals 118
Membership and identity 120
Discussion 123
References 125

H. Modelling the management of evolving operations

James S. Baldwin & Keith Ridgway

Introduction 131
Research methods and preliminary results 136
The simulation model 139
Simulation results and discussion 143
Closing remarks 149
References 150
Appendix 154

Θ. The evolution of industries in diverse markets: A complexity approach

Elizabeth Garnsey, Simon Ford & Paul Heffernan

Introduction	159
Industrial development in product-based industries	162
Industrial evolution	162
Variations in the product life cycle	163
Industrial development in instrumentation	165
Reflections on industrial development in instrumentation	168
Industrial development in semiconductors	169
Reflections on industrial development in semiconductors	171
Industrial development in microcomputers	172
Reflections on industrial development in microcomputers	174
Industrial development around the Network Computer	176
Reflections on industrial development around the Network Computer	177
Discussion: The organization of industrial activity	178
Complex systems and coevolution	178
Self-organization through interaction	182
Adaptation and attraction	184
Reflections on theory and evidence	185
Acknowledgements	186
References	186

I. The evolution of complex exchange dynamics in a pre-market economy: A case study from north-east Iberia

James McGlade & Mark Strathern

Introduction	191
Exchange processes in pre-market economies	192
Exchange and commodities	192
Prestige goods economies	193
A case study from proto-historic Iberia	194
The Iberian Iron Age landscape	194
Colonial encounters: The coming of the Greeks	195
Emerging polities: The new oppida landscape	199

The dynamics of socio-political relations 200
    Peer polity interaction 200
    Complex networks and emergent behavior 201
Elements of a model 204
    Model assumptions and boundaries 204
    Model description 206
Model results 209
    Population dynamics 209
    Disease vectors 209
    Prestige goods exchange dynamics 211
    Long-term versus short-term dynamics 212
Discussion 214
Conclusion 215
References 216
Appendix 219
    Key to parameters and variables 219

K. Complexity and economic evolution from a Schumpeterian Perspective

Stan Metcalfe

Introduction 223
Schumpeter's dynamics and complexity 225
Three Schumpeterian elements 228
    Enterprise and knowledge 228
    The competitive process 233
    The transience of the prevailing economic order 235
    Adaptation, growth rates and evolutionary change 236
    Fisher/Price dynamics 239
Reprise 242
References 242

Λ. Innovation systems, economic systems, complexity and development policy

Norman Clark

Introduction	247
Economic systems as complex systems	249
Innovation systems	251
Research into use programme	255
Some concluding points	259
References	259

M. Fixing the UK's economy

Bill McKelvey

Introduction	263
Cityscapes as self-organizing economies	265
Defining power-law distributions	265
Rank vs. Size of firms' market-capitalization values	265
PLs as indicators of self-organizing economies and industries	267
Lord Turner asks: "Is the city too large to be socially useful?"	272
Calls for a Tobin tax	272
Disagreements	273
How to fix the UK?	274
UK should switch from Pound to the Euro. Really!	275

Conclusion 277
References 278

N. Complexity and the evolution of market structure

Paul Ormerod

Introduction 283
Competition and market structure 285
Entry into a simple market 286
Methodological reflections 289
The evolution of market structure and competition 291
Conclusion 300
References 301

Index 305

II. Complexity and the evolution of market structure

Professor Peter Murray Allen

Encomium

Peter was born in Norbury, South London, on the seventh of February 1944 and this festschrift is to mark his 70th birthday and his more than 45 years of academic research, in which he is still engaged; actively still researching, publishing and lecturing. To merely give a dry outline of his many academic achievements would be to entirely miss the point. Peter is known both for his research insights. His inspiration and encouragement, and not least his warmth and generosity.

The dry outline of his academic achievements is that Peter is Emeritus Professor in Complex Systems at Cranfield School of Management. He is editor in Chief of the Journal Emergency: Complexity and Organization. He has a PhD in Theoretical Physics from Hull, was a Royal Society European Research Fellow 1969-71 and a Senior Research Fellow at the Universite Libre de Bruxelles from 1972-1987, where he worked with Nobel Laureate, Ilya Prigogine. Since 1987 he has run two Research Centres at Cranfield University. He also cofounded the Complexity Society that has organized courses and research seminars across the UK. For over 30 years Peter has been working on the mathematical modelling of change and innovation in physical, social, economic, financial and ecological systems, and the development of integrated systems models linking the physical, ecological and socio-economic aspects of complex systems as a basis for improved decision support systems. He has developed a range of dynamic integrated models has been developed in such diverse domains as industrial networks, supply chains, river catchments, urban and regional development, fisheries and also economic and financial markets and has been a leader in the development of evolutionary social modelling. Peter has written and edited several books and published well over 200 articles in fields ranging through ecology, social science, urban and regional science, economics, systems theory, and physics. He has been a valued consultant to many organizations including: Cambridge Econometrics, the Asian Development Bank and a number of City of London financial institutions, the Canadian Fishing Industry, Elf Aquitaine, BT, GlaxoSmithKline, DERA, DSTL, the United Nations University, and the European Commission.

But as I have said it is not the dry outline that defines Peter or his contribution. I joined Peter at Cranfield in 1988 since then I have worked on many modelling projects with him. He has also helped and encouraged me through a far from straight forward academic progression. He took it entirely within his stride when I arrived at Cranfield, a post-graduate institution, without a first degree and inspired and encouraged me through to my PhD. There were many hiccups but they were mine, and throughout he was unwavering in the generosity of his inspiration, support and help.

Peter is famous for his modelling. And typical of him this is not confined to mere dry academic expression. Perhaps his earliest brush with modelling was his youthful enthusiasm for making model boats, "a radio controlled Veron Police Launch was the zenith of (t)his modelling career" according to his brother Michael. But his modelling talents were also expressed in a number of other slightly unusual forms. There is his 1982 paper "Origami and Management in an Evolving World" (Allen, 1982) but also, with his Brussels colleagues, his early modelling movie "Inter-urban dynamics" using stop frame and Lego! (Allen, avi)

His early years were spent in south London, first Norbury then in 1947, Norbury Cross. He went to Norbury Manor Infants and Junior school 1949-1955. And then John Ruskin Grammar School 1955-1962. Here the school promoted boys with potential from the first form to a special third year form called 3u. Peter was one of this class and they took their O levels a year early, giving the possibility of 3 years in the sixth form, his potential had already been spotted. These were good years, indeed his mother called him "happy Peter".

After John Ruskin he went up to Hull University to read Physics. He lived in digs at Anlaby Road, Hull, while at Hull University, where he reputedly had a landlady with a distinctive personality that made an impression on all. Whilst at Hull, according to his brother **Micheal** again,

> He had a few adventures with motorbikes and cars. One early journey back to Hull on a 200cc Zundapp motorcycle was curtailed in North London when he dismounted thanks to a wet manhole cover, lightly damaging the bike and his ego.

Another attempt to reach Hull by motorcycle was more successful. It was achieved on a 1953 BSA Golden Flash 650cc A10. A bike, which had been stored carefully for some years because the owner was a bit nervous of its power and speed. Just the steed for Peter, it was in very good condition and was a tasteful subdued gold color. When not that many miles from Hull an oil return pipe slipped off and squirted oil over the rear of the bike including the tyre. This caused the road holding to become interesting. He never the less made it but the engine having been starved of oil was damaged.

As he had a doctorate in his sights he was keen to have the words 'Flying doctor' on the back of his motorcycle jacket. But he swiftly moved on to cars, and a 400cc Goggomobil coupe became the apple of his eye. He most industriously resprayed the car a light silver blue, using several aerosols. He was proud of the cars preselector gearbox, which worked surprisingly well.

A Ford 1172cc estate car followed, having been converted from a van, and this served him well until he moved up to a motor caravan. He invested in a Bedford Dormobile, which had been converted from a van by a man who claimed to be a ships carpenter. Peter likes to see the best in people and accepted this bold claim, but this seemed rather unlikely, considering the amateurish elevating roof and fitments. This home on wheels allowed him to escape Hull on fine weather days and complete his PhD in the surrounding countryside.

He arranged for a cheap rebore in order to ready it for an adventure through Spain to Morocco and Algeria. He had had experience of continental motoring earlier when a group of them went to Spain in an elderly VW minibus. They reached North Africa and holidayed. They were mistaken for wealthy British tourists and highjacked. He managed to escape and drive home. He had arranged breakdown insurance to cover any problems on the continent.

However, the van must have been aware of this and groaned to a halt shortly after having landed at Dover, where the insurance cover did not apply.'

His wife, **Francine** tells us:

After several years of hard work at Hull University: studying by day, by night playing the piano in bars, playing card games and smoking till the early hours of the morning—a tough life and not for the faint-hearted—Peter got his Ph.D. in Physics. Subsequently, Professor Cole suggested that he'd go to Brussels to work with Professor Ilya Prigogine at the Free University of Brussels. When, in 1977, Ilya Prigogine was awarded the Nobel Prize for Chemistry (Peter's work on Cities was specifically mentioned), there followed a constant flow of visitors from all over the world: including Chinese, Japanese, East Europeans, Americans... further enriching the pool of ideas. Those were exhilarating years. In 1987, he was offered a chair at Cranfield University. Feeling the time had come for him to get back to England, he accepted the post.

Again when asked about the time in Belgium which he spent in a caravan **Francine** added,

Humm... the gipsy phase (caravan) was before 'my' time, though he still had it when we got together. Should have mentioned that, because he could not have a piano in there (!) he bought himself a guitar to have something to practise playing music on.'

Adding:

He did actually meet a Belgian woman—in a vegetarian/macrobiotic restaurant in Waterloo. She was running a health food shop at the time.

Ever fond of good food as Peter was, I think it was the way she had with spinach that made him decide to marry her: he had hitherto disliked boiled spinach, but with cream etc they were more than palatable. If she could do that to spinach, he thought he was onto a good thing here. And so, in 1975, after living in sin and testing more of her cuisine, it all happened. On the day, his wife, unbeknownst to her, got bitten by a 'claig'. The newly weds took the night train to Scotland. In the morning, she could not walk on that foot. So, the honeymoon got spent with him supporting his wife who hopped along on t'other foot for the duration.

As far as foreign visitors where concerned, we had some memorable dinners parties, sometimes being up to 15 people round our table. Wonderful memories.'

But it is not just Peter's family who appreciate him; **Rosemary Cockfield** who administered the Nexsus project says:

Great memories of working with Peter—in particular my time as administrator for the Nexsus project. Thank you Peter for making going to work such a pleasure.

And he has stimulated many, including **Paul Ormerod**:

Peter and I met shortly after I published the Death of Economics *in 1994. I was captivated by his model of fishing fleets. Here was someone who not only shared my criticisms of rational agent, rational equilibrium economics, but was successfully working with the new paradigm of model building. This was a great inspiration to me. Twenty years ago, the world was a very lonely place for critics of mainstream economics, and Peter shone as an intellectual beacon.*

His fishing fleet model was not only intellectually appealing, for someone like me who had been actively engaged in running a company, it struck an immediate chord. Yes, this was what business was actually like. Just like the fishing boat crews in Peter's model, we didn't sit around trying to discover our marginal cost curve and trying to work out optimal strategies. We acted, took decisions which seemed reasonable. Sometimes they worked, sometimes they did not. We could never tell in advance, no matter how hard we thought about it, which it would be. In my contribution to the volume, I have focused on the evolution of markets, inspired by the fishing fleet model.

He has converted and inspired many students to the complexity viewpoint. **Liz Varga**, now head of the Complexity group which he founded in the Cranfield School of Management:

A chance visit to Peter's Complexity elective during my MBA at Cranfield was the spring-board to the PhD under his supervision. It was a defining moment in my career: the fog suddenly cleared and a warm feeling of meaning and empathy sparked, fuelling a hunger for more Complexity (with a capital 'C'). The MBA no longer runs the Complexity elective as we've become lean like so many other Schools of Management, which means these chance stimuli will occur only at conferences and events where potential learners have already selected themselves into the fold. However the burgeoning colloquial use of complexity (with lower-case 'c') is the first step on the journey to Complexity science and many fields are dipping toes into this field of diversity, creativity and emergence. And what a journey Peter had. With the privilege of working for 15 years for Ilya Prigogine, Nobel Prize-winner for Chemistry, he feasted on the secular bent of the Université Libre de Bruxelles and the competition with Université Catholique de Louvain. Of course many scientists are agnostic, or even sympathetic to Richard Dawkins's atheism, but Peter is foremost a pragmatist. He gives priority to experimentation, primarily in silica, to represent the interpretations of the many and to accept the plurality and diversity of whole. It's been a privilege to spend 10 years exposed to his wisdom.'

And his effect was equally deep on his colleagues at Cranfield, **Mark Lemon**:

I have worked with Peter Allen for over twenty years both on EU research projects and as a member of the IERC which he headed for a number of years.

Peter has managed to combine an academic humility with an openness and inquisitiveness that is both stimulating and fun. He has never lost the ability to pursue that academic endeavour with an equivalent appetite for good food and drink and conversation; indeed linking the three in various Mediterranean locations has been a highlight of my academic career. On numerous occasions those conversations have re-emerged in a modified form having been subtly, and sometimes dramatically, reshaped following Peter's subsequent deliberations. I am indebted to Peter because he made it acceptable not to have answers but to generate questions and hopefully produce insight that helped people think differently about problem contexts and potential futures. I was always intuitively of a mind-set that accepted uncertainty, Peter added an academic weight and rigour to that position; it remains somewhat uncomfortable but, at the same time, honest and stimulating.

And of course with his project colleagues, **Elizabeth Garnsey**:

Working with Peter was a delightful but challenging experience. He is always very friendly, very accessible and full of interesting ideas and suggestions. He was great fun and had a fund of stories about the early days of complexity studies in Brussels and the US. His range of knowledge seemed to extend in every direction as a result of the many projects on which he had worked. But there was always a translation challenge in presenting material to him as he would re-conceptualise it on his own terms and one had to do mental gymnastics to keep up with his own intellectual acrobatics. He was very kind and prepared to go miles out of his way, literally, to help out with students or colleagues. Working with Peter was thoroughly enjoyable and he was at the heart of very collegial groups where competition was at a minimum and intellectual exchange and progress the centerpiece of the activity. This was manifested in the last project on which I worked with him, the inter-university ESRC NEXSUS project.

Also his effect has been profound even on those he consulted for, **Hans-Peter Brunner**:

When I met Peter first at the Smithsonian Institution in Washington, I was positively taken aback by his advanced visual modeling of economic phenomena of structural change. Since I am still working in an institution that deals with economic development, I thought this man is pointing the way structural change will be modeled in the future. Peter was on the right track then, and he still is a beacon for those who use his work in helping institutions like the Asian Development Bank, or the World Bank. Today his work still points to solutions which are not yet pursued, and his work still is very timely when we try to deal with changes in economic structure in the emerging economies of Asia. I got to know Peter as a brilliant and also very friendly, humane person when I asked him to work with me at the ADB on issues of economic structural change. A result of our cooperation was our coauthored book on Productivity, Competitiveness and Incomes in Asia—An Evolutionary Theory of International Trade (Ed Elgar 2005), which features among others Peter's work using dynamic land-use maps of Puerto-Rico. Well, this is work from which ADB and World Bank could learn today. Peter is a great visionary of economic geography and geosimulation work to come—he started this in the 1980s when the technical means to do this work, were much inferior compared to now.

Mark Strathern
Cranfield, 2014

References

Allen, P.M. (1982). Origami and Management in an Evolving World, Intermediaires, Brussels

Allen, P.M., avi, https://emergent.blob.core.windows.net/temp/urban.avi.

Four colleagues wished to pay particular respect:

Sander Van de Leeuw: Homage to Peter Allen

Time has prevented me from writing a scientific contribution to this volume. But I would like, nevertheless, to address Peter Allen on this day in admiration and gratitude for what he meant to a whole generation of early 'complex systems' devotees, including myself.

Recently, I had breakfast in Beijing with one of his early admirers, Zhangang HAN, now head of an important systems science department at Beijing Normal University. That meeting brought back a flood of memories about the days in the 1980s when Peter and I met, and in the 1990s when we worked together on the ARCHAEOMEDES Project, the world's first sustainability project that explicitly used a complex systems approach.

Our first meeting was at a conference organized by my -then future- boss Colin Renfrew in Southampton in late 1980. I was somewhat familiar with Prigogine's work, and jumped on the occasion to meet with one of his collaborators. The meeting, in a—typically British—semi-dark backroom full of chairs, behind the scenes of the conference, led to twenty years of learning a different outlook on 'things', and to years of working together that, for me, changed my personal and academic life and ultimately led me astray from archaeology into sustainability science. I owe Peter a huge debt.

Peter introduced me a few years later, around 1986, to James McGlade, a professor of Fine Arts in Canada, who wanted to study archaeology and had caught the complex systems 'bug'. He became the first archaeologist anywhere to get a PhD in Complex System and an early partner in the (then) crime of using models to make archaeology come to life as a discipline studying past dynamics, and not only statics. Peter also introduced me to Denise Pumain and her team, with whom I am still working together on urban issues. Then, beginning in 1990, Peter introduced me to his group in Cranfield, which led to yet other intellectual discoveries and many friendships that I remember with great fondness, and many long drives between Cambridge and Cranfield over narrow roads. But above all, with great generosity, Peter was the person who initiated ARCHAEOMEDES and actively participated in it until the end, in 2000, always pushing us to go beyond the typical social science approach of individual cases to a more general kind of thinking about the worlds we were studying.

How did that happen? Peter had been asked, by one of the European Commission's Research Directorate Officers, Roberto (?) Fantecchi, to bring together a team to develop a project based on complex systems thinking. After an initial meeting in Brussels, for several months we did not hear anything. When the request came to formally submit a proposal, Peter was in Japan. In his absence, but with his blessing, his Cranfield team, and James and myself in Cambridge developed a project that was very different from the kind of project he would have submitted. When we, somewhat unexpectedly, were awarded the (for that time very substantive) grant, Peter, with his usual great generosity said to me on the way to Brussels: "as you led the team that wrote the proposal, you might as well direct the project ..." and thus he further shaped me into the person I am today by allowing me to assemble a team of scientists from all over Europe and covering all corners of the disciplinary map—in effect allowing me to get another academic education from people I could choose myself.

I always look back on those years, especially 1990-1999, as a time of great fun, very intense discussions, many field trips and pleasant dinners together in which we became the core of a team of some 60 scientists and graduate students that were experimenting with this new outlook on society and its environment, as well as an early form of intellectual fusion between a wide range of disciplines. Of course this led to many debates—always stimulating and intellectually generous on his part. After all, this kind of trans-disciplinarity is the art of constructively upsetting existing disciplines and shaping new academic identities through (sometimes) strong disagreements.

Later, after I moved to France and then to the US, we only saw each other sporadically, but I always think of Peter with great respect, admiration and fondness as well as gratitude for the role he played in shaping my life as a renegade archaeologist...

James McGlade: Peter brings the gospel to the natives in Nova Scotia

I first met peter more than 25 years ago while I was living in Canada. Peter, in those days, travelled as a 'missionary' across Canada and the USA trying to convert the faithful to what was then a radically new creed, with a bunch of abstruse doctrines with strange names like 'self-organization', 'far-from-equilibrium' and weirdest of all, the 'Brusselator'. This was my introduction to Peter's world of complexity and I soon became a convert, aided by prolonged initiation ceremonies involving copious amounts of malt whisky!

In those days, Peter would roll up at the airport in Halifax always with the heaviest suitcase imaginable—it turned out that the hernia-inducing weight was due—not to the smuggling of gold bars—but rather to the hundreds of acetate overhead projector slides...yes, this was a world before the advent of Power Point (to say nothing of PCs!).

When it comes to convincing people of the need to take on the complexity 'gospel', Peter is a master and this is simply because he is one of the most engaging speakers any audience will ever hear. However, his skills were severely tested by the fisheries fraternity at the Bedford Institute of Oceanography in Canada—not only was Peter trying to give them a dose of complexity thinking, but most difficult of all, he was trying to convince a rather skeptical group of statisticians that the brave new world of simulation modeling was the only possible future for them. I have always thought that this was one of Peter's most difficult missions—I think they all thought it was 'smoke and mirrors' and somehow, not real science!

Canadian reminiscences would not be complete without recalling one memorable evening—or was it a whole day?—of a malt whisky tasting party...as I recall (with difficulty) everyone brought a bottle, so we had 16 different malts to sample! I can still see Peter thrashing out some first class pub music on the piano accompanied by myself on snare drum and brushes. Embarrassingly, as the night wore on, I think there was some singing of Scottish folk songs! Actually, this was nothing new for Peter—he was in his natural habitat, for I don't think it's widely known that Peter put himself through university by entertaining the punters in the pubs of Hull!

In those days, as an "exile in the colonies" as Peter liked to refer to it, I hadn't yet made the transition to a career in Archaeology and I was still working as an artist in Nova Scotia: I have a wonderful memory of installing one of my more radical installations which involved covering the gallery floor with tons of earth and a variety of plough-shaped artifacts... I can still see Peter, acting as my assistant wielding wheel barrow after wheel barrow of earth—I think he thought I was just a little odd, if not slightly bonkers, but he had the good grace not to say what he was really thinking!

So you see, contingency is a great thing,—it was Peter's missionary work in Canada that proved to be my introduction into the mysterious territory of nonlinear dynamics and evolutionary structuring, and which, thanks to those early encounters with Peter, continues to form the basis of my research life today.

Kurt Antony Richardson: DERA and beyond

I was first exposed and inspired by "complexity" in 1998 when I was put in charge of the Operations Analysis for Information Warfare (OA4IW) project when working for the Defence Evaluation and Research Agency (DERA) on Portsdown Hill in Hampshire. Those of us involved in the project quickly realized we'd bitten off more than we could chew—the information warfare domain is vast, claiming the sociological, psychological, technological, economical, theological, etc. domains as legitimate concerns. We soon stumbled on the work of Peter, esp. his Fisheries Modeling, and there found a branch of 'physics' (nonlinear dynamics) that was not only very cool (I did a PhD in SUPERconductivity, and was only interested in areas of research that were 'super' or very cool!), but also gave us a way to approach the vast networked mess that we saw as information warfare, which was really an effort to understand all of society at its many levels and understand how an actor might intervene with semi-predictable outcomes. We were very fortunate indeed to have Peter visit us on the hill to run a small workshop, giving us the opportunity to hear more about complexity science from the guy who'd written the few papers we'd had access to. The workshop was so successful in fact that most of those in attendance realized the research they were doing was too dull to be spending our valuable time on, that a government consultancy was not the place to develop our interest in the field much further, and within a year or so we'd left for pastures new to explore complexity more freely. (We did of course meet our commitments to the customer regarding the OA4IW project, which benefitted enormously from our day with Peter—at least the customer was pleased!)

I ended up in Boston in January of 1999 with the New England Complex Systems Institute (NECSI) doing a Postdoc in Complexity and Management. Peter's work continued to influence my own once I got state-side, and we were lucky enough that he was willing to contribute papers to the fledgling journal, *Emergence* which was launched in 1999. NECSI became ISCE (Institute for the Study of Coherence and Emergence headed by Michael Lissack), or at least the part I worked for did. Unfortunately, the publisher of *Emergence*, Lawrence Erlbaum, decided to drop the journal in 2003 after only five volumes. We never had any problems getting content for the journal, the problem was simply that there were not enough 'complexity' researchers to drive sufficient subscription sales to please the bottom-line thinkers at Lawrence Erlbaum.

In 2003 I founded Emergent Publications for the sole purpose (initially) to take over the publication of *Emergence*, which was relaunched as *Emergence: Complexity & Organization* with the start of Volume 6. I was very pleased and very fortunate to have as one of the Editors-in-Chief, Peter (and have The Complexity Society as a sponsor). Not only is Peter's long and ongoing contribution to the study of complexity well recognized by everyone in the field, which is great 'branding' for a small journal such as *E:CO*, but the conscientiousness with which he has approached his role as an Editor-in-Chief has also been essential and central to *E:CO* making it to its 16th volume this year (2014). Getting a journal issue ready for publications is always a crazy rush—all of those involved have 'real' jobs that take priority, and the appearance of each new issue is a testament to those who willingly and generously give their time to keep *E:CO* going. I find it easy to acknowledge that especially without Peter, and Jeffrey Goldstein, *E:CO* would not have made it this far.

So thank you Peter for inspiring me to get stuck into a field—for which I still have titles for 100s papers and books in my mind that I wish I had the time to research and write—, and for your professionalism (i.e., always dealing with my unreasonably late requests for editorials and commentary so well!) in working with me to keep an important publication avenue for complexity researchers alive and well and now in its 16th year. My only regret is that being in the US has meant that I have missed out on being able to just hang out with you from time to time ;-)

On Peter Allen by Mike Batty

I first met Peter in the temporary one storey buildings, the so-called 'Terrapins', that was my home in the Geography Department at Reading University in late 1978 or early 1979. It was the winter time, I know, with the sun low on the horizon for most of the day and Peter arrived with a large group of researchers from the Free University in Brussels. He had been to see Peter Haggett in Bristol who I remember had advised him on who to go and see concerning his interest in building urban simulation models of city systems. In Reading I was still in the throws of building such aggregate models as were my colleagues Dave Foot and Erlet Cater and although the first wave of interest had passed, I was part of the residue of researchers who remained wedded to these ideas who had begun to mature their science. I really can't remember what we talked about but we must have talked about urban models, and of course about his interest in developing his model of Brussels, and his focus on building in temporal dynamics which would lead to bifurcating and surprising changes in direction for such systems. By then ideas from nonlinear dynamics and catastrophe theory had permeated our field, driven hard in the UK by Alan Wilson, so it was easy to see where Peter was coming from but he and his group brought a new view of systems into this whole debate. This of course was the view based on systems where entropy was being reduced as more organization was imported into such systems and it seemed to fit rather well the idea that cities were getting more complex, something which would gather pace in complexity theory in the next 20 years.

Somehow we saw a lot of each other over the next 3 or 4 years, meeting several times at conferences, in particular at the meeting at New College Oxford in 1980 which was organized under the auspices of a NATO Advanced Research Institute. Peter made a big impact on our field. His infectious enthusiasm for variety and diversity in city systems and his quest to model and predict such change contrasted with the more routine development of the field and it projected us into an era when temporal dynamics became central to such modelling efforts. I have sort of forgotten that during most of the 1980s he was still working in Brussels and in 1990 when I left to work in America he was transplanting himself to Cranfield. My own interests by then were firmly in complexity theory and less in urban models and it was not until the mid-1990s when I returned to University College London that we really began to interact.

I examined several PhDs that Peter was advising at Cranfield and I always remember that the convenor of the oral would tell us the ground rules that the examiners could ask questions but the advisor (who was often present) had a non-speaking observatory role. Of course as soon as I or the internal examiner asked the first question, Peter answered it and so it continued. During the 1990s, Peter 'received' several PhDs this way. But his infectious enthusiasm for the work and his students shone through all of this and it was impossible not to be excited by the research that Peter pioneered and espoused.

By the year 2000, complexity theory was really in the air in Britain and Peter put together one of the most interesting and fruitful groups of people I have had the privilege to interact with using funding from the Economic and Social Research Council. This was NeXSUS—which he defined as the Network on compleXity and SUStainability, many of whose members are writing in this Festschift. It was during the early 2000s that I learnt from Peter of the amazingly whacky setup that his group seemed to have been catapulted into at Cranfield. These were years when we were all being forced to write research grants and as usual, those who ran the show—not us the academics and researchers—had little clue as to how hard it is to compete. Peter seemed to move incessantly at Cranfield and eventually ended up in the Management School. I could go on but we seem to have now entered an era where what you have done counts for little and those who shout loudest and longest are able to get the goods. In all of this, Peter has remained clear and focussed. He stands above it and his great strength is his undying enthusiasm for how we might explain the variety, diversity and heterogeneity that comprise complex systems. Long may his influence continue.

Chapter A

Knowledge, complexity and understanding[1]

Paul Cilliers

Knowledge is often seen as some kind of abstract thing which can be stored, manipulated and managed. In this paper it will be argued that knowledge, as opposed to information, is contextual and historically situated. The matter is complicated when we deal with complex phenomena. Since we cannot know complex things in their complexity, we have to reduce the complexity in order to say something at all. This process of "framing" necessitates a normative element which forms part of our understanding of something complex.

Introduction

The strange thing about television is that it doesn't tell you everything. It shows you everything about life on earth, but the mysteries remain. Perhaps it is in the nature of television.

Thomas Jerome Newton in *The Man Who Fell to Earth*

During most events concerned with knowledge management someone starts a presentation by saying that they will not revisit the problem of the distinction between knowledge and data. Usually a sigh goes through the audience, seemingly signifying relief. But why relief? Is it because they will not be bored with an issue that has been resolved already, or because they are glad that they will not be confronted with these thorns again? I suspect that they want to believe that the first reason is the case, but that in fact it is the second. In what follows I therefore want to problematize the notion of "knowledge". I will argue that when talking about the man-

---

1. This paper (apart from some small alterations) first appeared in *Emergence*, 2(4), 7-13. It also formed the basis for a paper delivered at the International Workshop on *The Limits of Scientific Knowledge*, held at the Central European University, Budapest in June 2001 to which I was invited by Peter Allen. Since then we have collaborated at several workshops and conferences. We have also mutually contributed to publication projects, supported each other's students and spent many hours discussing the deeper implications of thinking about Complexity with integrity. In this he was always incisive, perceptive, creative, human and amusing.

agement of "knowledge", whether by humans or computers, there is a danger of getting caught in the objectivist/subjectivist (or fundamentalist/relativist) dichotomy. The nature of the problem changes if one acknowledges the complex, interactive nature of knowledge. These arguments, presented from a philosophical perspective, should have less influence on the practical techniques employed in implementing knowledge management systems than on the claims made about what is actually achieved by these systems.

The traditional trap

The issues around knowledge—*what* can we know about the world, *how* do we know it, what is the *status* of our experiences—have been central to philosophical reflection for ages. Answers to these questions, admittedly oversimplified here, have traditionally taken one of two forms. On the one hand there is the belief that the world can be made rationally transparent, that with enough hard work knowledge about the world can be made objective. Thinkers like Descartes and Habermas are often framed as being responsible for this kind of attitude. It goes under numerous names including positivism, modernism, objectivism, rationalism and epistemological fundamentalism. On the other hand, there is the belief that knowledge is only possible from a personal or cultural-specific perspective, and that it can therefore never be objective or universal. This position is ascribed, correctly or not, to numerous thinkers in the more recent past like Kuhn, Rorty and Derrida, and its many names include relativism, idealism, post-modernism, perspectivism and flapdoodle.

Relativism is not a position that can be maintained consistently[2], and of course the thinkers mentioned above have far more sophisticated positions than portrayed in this bipolar caricature. There are also recent thinkers who attempt to move beyond the fundamentalist/relativist dichotomy[3], but it seems to me that when it comes to the technological applications of theories of knowledge, there is an implicit reversion to one of these traditional positions. For those who want to computerise knowledge, knowledge has to be objective. It must be possible to gather, store and manipulate knowledge without the intervention of a subject. The critics of formalized knowledge,

---

2. If relativism is maintained consistently, it becomes an absolute position. From this one can see that a relativist is nothing else but a disappointed fundamentalist. However, this should not lead one to conclude that everything that is called post-modern leads to this weak position. Lyotard's seminal work, *The Postmodern Condition* (Lyotard, 1984), is subtitled *A Report on Knowledge*. He is primarily concerned with the structure and form of different kinds of knowledge, not with relativism. An informed reading of Derrida will also show that deconstruction does not imply relativism at all. For a penetrating philosophical study of the problem, see *Against Relativism* (Norris 1997).
3. The critical realism of Bhaskar (1986) is a good example.

on the other hand, usually fall back on arguments based on subjective or culture-specific perspectives to show that it is not possible, that we cannot talk about knowledge independently of the knowing subject.

I am of the opinion that a shouting match between these two positions will not get us much further. The first thing we have to do is to acknowledge the complexity of the problem we are dealing with. This will unfortunately not lead us out of the woods, but it should enable a discussion that is more fruitful than the objectivist/subjectivist debate.

Complexity and understanding

An understanding of knowledge as constituted within a complex system of interactions[4] would, on the one hand, deny that knowledge can be seen as atomized "facts" that have objective meaning. Knowledge comes to be in a dynamic network of interactions, a network that does not have distinctive borders. On the other hand, this perspective would also deny that knowledge is something purely subjective, mainly because one cannot conceive of the subject as something *prior* to the "network of knowledge", but rather as something constituted *within* that network. The argument from complexity thus wants to move beyond the objective/subjective dichotomy. The dialectical relationship between knowledge and the system within which it is constituted has to be acknowledged. The two do not exist independently, thus making it impossible to first sort out the system (or context), and then to identify the knowledge within the system. This co-determination also means that knowledge and the system within which it is constituted is in constant transformation. What appears to be uncontroversial at one point may not remain so for long.

The points made above are just a restatement of the claim that complex systems have a history, and that they cannot be conceived of without taking their context into account. The burning question at this stage is whether it is possible to *do* that formally or computationally. Can we incorporate the context and the history of a system into its description, thereby making it possible to extract knowledge from it? This is certainly possible (and very useful) in the case of relatively simple systems, but with complex systems there are a number of problems. These problems are, at least to my mind, not of a metaphysical, but of a practical nature.

The first problem has to do with the non-linear nature of the interactions in a complex system. From this it can be argued (see Cilliers, 1998: 9-10; Richardson *et*

---

4. Complex systems are discussed in detail in Cilliers (1998).

al., 2000) that complexity is incompressible. There is no accurate (or rather, perfect) representation of the system which is simpler than the system itself. In building representations of open systems, we are forced to leave things out, and since the effects of these omissions are non-linear, we cannot predict their magnitude. This is not an argument claiming that reasonable representations should not be constructed, but rather an argument that the unavoidable limitations of the representations should be acknowledged.

This problem—which can be called the problem of boundaries[5]—is compounded by the dynamic nature of the interactions in a complex system. The system is constituted by rich interaction, but since there are an abundance of direct and indirect feedback paths, the interactions are constantly changing. Any activity in the system reverberates throughout the system, and can have effects that are very difficult to predict—once again as a result of the large amount of non-linear interactions. I do not claim that these dynamics cannot be modelled. It could be possible that richly connected network models can be constructed. However, as soon as these networks become sizeable, they become extremely difficult to train. It also becomes rather hard to figure out what is actually happening in them. This is no surprise if one grants the argument that a model of a complex system will have to be as complex as the system itself. Reduction of complexity always leads to distortion.

What are the implications of the arguments from complexity for our understanding of the distinction between data and knowledge? In the first place it problematizes any notion that data can be transformed into knowledge through a pure, mechanical and objective process. It however also problematizes any notion that would see the two as totally different things. There are facts that exist independently of the observer of those facts, but the facts do not have their meaning written on their faces. Meaning only comes to be in the process of interaction. Knowledge is interpreted data. This leads us to the next big question: what is involved in interpretation, and who (or what) can do it?

Knowledge and the subject

The function of knowledge management seems to be either to supplement the efforts of a human subject who has to deal with more data than is humanly possible, or to free the subject up for other activities (perhaps to do some thinking for a change). Both these functions presuppose that human subject can manipulate knowledge. This realisation leads to questions in two directions. One could

---

5. The problem of boundaries is discussed in more detail in Cilliers (2001).

question the efficiency of human strategies to deal with knowledge and then attempt to develop them in new directions. This important issue will not be pursued further here. There is another, perhaps philosophically more basic question, and that has to do with how the human subject deal with knowledge at all. Given the complexities of the issue, how does the subject come to forms of understanding, and what is the status of knowledge as understood by a specific subject? This issue has been pursued by many philosophers, especially in the discipline known as hermeneutics. However, I am not aware that this has been done in any depth in the context of complexity theory[6]. How does one perceive of the subject as something that is not atomistically self-contained, but is constituted through dynamic interaction? Moreover, what is the relationship between such a subject and its understanding of the world. A deeper understanding of what knowledge is, and how to "manage" it, will depend heavily on a better understanding of the subject. This is a field of study with lots of opportunities.

Apart from calling for renewed effort in this field, I only want to make one important remark. It seems that the development of the subject from something totally incapable of dealing with the world on its own into something that can begin to interpret —and change—its environment is a rather lengthy process. Childhood and adolescence are necessary phases (sometimes the only phases) in human development. In dealing with the complexities of the world there seems to be no substitute for experience (and education). This would lead one to conclude that when we attempt to automate understanding, a learning process will also be inevitable. This argument leads one to support computing techniques which incorporate learning (like neural networks) rather than techniques which attempt to abstract the essence of certain facts and manipulate them in terms of purely logical principles. Attempts to develop a better understanding of the subject will not only be helpful in building machines that can manage knowledge, it will also help humans to better understand what they do themselves. We should not allow that the importance of machines (read computers) in our world leads to a machine-like understanding of what it is to be human.

Implications

In Nicholas Roeg's remarkably visionary film *The Man Who Fell to Earth* (1976), an alien using the name Thomas Jerome Newton (superbly played by David Bowie), tries to understand human culture by watching television—usually a whole bunch of screens at the same time. Despite the immense amount of data available to him, he

---

6. An important contribution was made by reinterpreting action theory from the perspective of complexity (Juarrero 2000). Some preliminary remarks, more specifically on complexity and the subject, are made in Cilliers and De Villiers (2000).

is not able to understand what is going on directly. It is only through the actual *experience* of political complexities, as they unfold in time, that he begins to understand. By then he is doomed to remain earthbound.

I am convinced that something similar is at stake with all of us. Having access to untold amounts of information does not increase our understanding of what it means one bit. Understanding, and therefore knowledge, follows only after interpretation. Since we hardly understand how humans do it, we should not oversimplify the problems involved in doing knowledge management computationally. This does not imply that we should not attempt what we can—and certain spectacular advances have been made already—but that we should be careful in the claims we make about our (often still to be finalised) achievements. The perspective from complexity urges that, amongst others, the following things should be kept in mind:

- Although systems which filter data enable us to deal better with large amounts of it, it should be remembered that filtering is a form of compression. We should never trust a filter too much.

- Consequently, when we talk of mechanized knowledge management systems, we can (at present?) only use the word "knowledge" in a very lean sense. There may be wonderful things to come, but at present I do not know of existing computational systems that can in any way be seen as producing "knowledge". Real breakthroughs are still required before we will have systems that can be distinguished in a fundamental way from database management. Good data management is tremendously valuable, but cannot be a substitute for the interpretation of data.

- Since human capabilities in dealing with complex issues are also far from perfect, interpretation is never a merely mechanical process, but one that involves decisions and values. This implies a normative dimension to the "management" of knowledge. Computational systems which assist in knowledge management will not let us escape from this normativity. Interpretation implies a reduction in complexity. The responsibility for the effects of this reduction cannot be shifted away onto a machine.

- The importance of context and history means that there is no substitute for experience. Although different generations will probably place the emphasis differently, the tension between innovation and experience will remain important.

- We should manage complex systems with respect for the diversity they contain. Knowledge emerges from form the interaction between many different components. Thus, if we can maintain a rich diversity, the resources in the system will be richer. Peter Allen (2001) makes this point even stronger by calling for "excess diversity". A system should not only have the "requisite variety" it needs to cope with its environment (Ashby's law), it should have *more* variety. Excess diversity in the system allows the system to cope with novel features in the environment. What is more, if a system has more diversity than it needs in order to merely cope with its environment, it can experiment internally with alternative possibilities. These creative capabilities should not "managed out"

These considerations should assist in developing an understanding of knowledge management which could be called "organic", but perhaps also "ethical". The act of management cannot be divorced from acknowledging the complexity of that which is managed, and if this is the case, we cannot formalise the process of management fully. Management implies choice, interpretation and decision. Since these are based on knowledge which cannot be perfect, the manager cannot escape assuming responsibility for what he does.

References

Allen, P.M. (2001). "A complex systems approach to learning, adaptive networks," *International Journal of Innovation Management*, ISSN 1363-9196, 5(2): 149-180

Bhaskar, R. (1986). *Scientific Realism and Human Emancipation*, ISBN 0415454956 (2009).

Cilliers, P. 1998. *Complexity and Postmodernism: Understanding Complex Systems*, ISBN 0415152879.

Cilliers, P. (2001). "Boundaries, hierarchies and networks in complex systems," *International Journal of Innovation Management*, ISSN 1363-9196, 5(2): 135-147.

Cilliers, P. and De Villiers, T. (2000). "The complex 'I'," in W. Wheeler (ed.), *The Political Subject*, ISBN 0853159149.

Juarrero, A. (2000). *Dynamics in Action: Intentional Behavior as a Complex System*, ISBN 0262100819.

Lyotard, J.F. (1984). *The Postmodern Condition: A Report on Knowledge*, ISBN 0719014506.

Norris, C. (1997). *Against Relativism: Philosophy of Science, Deconstruction and Critical Theory*, ISBN 0631198652.

Richardson, K. Cilliers, P. and Lissack, M. (2000). "Complexity science: A 'grey' science for the 'stuff in between'," *Proceedings of the First International Conference on Systems Thinking in Management*, Geelong Australia, pp. 532-537.

Paul Cilliers studied at the Stellenbosch University from mid-1980s to 1994. There he received his BA in Electronic Engineering in 1980, his BA cum laude in Political Philosophy in 1987, and his MA cum laude in Philosophy in 1989. He received his PhD in 1994 from the Stellenbosch University and Cambridge University under supervision of Johan Degenaar and Mary Hesse. In 1998 he published his seminal work *Complexity and Postmodernism: Understanding Complex Systems*.

Cilliers started his academic career at the Stellenbosch University, where in 1993 he became a lecturer in philosophy. Since 2003 he was Professor of Complexity and Philosophy at the University of Stellenbosch. In the year 2008 he was Visiting Professor at the University of Humanistic Studies in Utrecht in the Netherlands.

In 2006 Cilliers was awarded the Harry Oppenheimer Fellowship Award in recognition of his outstanding achievements in developing a general understanding of the characteristics and nature of complex systems. In 2008 the National Research Foundation of South Africa awarded him an A-rating, and in 2010 he was elected fellow of the Royal Society of South Africa

Very sadly Paul Cilliers died whist this book was being prepared. Fortunately we received his contribution prior to this with a covering note saying: "I will gladly participate in this project. I have the greatest respect for Peter."

Paul's thought provoking contribution clearly shows how much his incisive insights into complexity theory will be missed.

Chapter B

Complexity science: The integrator

Liz Varga

Complexity Science sheds new light upon real-world phenomena, explaining the integration of environments, people, and society, at multiple scales and of great diversity, and with unforeseen and emergent properties. This chapter considers the integrative role of complexity in the physical and meta-physical domains, and brings together the broad spectrum of Allen's contribution to complexity as an integrator. Complexity science straddles multiple disciplines and embraces pluralism of world views and ways of knowing. Real-world phenomena of coevolution, emergence and path-dependency are explained by complexity science notions of diversity, feedback, and non-linearity. Complexity catalyses learning and adaptation by employing mixed methods to explore, explain and understand the world, making it highly relevant for practice and reinforcing the unpredictability of integrated reality.

Introduction

Of all the discourses held and narratives defended in the Complex Systems Research Centre since my arrival in 2003 the one which represents complexity science most thoroughly is that of the integrator: of real-world systems in real-world contexts and of knowledge and knowing in our representations of real-world systems. In its role as integrator, complexity connects practice and theory so fundamentally that they are inseparable.

As a consequence of its integrative role, complexity science hits all of the buzz words of the 21st century, such as multi-disciplinary, integration management, resource management, stakeholder management, context (political, economic, social, technological, environmental) sensitivity, and integrated decision-support tools. It is no wonder that complexity science addresses highly integrated topics of practical relevance such as climate change, sustainable development, infrastructure systems management, strategy and policy, urbanization and industrialization, ecological innovation, economic development, urban planning, and so on.

The reason for its integrative leadership is that complexity is focused on how the real-world interactions of many diverse individuals create structures (and are influenced by the structures they create) and which persist and coevolve within an environment which provides resources. Complexity science sheds a new light upon real-world phenomena, of multiple scales, of great diversity, with emergent properties providing better explanations of these phenomena than extant theories which ignore this integrative perspective.

By admitting theories and heuristics developed from a wide range of disciplines from mathematics through to medicine, networks to nanotechnology, psychology to physics, complexity science adopts a pluralistic view of the world, thereby permitting multi-paradigmatic views of the world, from the post-positivist to the post-modernist. This pluralism is able to inform learning and adaptation through its capability to embrace multi-ontological and multi-epistemological perspectives which represent real-world systems, and employ mixed methods (quantitative and qualitative) to explore, explain and understand real-world phenomena.

With a focus on integrating in real-world systems, namely, individuals, socioeconomic units, and the environment, with metaphysics, namely, ontology and epistemology, methods and learning and adaptation, as shown in Figure 1, this paper highlights various outcomes and impacts generated during Peter Allen's incumbency, as Professor of Evolutionary Complex Systems and Director of the Complex Systems Research Centre, which have made a significant contribution to the field of complexity science in its capacity as integrator.

Socioeconomic structures: How systems emerge from the interactions of individuals

Complexity science targets a sub-set of all systems; a sub-set which is abundant and is the basis of all novelty; a sub-set which is evidenced in biology, chemistry, physics, social, technical and economic domains; a sub-set which coevolves with its environment; a sub-set from which structure emerges. That is, self-organization occurs through the dynamics, interactions and feedbacks of heterogeneous components (Allen, 2007). This sub-set of all systems is known as complex systems. It follows that not all systems are complex. And in fact, we observe that in practice some complex systems are constrained, or locked-in, such that they cannot evolve or change (Unruh, 2002).

Figure 1 *Interactions*

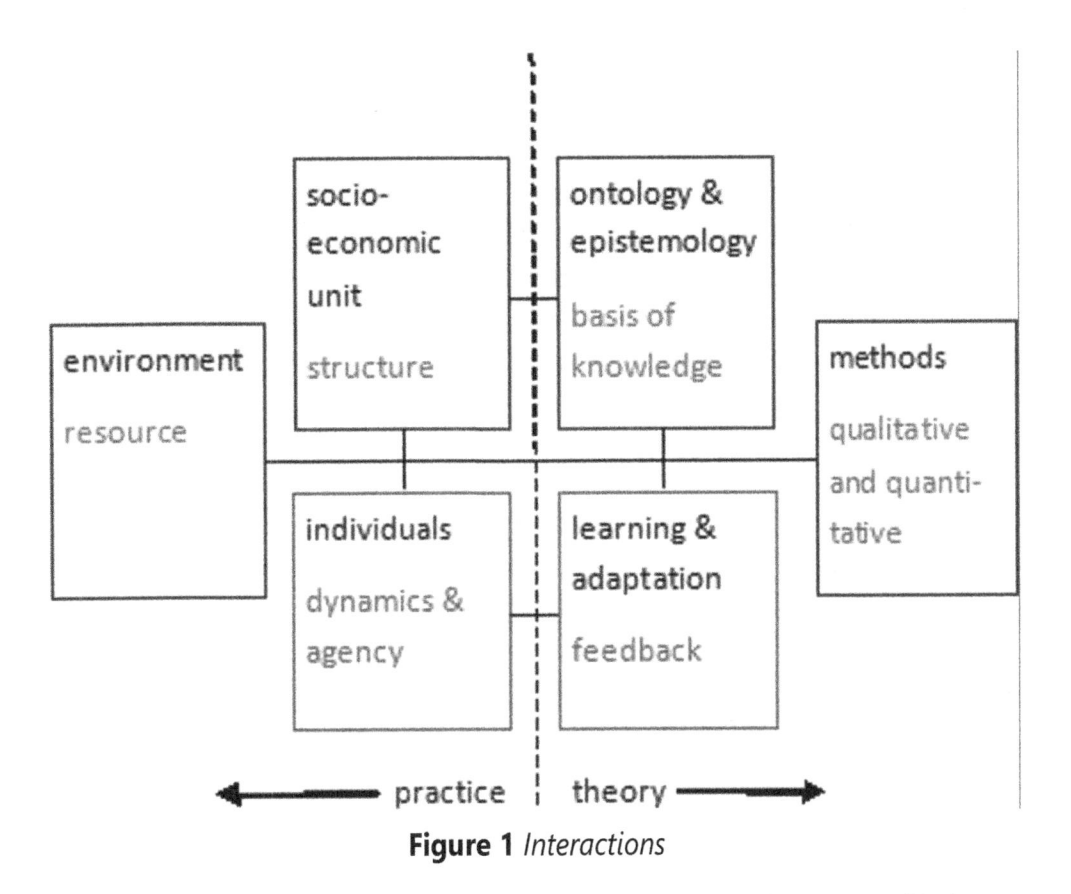

Complex systems are evolutionary and generate structural diversity. This is most fundamentally seen through biology and the nature of man, having a structurally evolved phenotype from his ancestors, however, being constituted from genetic variety via neo-Darwinian processes of variation, selection and retention. Socio-technical-economic (co-)evolution works in exactly the same way; by innovations of objects and evolving parts within the system, such as routines (Nelson & Winter, 1982), which are attempted and sometimes retained, and may lead to new structural attractors which in practice demonstrate qualitatively different characteristics (Allen *et al.*, 2006a).

However, in socio-technical-economic systems, we also note the operation of Lamarckian evolution in which the acquisition of objects and evolving parts occurs. Acquired objects and evolving parts might have arisen from outside the complex system, such as new people with different knowledge, ignorance and perspectives, and technologies promising transformational capability. It follows that the possibilities created by Lamarckian evolution create even more potential for diversity.

The process of change in socio-technical-economic systems occurs non-linearly. Evolutionary processes unfold in characteristic patterns as waves of Creative Destruction (Schumpeter, 1994). Markets have periods of comparative quiet, when firms have developed superior products, technologies or organizational capabilities which earn positive economic profits. Quiet periods are punctuated by shocks or discontinuities that destroy old sources of advantage and replace them with new ones. These are phase transitions and replace 'gradualist evolution' with punctuated equilibrium (Eldredge & Gould, 1972) and arise as new structural attractors are established through multiple interactions of diverse components (Allen et al., 2006a). Existing structures are the consequence of irreversibility (Prigogine & Stengers, 1997) and path-dependency, the past coproducing the present and future system states following what Eddington (1930) called the 'arrow of time'.

As systems coevolve, our language to describe them changes. The qualitative description of system states often corresponds to a rational articulation of an overall aim, but more often is simply the result of an accretion of events and instabilities that have marked the system over time (Allen et al., 2006b).

The emergence of successive dynamical systems, each exhibiting an ephemeral stability provides an overarching perspective of succession and emergence (Allen, 2008) over time which is incapable of explanation by traditional approaches. Two examples follow of the evolution of industries.

Under the ESRC Nexsus project (Allen, 2004) a survey was completed by 73 manufacturers regarding the interaction of 53 automotive practices (McCarthy et al., 1997). A matrix of each pair of 53 by 53 practices was constructed and the relationship between each pair quantified by an increasingly positive value if positively synergetic or reinforcing, and increasingly negative if conflicting. This interaction matrix allowed potential structures to be investigated in which new practices are adopted at random and showed that successful organizational forms are those whose constituent parts are internally coherent (Allen, 2001b). This explains the relationship between the identity of a firm with its emergent capabilities and attributes, and its constituent parts. The interaction between parts and structure thus is related to both the parts and their interactions.

In aerospace, about 18 distinct forms of aerospace supply chains were identified based on the interpretation of the 78 character states (or characteristics) determined through literature review and interview (Rose-Anderssen et al., 2009a). The classification system, in particular the emergence of new character states and distinct forms which arise at times of major change in the national and global environment, such as the world wars, help explain the history and path-dependency of supply chains as complex systems, and also suggest how changing present structures and creating new ones could create competitive advantage. It indicates which new characteristics are likely to be suitable for a supply network, complex system or firm by considering the synergy of a new characteristic with its existing characteristics. The dynamics of a firm are such that characteristics might also be lost during their evolution, particularly if the loss of a characteristic aids the adoption of one or more new characteristics which make the firm more synergetic in its manifestation. In this way, we can also assess the overall health of a firm by the extent of synergy between its interacting characteristics (Allen, 2004).

Philosophical perspectives

Whilst we may arrive at a consensus embodied in a qualitative description of a complex system and its emergent properties, it will undoubtedly be incomplete and may favour the motives of the stakeholders. This occurs because our representations of reality are not reality itself, typically demonstrated by Magritte's drawing of a pipe entitled "Ceci n'est pas une pipe" highlighting the fact that the representation is just that and not a real pipe.

Representations are not always as clear as reality. We strive for parsimony in our representations but must provide sufficient richness and uniqueness of description to identify and differentiate a new form. Representations with greatest clarity are most useful and examples include global industrial waves, cultural norms, and industrial types, examples respectively are Kondratiev waves, civilizations, and economic types.

Kondratiev's macroeconomic cycles describe the dominant or standard ways of economic development over different time periods (Kondratiev, 1984). Our qualitative descriptions today recognize them as the age of steam, the age of steel, the age of oil, and the age of information technology.

The evolution (and destruction) of civilizations is charted and ascribed to cultural changes (Ehrlich, 2002). Jablonka and Lamb (2005) describe the process of symbolic inheritance to represent cultural evolution and which encompasses cognition, communication, language and other types of symbol. Examples of civilizations include Bronze, Minoan, Mongolian, and Modern.

Archetypes of organizational forms were described by Mintzberg (1980) as Simple Structure, Machine Bureaucracy, Professional Bureaucracy, Divisionalized Form and Adhocracy. Regardless of our descriptions, evolution fits within the wider milieu of the social, cultural, environmental and technological history: the "eco-historical regime" (Garnsey & McGlade, 2006).

Inevitably our descriptions exclude the apparently inconsequential, invisible, and informal, which themselves may be the most significant component of the stability and reinforcement of a new structure. Complexity science recognizes this ambiguity and accepts the pluralism of reality and knowledge. Different stakeholders will describe reality differently based on how they know their reality. Complexity science integrates this pluralism of reality (ontology) and knowledge (epistemology) (Allen & Varga, 2006b): "It follows that complexity science concerns both the nature of the system and the nature of its interpretation. It is about both ontology and epistemology." (Allen & Varga, 2006a: 232). The metaphysical issues of ontology and epistemology are dealt with by recognizing that truth and knowledge are extracted along a continuum which evolves over time (Maguire et al., 2006).

Because of its integrative capacity, complexity may be used by researchers in all areas of ontological and epistemological realms. Snowden and Stanbridge (2004) propose four quadrants based on Ontology (ordered and un-ordered) and Epistemology (rules and heuristics). In the ordered/rules quadrant we find the view that the organization is mechanical and for many organizations and industries which are stable and subject to significant regulation, negative feedback, and controls, the organization's behavior can be predictable to a large degree. This is largely the field of traditional economics and operations management in which optimization and macro imposed structures are used. In the ordered/heuristics quadrant, the roots are in systems thinking with a focus on values, competencies, and learning (see for example Senge, 1990). In the unordered/rules quadrant, called mathematical complexity, the idea is of discovering or designing simple rules at the agent level in order to aid decision making to optimize complex phenomena. In the final quadrant, un-ordered/heuristics, the emphasis is on socially constructed meaning (see for example Stacey, 1996; Cilliers, 1998) and distancing social systems from biological, physical and chemical systems.

Allen's work integrates all four of these quadrants: in ordered/rules we find many of Allen's systems dynamics models (see for example Allen et al., 1986); in ordered/heuristics we find Allen's work on learning and knowledge (see for example Allen, 2001b, 2001a); in unordered/rules, Allen's work on emergent behaviors is found (see for example Allen et al., 2013); and finally in unordered/heuristics we find Allen's work on social identity (see for example Allen et al., 2010).

Integrated methods

Complexity science highlights the value of absorbing complexity and keeping simplification to a minimum (Boisot & Child, 1999). As more assumptions are made, the representation of a complex system become more simplified such that the diversity, dynamics and distribution of the components of the system may be lost (Allen & Varga, 2006b).

By using both qualitative methods and quantitative methods, in a logical research design (see for example Creswell & Plano Clark, 2010), complexity science integrates representations of the real-world using both descriptions and numbers. Language is itself an evolutionary system: our descriptions of the real world emerge as we observe changes in the real world and create new words and meanings for our observations. Our world is constantly 'becoming' (Tsoukas & Chia, 2002).

Computational methods and information systems also coevolve with our increasingly rich representations of the real-world. Traditional quantitative methods are increasingly superseded by methods which make fewer assumptions thus embracing the complexity of the real-world and making use of multi-processor power to process big data and more variables.

Agent based modeling holds a particular value for complexity science, because it integrates bottom up interactions with the potential for emergent structures whilst also using qualitative methods to describe behavior and structure. Agent based modeling does not assume structure, thereby allowing structures to emerge, together with qualitative descriptions of these structures, from the quantitative models. This integration of quantitative understanding of physical energy systems with more qualitative understanding of the social aspects of these systems is an important integrated need (Smil, 2010).

Modeling is a core research method in the Complex Systems Research Centre. Models are used to explore, imagine and experiment upon different rules, parameters such as policy settings, and contingent connections and relationships for the interacting elements and agents that make up virtual firms and virtual networks. A model can consider systematically all possible sets of rules and responses, something that no real complex system can perform, and so assists with risk reduction (Baldwin *et al.*, 2006).

A very recent example is the creation of a UK energy model (Allen *et al.*, 2013) which, rather than simulate the current operation of the UK energy demand and supply systems, explores possible futures to find those which meet multiple criteria. The current configuration and location of power generation, inter-connectors, and demand are used as initial conditions for the model. Near term plans to decommission and build replacement plants in different UK locations are used as fixed decisions in the early years of evolution. In medium and long-term decision making, user preferences can include avoidance of on-shore wind, or fossil fuel plants of any sort. The model will determine the effect on carbon emissions with the aim of reducing these by 80% whilst accommodating growing demand. The model self-organizes supply side investments based on the preferences of the user and the rules for the attractiveness of different energy supply side investments creating different possibilities for generation in different locations, such as building nuclear only where it has been historically located. The model therefore finds alternative not pre-determined structures (combinations of energy generation mix and location) that meet energy demand, user choices (such as minimizing coal fired plants), and carbon emissions targets. As new technologies, new policies, and changes in demand occur in real systems, so too can the representation be updated in the model and new future pathways generated to 2050.

Such models facilitate dialogue between stakeholders: industry, consumers, investors, policy makers, and so on. An example contained in "Innovation Systems, Economic Systems, Complexity and Development Policy" (Clark, N., this volume) describes how Allen's aim for the Senegal model was to provide a planning tool "to integrate policy discussion across professional groups" and "to achieve new levels of integrated verisimilitude" thereby recognizing the plurality of views.

Environments

Context dependency is found in many theories of socioeconomic systems. For example, Hannan and Freeman (1977) introduced us to Population Ecology, Lawrence and Lorsch (1967) to structural contingency theory, Meyer and Rowan (Meyer & Scott, 1983) to institutionalization theory, and Pfeffer and Salancik (1978) developed resource dependence theory.

Complexity science takes a couple of steps forward in this respect: 1. The systems created by humans are inter-dependent with the environment: climate change is a prime example, 2. The environment is just one layer or level in at least three layers of all systems: the individual elements, the connected elements (organizations, institutions for example) and the environment (Allen, 2000). This context inter-dependency is made explicit for example in formal models which "represent, in a detailed and explicit manner, the major assumptions, relationships, variables and decision parameters about the 'real-world' domain which they cover" (Connor & Allen, 1994: 97). Policy interventions can be seen in Connor and Allen's paper to effect changes in the environment in sub-Sahara such as water-resources, migratory behavior and so on.

Alternate representations of possible future environments are easily integrated into models usually through the use of scenarios (see for example Hughes & Strachan, 2010). This allows testing of the model assumptions against different settings, and to see if policy interventions create different outcomes under different settings. In energy modelling, energy projections are used for policy decision-making and in policy debates particularly in respect of low carbon technologies (Strachan, 2011).

However, all models are invariably inaccurate. In social systems emergent structural attractors can occur simply because there is demand for their particular emergent capabilities. Fashion, lifestyles, art, artefacts and communities of practice can emerge and survive providing that there is a clientele for them. They are not about being "true or false" but simply about whether there is a "market" for them. These possibilities are very difficult to foresee and demonstrate that socioeconomic systems are not deterministic. Prediction is not possible and the future is not given (Prigogine, 2000). This does not undermine the value of modeling which is primarily in the exploration of assumptions about the system, the individuals, the context and its futures.

Learning and adaptation

Complex systems have unique histories and exist in particular environments. They are path-dependent, with the past coproducing the present and the future (Maguire *et al.*, 2006).

Individuals in the real-world attempt to learn about the systems in which they live and work, but ambiguity and uncertainty (Allen *et al.*, 2010) occur due to the limits to knowledge (Allen, 2001b, 2001a, 1994, 1993). All knowledge is a reduction of reality of some kind, and evolution in human systems is a continual, imperfect learning process driven by differences between expectations and experience that rarely provides complete understanding. Ignorance is the outcome but also the motor of evolution in complex systems (Allen, 2011).

If a system is structurally stable then prediction, either dynamic or static, is possible even if probabilistic. The statistics will probably not be "normal," but "present future value," optimization, and so on can be calculated. But structural stability means that no real innovations occur and probable extinction exists for the system in a changing environment. Evolution and learning are structural instabilities, implying:

- Unpredictability; the future does not exist—we will help to create it; learning and exploration strategies are therefore most effective;
- Small things (the non-average) can have big effects: all large things were once small, and statistical distributions are part of the evolution;
- Multiple-levels of behavior: environment, system configuration and coordination, internal knowledge, beliefs, aims of people, all result in behaviors at different levels;
- If there is no internal diversity within different functional types, there can be no learning or evolution, and;
- New capabilities, features and qualities emerge in ways that are not foreseen—strategy needs to emerge and be honestly evaluated.

(Allen *et al.*, 2009)

The multiple, interlocking heuristic rules that constitute a firm's current operations have been derived from a "satisficing" principle whereby only behavioral rules which do not work can be eliminated while rules only have to be "good enough" in order to survive. In other words, most organizations are made up of connected behavior that has worked sufficiently well up to now, but will not necessarily be able to deal with tomorrow's conditions. Behavioral rules are all coupled, feed back on each other and have time delays and logistical issues which mean that if everybody tries to experiment with their behavior, nobody can learn.

An individual's perspective of reality is constructed and limited by their knowledge and experience. Whilst we might strive to find the truth, we are limited by our cognition to see the truth (Allen, 2001a, 2000).

Knowledge and adaptation is integrated with the systems in which we exist; as we learn and experiment we discover there is yet more to learn and adapt. Complexity Science tells us that evolution/transformation over time selects for systems that retain noise (fluctuations, disturbances, and small scale events), micro-diversity (people with different experience, views and values) and people and groups that possess exploratory, innovative capacities which are micro-diversity generating mechanisms within them (Allen *et al.*, 2007). This means that firms will be eliminated over time if they do not retain the "learning" mechanisms resulting from the dynamic selection between different pathways suggested by diverse individuals, including their different connections and channels into the outside world. If a firm is to survive it must do experiments, which can only be imagined by people with diverse, incommensurable, paradoxical views.

This is not about "official diversity" (gender, ethnic, etc.) It is about individual's different histories, origins, and views of the problem (Allen & Andriani, 2007).

In considering the evolution of knowledge in whatever domain over time, we see that knowledge involves articulated words, concepts and variables that inhabit certain dimensions and which together are deemed to provide some emergent capability. Over time the knowledge cluster evolves as new connections are made involving new variables and dimensions. These either run in parallel, catering to different parts of the "market" or actually replace them as being the "improved theoretical view". Structural attractors correspond to these emergent clusters: for products it is bundled technologies; for markets it is bundles of coevolving firms; for organizations it is bundles of coevolving practices and techniques; for knowledge more generally it is bundles of connected words, concepts and variables that emerge for a time.

Conclusion

This chapter has highlighted the works of Peter Allen which underpin the area of Complexity Science as an integrator. This integrative role of Complexity Science makes it highly relevant for practice, explaining the integration of the individual with social and economic structures, and with the environment. Through notions of diversity, feedback, and learning and adaptation, the phenomena of coevolution, emergence and path-dependency have been explained. The integration of real-world systems with meta-physical positions has also been examined, showing Peter Allen's capability in major segments of ontology and epistemology. Finally, methods of knowing complex systems and ways in which complex systems are represented have shown how Complexity Science accepts mixed quantitative and qualitative methods, reflecting the true diversity of real-world systems.

References

Allen, P.M. (1993). "Evolution: Persistent ignorance from continual learning," in R.H. Day and P. Chen (eds.), *Nonlinear Dynamics and Evolutionary Economics*, ISBN 0195078594, pp. 101-112.

Allen, P.M. (1994). "Coherence, chaos and evolution in the social context," *Futures*, ISSN 0016-3287, 26(6): 583-597.

Allen, P.M. (2000). "Knowledge, ignorance and the evolution of complex systems," in J. Foster and S. Metcalfe (eds.), *Frontiers of Evolutionary Economics: Competition, Self-Organization and Innovation Policy*, ISBN 3540237739.

Allen, P.M. (2001a). "A complex systems approach to learning, adaptive networks," *International Journal of Innovation Management*, ISSN 1363-9196, 5(2): 149-180.

Allen, P.M. (2001b). "What is complexity science? Knowledge of the limits to knowledge," *Emergence*, ISSN 1525-3250, 3(1): 24-42.

Allen, P.M. (2004). "Nexsus: The first priority network of the ESRC—Sustainability of socioeconomic systems," http://www.som.cranfield.ac.uk/som/dinamic-content/research/csrc/Nexsus.pdf.

Allen, P.M. (2007). "Self-organization in economic systems," in H. Hanusch and A. Pyka (eds.), *Elgar Companion to Neo-Schumpeterian Economics*, ISBN 1848447027, pp. 1111-1148.

Allen, P.M. (2008). "Complexity and emergent temporal structure," in S. Vrobel, O.E. Roessler, and T. Marks-Tarlow (eds.), *Simultaneity: Temporal Structures and Observer Perspectives*, ISBN 9812792414, pp. 150-181.

Allen, P.M. (2011). "Complexity, evolution and organizational behavior," in S.J. Guastello, M. Koopmans, and D. Pincus (eds.). *Chaos and Complexity in Psychology: The Theory of Nonlinear Dynamical Systems*, ISBN 1107680263, pp. 452-474.

Allen, P.M. and Andriani, P. (2007). "Diversity, interconnectivty and sustainability," in J. Bogg and R. Geyer (eds.), *Complexity, Science and Society*, ISBN 184619203X, pp. 11-32.

Allen, P.M. and Varga, L. (2006a) "A coevolutionary complex systems perspective on information systems," *Journal of Information Technology*, ISSN 0268-3962, 21(4): 229-238.

Allen, P.M. and Varga, L. (2006b). "Complexity: The coevolution of epistemology, axiology and ontology," *Non-Linear Dynamics, Psychology and Life Sciences*, ISSN 1090-0578, 11(1): 19-50.

Allen, P.M., Engelen, G. and Sanglier, M. (1986). "Self-organizing systems and the 'Laws of Socioeconomic Geograhy'," *European Journal of Operational Research*, ISSN 0377-2217, 25(1): 127-140.

Allen, P.M., Strathern, M. and Baldwin, J. (2006a). "Evolution, complexity and organization," in E. Garnsey and J. McGlade (eds.), *Complexity and Coevolution: Continuity and Change in Socio-Economic Systems*, ISBN 184542140X, pp. 22-60.

Allen, P.M., Strathern, M. and Baldwin, J. (2006b). "Evolutionary drive: A new understanding of change in socioeconomic systems," *Emergence Complexity & Organization*, ISSN 1525-3250, 8(2).

Allen, P.M., Strathern, M. and Baldwin, J.S. (2007). "Complexity and the limits to learning," *Journal of Evolutionary Economics*, ISSN 0936-9937, 17(4): 401-431.

Allen, P.M., Strathern, M. and Varga, L. (2010). "Complexity: the Evolution of Identity and Diversity," in P. Cilliers and R. Preiser (eds.). *Complexity, Difference and Identity: An Ethical Perspective*, ISBN 9400732708, pp. 273.

Allen, P.M., Varga, L. and Strathern, M. (2009). "Complexity, innovation and organizational evolution," *International Journal of Projectics*, ISSN 2031-9703, 1: 31-49.

Allen, P.M., Varga, L. and Strathern, M. (2010). "The evolutionary complexity of social and economic systems: The inevitability of uncertainty and surprise," *Risk Management*, ISSN 1460-3799, 12(1): 9-30.

Allen, P.M., Varga, L., Strathern, M., Savill, M. and Fletcher, G. (2013). "Exploring possible energy futures for the UK: Evolving power generation," *Emergence Complexity & Organization*, ISSN 1525-3250, 15(2): 38-63.

Baldwin, J.S., Rose-Anderssen, C., Ridgway, K., Allen, P.M., Alvaro, L., Strathern, M. and Varga, L. (2006). "Management decision-making: risk reduction through simulation," *Risk Management*, ISSN 1460-3799, 8(4): 310-328.

Boisot, M. and Child, J. (1999). "Organizations as adaptive systems in complex environments: The case in China," *Organization Science*, ISSN 1047-7039, 10: 237-253.

Cilliers, P. (1998). *Complexity and Postmodernism: Understanding Complex Systems*, ISBN 0415152879.

Connor, S.J. and Allen, P.M. (1994). "Policy and decision support in sustainable development planning: a 'complex systems methodology' for use in sub-Saharan Africa," *Project Appraisal*, ISSN 0268-8867, 9(2): 95-98.

Creswell, J.W. and Plano Clark, V.L. (2010). *Designing and Conducting Mixed Methods Research*, ISBN 1412975174.

Eddington, A.S. (1930). *The Nature of the Physical World*, Macmillan, London.

Ehrlich, P.R. (2002). *Human Natures: Genes, Cultures, and the Human Prospect*, ISBN 0142000531.

Eldredge, N. and Gould, S.J. (1972). "Punctuated equilibria: An alternative to phyletic gradualism," J.M. Schopf (ed.), *Models in Paleobiology*, ISBN 0877353255, pp. 82-115.

Garnsey, E. and McGlade, J. (2006). *Complexity and Coevolution: Continuity and Change in Socio-Economic Systems*, ISBN 184542140X.

Hannan, M.T. and Freeman, J. (1977). "The population ecology of organizations," *American Journal of Sociology*, ISSN 0002-9602, 82: 929-964.

Hughes, N. and Strachan, N. (2010). "Methodological review of UK and international low carbon scenarios," *Energy Policy*, ISSN 0301-4215, 38(1): 6056-6065.

Jablonka, E. and Lamb, M.J. (2005). *Evolution in Four Dimenions: Genetic, Epigenetic, Behavioral, and Symbolic Variation in the History of Life*, ISBN 0262600692 (2006).

Kondratiev, N. (1984). *The Long Wave Cycle*, (translated from The Major Economic Cycles, originally published in Russian in 1925), ISBN 0943940079.

Lawrence, P.R. and Lorsch, J.W. (1967). *Organization and Environment*, ISBN 0875841295 (1986).

Maguire, S., McKelvey, B., Mirabeau, L. and Õztas, N. (2006). "Organizational complexity science," in S. Clegg, C. Hardy, T. Lawrence, et al. (eds.), *Handbook of Organizational Studies*, ISBN 0761949968, pp. 165-214.

McCarthy, I., Leseure, M., Ridgway, K. and Fieller, N. (1997). "Building a manufacturing cladogram," *International Journal of Technology Management*, ISSN 0267-5730, 13(3): 2269-2296.

Meyer, J.W. and Scott, W.R. (1983). *Organizational Environments: Ritual and Rationality*, ISBN 0803920814.

Mintzberg, H. (1980). "Structure in 5's: A synthesis of the research on organization design," *Management Science*, ISSN 0025-1909, 26(3): 322-341.

Nelson, R.R. and Winter, S.G. (1982). *An Evolutionary Theory of Economic Change*, ISBN 0674272285.

Pfeffer, J. and Salancik, G.R. (1978). *The External Control of Organizations: A Resource Dependence View*, ISBN 080474789X (2003).

Prigogine, I. (2000). "The future is not given," *New Perspectives Quarterly*, ISSN 1540-5842, 17(2): 35-37.

Prigogine, I. and Stengers, I. (1997). *The End of Certainty: Time, Chaos and the New Laws of Nature*, ISBN 0684837056.

Rose-Anderssen, C., Baldwin, J. S., Ridgway, K., Allen, P.M., Varga, L. and Strathern, M. (2009a). "A cladistic classification of commercial aerospace supply chain evolution," *Journal of Manufacturing Technology Management*, ISSN 1741-038X, 20(2).

Schumpeter, J.A. (1994). *Capitalism, Socialism and Democracy*, originally published in 1942, ISBN 0415107628.

Senge, P.M. (1990). *The Fifth Discipline: The Art and Practice of the Learning Organization*, ISBN 071269885X.

Smil, V. (2010). *Energy Transitions: History, Requirements, Prospects*, ISBN 0313381771.

Snowden, D. and Stanbridge, P. (2004). "The landscape of management: Creating the context for understanding social complexity," *Emergence: Complexity & Organization*, ISSN 1525-3250, 6(1-2): 140-148.

Stacey, R. (1996). "Management and the Science of Complexity: If Organizational Life is Nonlinear, Can Business Strategies Prevail?" *Research Technology Management*, ISSN 0895-6308, 39(3): 8-10.

Strachan, N. (2011). "UKERC energy research landscape: Energy systems modelling," UKERC, http://ukerc.rl.ac.uk/Landscapes/Modelling.pdf.

Tsoukas, H. and Chia, R. (2002). "On organizational becoming: Rethinking organizational change," *Organization Science*, ISSN 1047-7039, 13(5): 567-582.

Unruh, G.C. (2002). "Escaping carbon lock-in," *Energy Policy*, ISSN 0301-4215, 30(4): 317-325.

Liz Varga is Principal Research Fellow and Director of the Complex Systems Research Centre in the School of Management at Cranfield University. Liz has a number of research grants located within infrastructure systems, each of which has a strong modeling component with a focus on demonstrating the relevance of complex systems approaches to coevolutionary socio-economic technical systems. Existing research includes: European FP7 project EU-InnovatE: modeling sustainable user innovations within sustainable scenarios (SPREAD); International Centre for Infrastructure Futures (ICIF) modeling of inter-dependency risks and business models; EPSRC funded Multi-Utility Service Companies: modelling of business models. Recent projects include multi-modal freight logistics (ABIL), smart grids (CASCADE) and utilities' conversion points (TUCP). On-going projects include: *The Handbook for Complex Delivery: Managing Complexity in Projects and Programs* (2015); and multi-disciplinary projects in transport and complexity. Liz teaches research methods and perspectives from complexity science and she has a mix of doctoral students straddling engineering, applied sciences and management, connected by a shared view of infrastructure as complex systems in use. Her work makes a general contribution to the coevolution of infrastructure systems, the use of agent-based models for representing infrastructure systems, and to approaches that have historically rarely moved from the purely academic realm.

Chapter Γ

Levels of abstraction and cross-cutting skills: Making sense of context in pursuit of more sustainable futures

Mark Lemon, Paul Jeffrey & Richard Snape

In light of the audience for this text there is unlikely to be anything radically new in these statements; they do however express an ontological position that both questions how science might contribute to understanding the 'complexity' of human systems and the inseparability of those systems from their environment and, equally importantly, highlights an enthusiasm and an energy for the endeavour. It is this willingness to question the logic of mechanistic reduction and everything else being equal; to accept uncertainty rather than fight it; and to be stimulated by, and enjoy, the research process that has been the key feature of working with Peter Allen. This working relationship, based on this shared ontology, has stretched over twenty years for two authors (Lemon and Jeffrey); and for some three years for the other (Snape). It has also led to a stimulating and ongoing dialogue about the relationship between contextual narratives and their incorporation into multi-scalar models for exploring (un)sustainable futures; it is this dialogue that underlies the following chapter.

Introduction: A systematic approach to complex systems

If we accept that the world is, and our personal worlds are, increasingly inter-connected and complex we must also recognise that prediction and thereby planning are risky ventures. This calls into question the premature adoption of reductionist approaches that compartmentalise and bound processes and problems by discipline, institutional structure, role, cost centre etc. and then only respond within the confines of those often arbitrary boundaries. It also questions the assumption of everything else being equal which underpins much educational, experimental and scientific endeavour. In a dynamic and continually changing world everything else is never equal; indeed it is the recognition of this that guards against reliance upon prediction and encourages openness to uncertainty and exploration.

At the same time this acceptance does not absolve us of a responsibility to understand and manage change, nor does it allow us to fall into a position of fatalistic indifference. The incompressibility of complex systems (Richardson *et al.*, 2000) does not mean that we can deal holistically with that complexity, but that we need to recognize the 'artificiality' of the boundaries that we draw in order to make sense of them. The ability and willingness to see the links and interactions that constitute 'whole pictures', even with a low level of resolution and certainty, should form the basis for anticipating possible futures and identifying the skills and competencies necessary for responding to them. This may require both integrity and humility if the most appropriate response is not grounded in the disciplinary expertise, or the interest, of those who have scoped the potential futures.

We choose in this chapter to consider a particular complex concept: sustainability. What does the position we have begun to outline mean for how we investigate, communicate and pursue sustainability? Is sustainability a product of the way we engage with the natural environment or is it a holistic and complex process that links the ecological, social and economic along the lines of the Triple Bottom Line; albeit with the former being non-negotiable, as advocated in the Five Capitals model of Forum for the Future who are now focusing their strategy on an appreciation of complex systems (Madden, 2011)? If we think of sustainability as a set of complex and emergent interactions we have to question how confident we can be about what might constitute a sustainable future; indeed the question might be better framed as "What is likely to be unsustainable?". Within this frame, the pursuit of sustainability becomes one of identifying what skills and competencies can best equip us to avoid such unsustainable possibilities or respond and adapt to them? This approach, and the conceptual and computer based models that are derived from it, are less concerned with detailed description of what exists, although that is obviously very important, than with exploring possible futures and learning about how systems might transform over time (Allen, 2010). Such reframing also raises a set of additional issues that influence how we might become more 'literate' (Stone & Barlow, 2005) in terms of sustainability; or more practically how can that literacy become embedded in individual and collective action at the local, contextualized, level. A number of tentative observations can be made about the link between local1 context(s) and modeled abstractions (Eisenhart *et*

---

1. Local is not necessarily geographically constrained but refers to restricted, or bounded, contexts such as the board of the regional social housing group.

al., 2010) at different scales (time, space and organization, Lemon *et al.*, 1998, 1999)) and what this might mean for exploring the (un)sustainability of different futures.

- Complex and interconnected processes inevitably throw up unique problem contexts which seldom have clear cut 'sustainable' solutions;
- If the essence of sustainability is to increase our adaptive capability we need to be able to anticipate potential futures (i.e., that to which we have not yet been exposed) and to learn from them;
- Educational endeavour cannot be restricted to understanding the unique and contextual; tools, techniques and concepts must be provided which can utilize that understanding in new contexts. Learning becomes an iterative process moving between the contextual and the conceptual, supported by systematic, replicable and accessible methodologies (Flyvbjerg, 2001).

The initial premise of this chapter might appear paradoxical. On the one hand we are arguing that something is lost as systems are deconstructed or dismantled (Cilliers, 1998; Allen, 2007) and that something qualitatively new invariably emerges from systemic interaction—in other words we are continually dealing with the contextual and the unique albeit at different scales. At the same time we have to develop systematic approaches to understanding the contextual and how it might (re)configure into the future. The learning from this process, and the associated literacy, will be cumulative through engagement with multiple narratives, either directly as agent or vicariously as observer, but for it to be transferable we have to draw upon conceptual and methodological tools. This chapter will initially introduce some core concepts from the complexity literature and will consider how these might help us make sense of a complex, and ongoing, phenomenon—the retrofitting of residential dwellings. We will then consider how specific cross-cutting skills are of crucial importance, not only for understanding the process as researchers but for practitioners, including action researchers (Reason & Bradbury, 2001) who are engaged in implementing it.

Before we focus on some of the concepts that the complexity literature has developed, adopted and adapted to help us understand complex phenomena a number of background observations need to be made. Firstly we want to consider those concepts in the light of seeking consensus around what constitutes an issue, or problem, and how it might be addressed; secondly and related to this we would like to outline the process of abstraction that draws upon those concepts and more substantive descriptive themes. We will then introduce an ongoing case study on the retrofitting of social housing to explore firstly how an understanding of this might be enhanced by non linear, futures models and secondly how that must be informed by and integrated with insight gained from exposure to multiple and reflexive narratives.

Conceptual frame: Desperately seeking consensus

This paper, possibly in common with many of the other chapters in this book, will argue that the increasingly complex and interconnected nature of our world means that researchers are bombarded with demands to adopt multi (involving various disciplines), cross (collaborative and integrative working) and trans (cross-cutting expertise) disciplinary approaches without a clear steer as to how such an approach might be undertaken. At the same time the grounded and real world nature of many of the wicked (Rittel & Webber, 1973) problems researchers are wrestling with means that engagement with problem owners or stakeholders is also of increasing relevance and needs to be reflected in the methodologies adopted. This additional layer of complexity is enhanced when the researcher is also an agent in the process under evaluation.

The relevance of multiple perspectives will be discussed below but for the time being our concern is with the lure of consensus over what constitutes a problem or issue. The striving to achieve such a comfortable state in effect extends the longstanding and ongoing debates over the relative merits of positivist, reductionist models as compared with constructivist and holistic ones. Such distinctions and debates are premature and self-serving; the dismantling of meaning, or the mechanistic deconstruction of a technical process, is secondary to the need to generate consensus around the whole which we interpret as the coexistence of different views and perspectives about context. In other words we should not define what a problem is before we have 'mapped' out the range of perceived contexts within which it might be located. Consensus becomes agreement about the whole, which is constituted by multiple interpretations, rather than the particular. Figure 1 illustrates this idea, with each area representing the traditional boundaries of four disciplines' perspective on a given problem. Too often consensus is seen as the intersection where all points of view coincide (A) and almost inevitably where nothing new is to be learnt, whereas a more productive approach to consensus would be to agree on the whole picture of differing perspectives (dashed line). Disciplines in this context are taken to include the perspectives of researchers and stakeholders in the story.

Consensus is usually treated as finding and isolating the region A: we advocate a view of the process that defines the entire area enclosed by the dashed line and reflects the different perspectives on that process. Understandable concern is raised

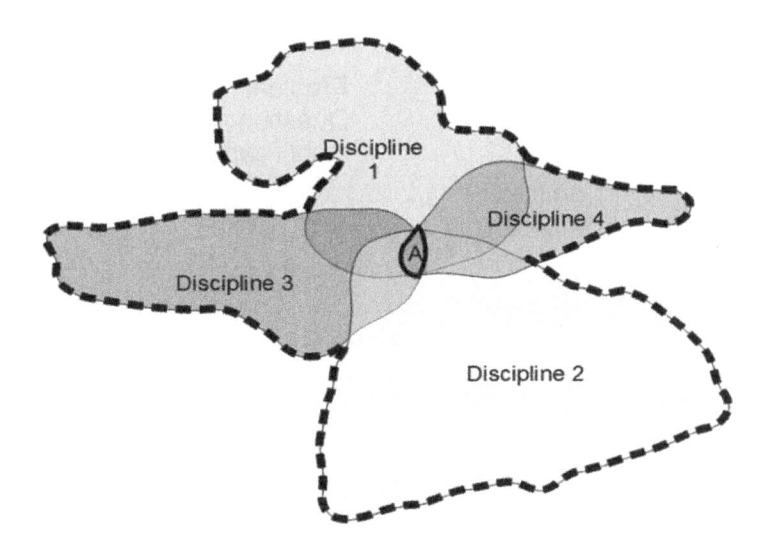

Figure 1 *Picturing the whole—consensus about difference.*

about the utility of such a 'messy' starting place; far better to use well defined disciplinary, or practitioner, paradigms to bound a problem and then to pursue the appropriate methodology to 'solve' it. We have some sympathy with such a concern and are not advocating the demise of the 'expert' or even rejecting reductionist responses; rather we are arguing that they are often employed too early. In prematurely deploying traditional experts to define the system, issue or problem at hand it is likely that we limit and prejudice the scope of that definition and, when applicable, the solution. It is especially important that this is explicitly recognized when considering complex systems where subtle interactions and individual actions can have far-reaching and unexpected effects.

Underpinning the establishment of consensus is the need to differentiate between an interconnected and complex environment or system (with multiple perspectives, emergent and irreversible behaviors, human and non-human actors) and the pursuit of 'solutions' often with a high level of complicatedness i.e., solutions that may require extremely high levels of expertise but which, with that expertise, can be deconstructed into their component parts. This suggests that our approach needs to explore the potential questions before advocating the requisite (often disciplinary) response; it also suggests that such a response, because it is reduced and bounded, can only be evaluated back in context. Thus, we have an iterative process between complex context—question / problem identification—potentially complicated response / solution and modified and new, and possibly unrecognizable, complex context (Figure 2).

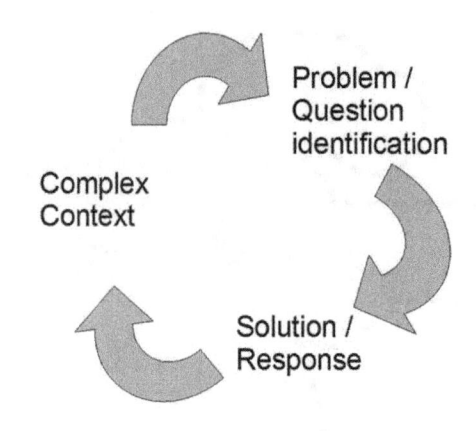

Figure 2 *Recursive problems, solution and context*

Are we talking about human centric phenomena and contexts necessarily? While we are undoubtedly grounding our ideas in complex systems that involve human actors and they may well be the most significant source of uncertainty; are people necessarily at the centre? We can explore the social and human components of a system and their interaction with other processes but the 'centrality' of that human agency may be an interesting question in the process of understanding, and modelling, of causality and emergence rather than a starting point. For example, while the urban environment is undoubtedly 'man-made' our responses to it—light, noise, comfort etc. are determined by the physical and natural phenomena as much as the social and organizational ones. In other words, if we are considering complex systems as multi-dimensional, irreversible and emergent it would seem problematic to assume that people are always at the centre even if we are assuming that they are invariably present somewhere in the exploration of what might constitute more sustainable 'human' systems.

The uniqueness of the contexts within which issues reside does indeed mean that they can provide us with interesting and diverse stories; they are constructed and we have suggested that consensus needs to relate to the totality of available constructions from which a specific research approach or intervention can be identified, and the implications of that approach clearly specified. Indeed, the concept of incompressibility (Cilliers, 1998) suggests that "it is impossible to have an account of a complex system that is less complex than the system without losing some of its aspects" (Richardson *et al.*, 2001: 533). However, our goal is the pursuit of a systematic method by which to understand complex systems and, as Lewis Carroll pointed out, a 1:1 map (or model) is useless or at least pointless[2]. While the methodology em-

---

2. And then came the grandest idea of all! We actually made a map of the country, on the scale of a

ployed should enable that complexity to be appreciated and potentially anticipated through the generation of scenarios there is also the need to compare unique and dynamic contexts in pursuit of cumulative knowledge. Richardson *et al.* (2001) briefly discuss this in terms of local knowledge and raise some interesting dilemmas about the adoption of context independent frameworks and related lexicons. However for the analyst and practitioner to make informed judgements about 'unique' and emerging contexts it is essential that conceptual devices are adopted and that 'knowledge' is generated through access to, and experience of, multiple contexts. These devices—such as incompressibility—provide us with a way of looking at complex phenomena, ideally supported by a systematic methodology; they are ways of providing a cumulative and comparative knowledge base that is not merely the accumulation of tacit and individualistic understanding. The paradox at the heart of complex contexts is to be able to learn from them and, systematically, to draw upon and advance that learning in the same 'location' over time or in different situations in time and space. The role of reflexive narratives and their integration into modeled explorations of the future is both fundamental and problematic. Uprichard and Byrnes (2006) suggest that such narratives can aid the conceptualization of complex models and or those models can themselves 'tell a story'. Elsewhere (Lemon *et al.*, 1998 & 1999) we have suggested that the role of qualitative understanding and narrative is both essential for providing the 'mess' (Law, 2004) of multiple contexts (from which an informed classification can be derived for modeling purposes) as well as making a fundamental contribution to interpreting the futures generated by those 'complex' models. We might visualize this as in Figure 3 below, although it is not suggested that all these levels are necessary for any given task.

There are two types of abstract conceptual devices that might help us create a unified understanding from unique, diverse and local experiences (Eisenhardt, 2010); firstly those that help characterize complex systems and underpin this view of the world, e.g., incompressibility, irreversibility and emergence, connectivity. Secondly, others provide a more descriptive and potentially 'measurable' framework for evaluating context and inevitably reduce complexity through categorization—for example describing context in terms of some or all of temporal and geographical scale, multiple perspectives, interaction and networks.

---

mile to the mile!" \ "Have you used it much?" I enquired. \ "It has never been spread out, yet," said Mein Herr: "the farmers objected: they said it would cover the whole country, and shut out the sunlight! So we now use the country itself, as its own map, and I assure you it does nearly as well. Lewis Carroll, Sylvie and Bruno Concluded (1893).

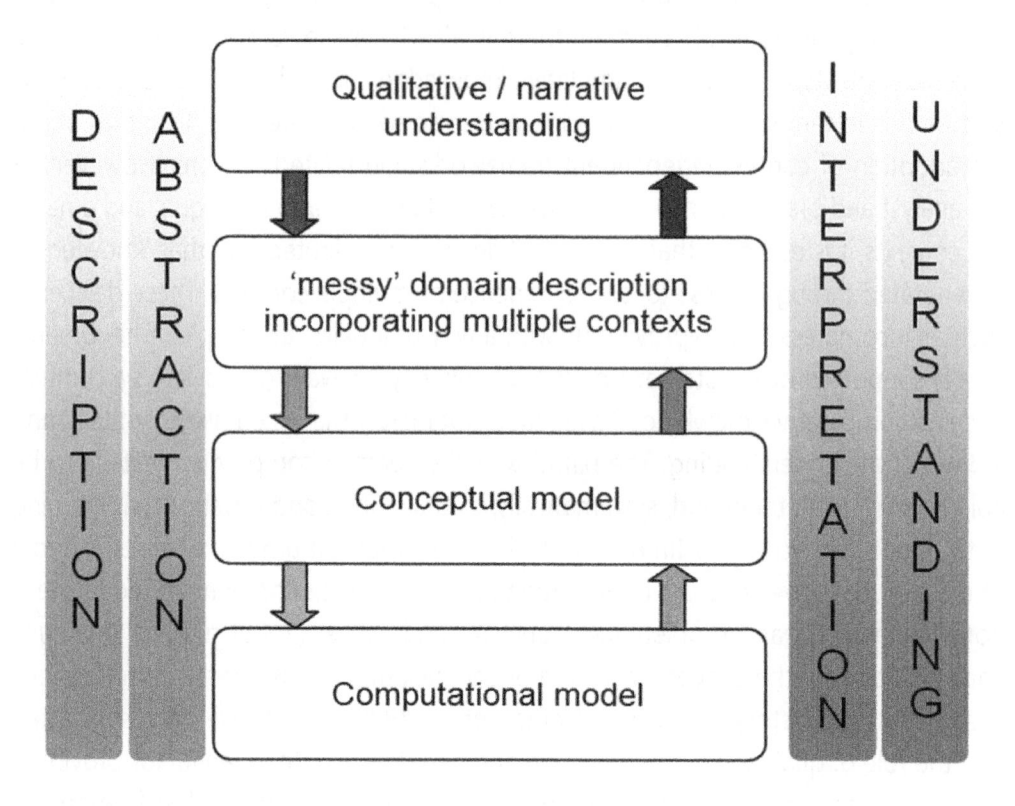

Figure 3 *Use of qualitative understanding and narrative for providing multiple contexts and interpreting modeled 'futures'*

We will now consider some of these devices that are central to Peter Allen's approach in the context of an ongoing study into the retrofitting of domestic social housing. The case is being used to exemplify the earlier discussion about conceptual devices being applied to a complex issue, it is not intended to convey a comprehensive picture of the retrofit process.

Case study: A story of retrofit

27% of UK carbon emissions are derived from housing and more than 60% of the 30 million+ houses that will be occupied in 2050 have already been built. Of that housing stock over 4.5 million homes are in the social housing sector which with the help of government schemes such as the Decent Homes Programme has become possibly the most energy efficient performer in the residential sector (TSB, 2009). That said even if the number of homes remains stable, those currently built will have to be retrofitted in line with overall carbon reduction targets (80% reduction by 2050) equating to one property being completed every few minutes in the intervening period. The practical implications of achieving such figures are significant to say the least; we begin to detect some of the characteristic features of complex systems described above. Multiple spatial and temporal scales interact (national policy and take-up, sectoral distinctiveness, household attitude and culture, building's physical characteristics) along with many traditional perspectives and disciplines (legislator, landlord, builder, tenant...; politics, economics, physics, engineering, sociology etc.).

In 2010 the Technology Strategy Board (TSB) through its Retrofit for the Future programme commissioned a number of refurbishment projects on social housing properties. The aim of this was to develop whole property solutions for achieving significant reductions in carbon through enhanced building efficiency and to generate a knowledge base that will support the effective scaling up of the retrofit process in the near future. De Montfort University (IESD) was engaged in several of these projects with Social Housing partners who had substantial portfolios of properties in the North East and East Midlands. One of these properties was a Victorian terrace in Leicester and we will concentrate on this in the following discussion.

A number of issues were immediately raised relating to the nature of the call and the practice of Social Housing. The housing association undertakes approximately one refurbishment a week out of its stock of over 15,000 dwellings; these are carried out in line with the requirements of the Decent Homes Programme to improve the housing conditions of vulnerable tenants. The work is undertaken by contractors who have developed considerable expertise in the effective (time, quality and economic) implementation of housing refurbishment (e.g., standard kitchen and toilet improvements). However, neither they, nor the housing provider had experience of retrofitting for carbon reduction. Core concepts in energy efficient dwellings such as air tightness were not part of the 'language' of either the association's property manager or the project manager for the contractors; this extended to the practical skills of the onsite work force.

I never thought of, or understood the relevance of, air tightness before, but now we will carry out tests on all our properties .....Initially the lads got fed up with me, e.g., over air tightness; but after time they got it and became obsessive themselves (Property Manager for the housing group).

It was recognized that the concepts—and multiple understandings of the same task—needed to be communicated before they could be incorporated into a new 'culture' of working. This did not relate necessarily to bad practice, e.g., the production of waste on site but to the changing of 'habitual' good practice. For example an insulation membrane, skirt, was trimmed to size rather than leave the excess to enable an airtight corner joint; also the properties of the materials were not clearly understood e.g., avoiding piercing or cutting into insulation panels. These knowledge issues i.e., understanding what is meant by low carbon interventions and the skills necessary for achieving that aim, were considered to be the main reason for the project over-running in time and budget.

This also highlighted a cultural difference between those team members with a background in manufacturing and those with one in construction—acceptable tolerances were far lower for the former who were keen to adopt an off-site approach to retrofit than for the latter who felt that the only practical way to deal with the 'variability' within buildings was to undertake the work on-site. This debate was exemplified by the decision to incorporate an additional room in the roof and the decision about how this should be facilitated (Figures 4 and 5). An offsite solution was decided upon; this was in large part due to confidence in the existence of manufacturers who would be able to produce a bespoke roof pod. This confidence was misplaced and it was only through the pursuit of a network of acquaintances that a small business was identified that could, and would, develop the product.

The roof pod was the product of a number of interconnected antecedents; firstly the retrofitting of the building to virtually a zero carbon specification had led to the extensive use of 'bulky' insulation materials and corresponding reduction in room size and available space. Secondly, many of the tenants of the housing associations had cultural requirements for two sitting rooms—male and female. This necessitated a significant redesign of the property (Figure 4) with the roof pod innovation.

A final set of issues can also be raised relating to this vignette. The Retrofit for the Future programme was introduced to identify ways forward for the social housing sector in reducing the carbon footprint of its housing stock. The extent of the problem outlined above (i.e., retrofitting 4.5 million homes) is exacerbated when one considers that the majority of properties are occupied, thereby either restricting what can be done, and how it is done, with householders in–situ or decamping residents with the associated practical and economic implications. Another feature of many social housing properties is that they are often disparately spread (pepper-potted) throughout communities and, when this is the case, do not lend themselves to the economies of scale available to contiguous dwellings.

This condensed, local and very 'specific'—one dwelling—story, or researcher narrative highlights a range of apparently disparate systems that interconnect and (re)

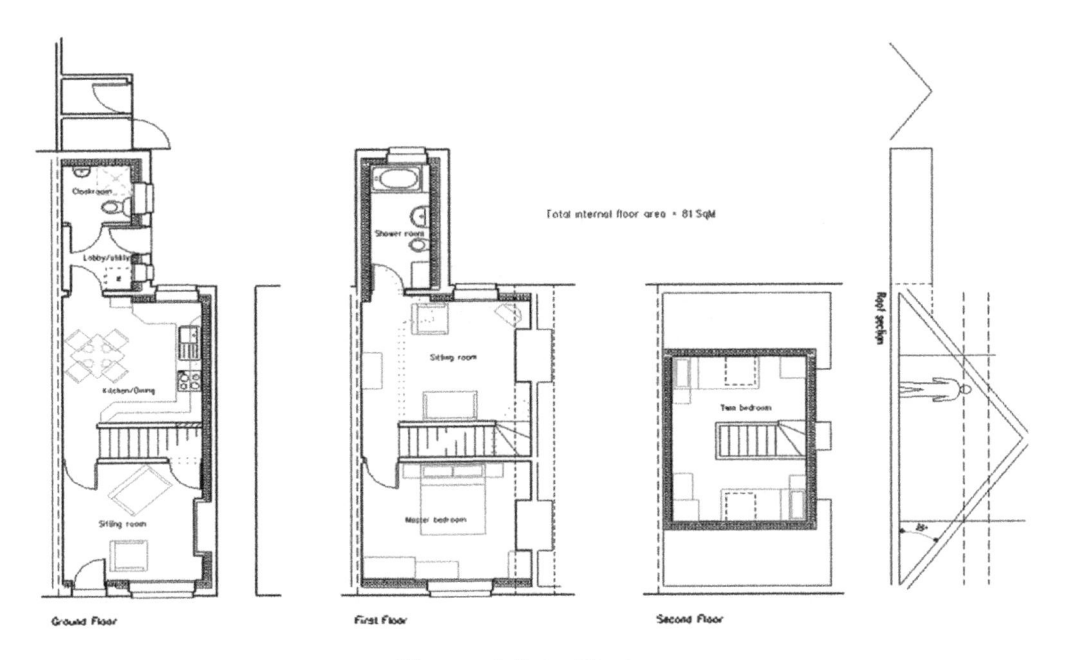

Figure 4 *Retrofit plans*

configure to generate that unique context. Within that set of interconnections there are multiple meanings and perceptions many of which do not overlap (e.g., as in Figure 1). The desire for a second lounge is unlikely to be related to the ability to produce a bespoke pod or to select, source and fix insulation effectively.

We will continue to refer to this story and consider how a selection of conceptual devices can help us make sense of such unique contexts in order to intervene more effectively in them, if that is required; more importantly however we want to explore how such unique stories can inform a modeling process designed to explore potential futures and how sustainable they might be.

Boundaries, scale and connectivity

If, as was the case outlined above, we bound the retrofit process at the perimeter of a Victorian terrace a number of legitimate questions can be asked about its effectiveness—did it come in under budget (no), on time (no), did it result in the anticipated reduction in energy use and thereby carbon (almost) and will it continue to deliver benefits over time (uncertain)? The project was in effect an experiment, bounded and protected by grant aid and the absence of tenants. The portfolio of the social housing group, however, comprises over fifteen thousand dwellings, almost all with tenants in-situ and significant variation in location, building characteristics, energy sources and occupant demographics. It is not possible to simply scale up the learning from this

Figure 5 *Installing the roof pod*

'unique' project to the rest of the group's portfolio let alone to the four million plus social housing dwellings that will need to be retrofitted before 2050. In other words fifty single property projects supported by the Retrofit for the Future programme are unlikely to give a clear direction for meeting carbon reduction targets. What they can do is to provide insight into how different materials might perform—e.g., wall and floor insulation, window and door seals etc and what knowledge gaps existed among site workers, estate managers, architects and occupants when they were involved in the project.

If the learning from the project is divided into independent components e.g., training of site workers and managers, decisions about the wider adoption of new materials and innovations then it is likely that incremental improvements will be made (in time, cost and carbon reduction) but questionable about whether these can be up-scaled to the necessary level. What can also happen is that the up-scaling can become 'locked–in' to particular modus operandi or technologies. For example, in the field of sustainability, emphasis on renewable electricity generation may well have been reinforced by Feed in Tarrifs whereas the Green Deal (implemented in 2012 following the onsite work) is focused on building efficiency. Investments in the technology for the former may have removed the finance and incentive to pursue the latter; one is tied into generating energy, albeit from renewable sources, the other to the introduction of the energy saving technologies available at present.

We have discussed the interconnectedness of economic (e.g., costs), ecological (e.g., carbon production), cultural (e.g., practices of on-site workers and requirements of ethnically diverse householders) technological (e.g., willingness and ability to trial innovative materials and procedures; see Jeffrey & Seaton, 2004) and political (e.g., policy interventions and targets) factors within a single retrofit project. These interconnections, or configuration of attributes, will 'play out' differently in another context with a different internal configuration of those attributes and relationship with its environment (Allen, 2010)—i.e., that which resides outside of the prescribed system boundary, in this case the retrofit building project. For example, the on-site team may have access to expertise in air-tightness or to a company wishing to trial thinner insulation materials. We can identify two distinct but, of course related, dimensions of connectivity; those that are evident within the 'system of interest' and those which impact on it from the environment. It can be suggested that the former connections (re)configure to form the 'messy' story or narrative whereas the latter constitute processes operating at multiple scales that can only be reproduced through conceptual or computer based models.

The use of the term scale in this discussion is meant to signify different indications of time and space as a way of bounding the process we are looking at. This helps with classification rather than being a characteristic of the system as for example with the hierarchical structures discussed by Holling (2001) or those of technological transition (Geels, 2002). Temporal scale is both the arrow of time (e.g., two hours to decide upon the use of a roof pod, one month searching for a supplier, four hours to install and, hopefully, twenty to thirty years in place) and the dynamic of a process (e.g., this was the first pod of its type to be installed in the UK, more have since been installed and it is now advertised on the web site of a large DIY chain indicating the possibility of rapid expansion). Spatial scale is both geographical (e.g., contiguous and or pepper potted housing in one city or across a region) and organizational structure (e.g., the on site management of properties by local managers through to the strategic management of the group across the region by senior executives).

It is interesting to consider the range of related multi-scalar modeling work undertaken by Peter Allen in the light of the case presented above. For example the restructuring of urban form (Allen & Sanglier, 1981)—in this case relating to social housing and the economic and demographic characteristics of it; the complexity of innovation processes and technology paths (Allen, 1989)—in this case relating to the take up of off-site manufacturing for the roof pod and the receptivity to state of the art insulation materials) and the inter-relation of technical and environmental phenomena (Allen, 1999)—in this case the impact of retrofit on carbon production. While these examples highlight the interconnected nature of phenomena within complex systems, and it would be interesting to see if other starting points would also lead to a range of relevant modeling activities, we are aware that the models provide aggregate representations of organizational, or hierarchical, scale whereas agent interactions are invariably interdependent and multi-level. Scale in this sense is a boundary condition; it is a practical modeling device that raises questions about how we can better represent nonlinear multilevel interactions (Uprichard & Byrne, 2006).

Agency, knowledge and multiple perspectives

The coevolution of individual agency and macro-structure is central to Peter Allen's work; latterly he has articulated the concept of 'structural attractors' that are configurations or ecologies of behavior, beliefs and strategies, clustered in a mutually consistent way and characterized by a mixture of competition and symbiosis (Allen, 2010: 20). Within this nested hierarchy there are a multitude of perspectives or subjectivities and motivations that influence micro-level behavior and are informed by past experience and (mis)information. For example, the motivation of the new tenant in the retrofitted property was for refurbished accommodation with low energy bills; it did not relate to the carbon reduction objective. Indeed some teething problems with the water heating were seen to be caused, without foundation, by the smaller boiler—a perception that if shared with other tenants could hinder subsequent take-up.

Allen (2010) continues that where the motivations or expectations are not met then behaviors will be modified and or confusion generated. This reflexive process generates micro-diversity which in turn leads to restructuring and uncertainty at different hierarchical scales; for example if the negative view associated with retrofit became the norm among any stakeholder groups then a form of pathological autopoeisis (Gregory, 2006) might occur. We have already suggested that this hierarchical structure is essentially fluid and futures models need to be able to reflect directly or indirectly micro-diversity and local context and the structural coupling (Leydesdorff, 2000) that may occur between social systems and between those systems and the environment (Gren & Zierhofer (2003). We would also suggest that this coupling is not just between social agents but that non-human actors, at different scales, have to be included in the analysis.

Let us initially take a human centric example, that of trust. Newell and Swan (2000) distinguish between competency based and relationship based trust. The discussion over whether the roof pod should be produced on site using traditional construction methods or off site using modern methods resulted in a decision to pursue the latter approach based largely on the project manager's off-site manufacturing experience and the trust of the other team members in him as a person and his expertise in modern methods of construction (MMC). This trust was compromised when no large scale pod manufacturers were interested in producing a bespoke product and a small and apparently unsophisticated company were found and commissioned to undertake the work.

We were impressed with them—even if they built it in a shed. Their enthusiasm and knowledge re-invigorated the project; can do attitude. Before this we had nothing—build from scratch, no pod or them. I did not believe the tolerances could be achieved—using a plumb line, in the hot loft for four hours—rafters out by 1.5 degrees on four metres—very impressive and confidence revived—albeit with some subsequent problems. Mad rush. (Property manager for the Social Housing Group)

Paradoxically the extensive and traditional skills of the manufacturer—e.g., using a plumb line rather than laser measurement and producing the pod in workshop conditions—enhanced trust in the decisions to incorporate it in the design and to produce it off-site even though the rationale was that this would be a more state of the art approach. As mentioned earlier, negotiations are now underway to mass produce the pod through an MMC process with associated implications for how retrofit might be undertaken into the future. While the purpose of the retrofit programme is to highlight technologies and practice that can be disseminated downwards from policy (landscape), to regime (sector / organizations / local government) and to local decisions this example highlights the two way and iterative nature of such a multi-level perspective (Geels, 2002).

We can argue that the pod has become a key actor not only in terms of aesthetics and the use of space, but through carbon reduction and energy efficiency within buildings and as the potential driver of a new manufacturing opportunity (customized manufacturing for retrofit) with associated implications for job creation and changes in the construction culture. It also provided a significant publicity opportunity for the project, the programme and the approach with television, radio and press coverage. The pod is therefore an actor, and an agent of change, at multiple and interrelated scales ranging from the individual retrofit building through to potential influence on how the retrofit process might be scaled up. In common with a wide range of other materials and processes it has a dynamic relationship with human actors and, in common with those other physical and technological actors in the narrative, it is not therefore a static object waiting to be 'done to' (Lemon et al., 1998). In the same way that physical and technological actors mirror human ones in their variety of interconnections and causal relations they also reflect them in the way we associate and interrelate with them—e.g., do we trust them to perform in the way we expect, 'do they do what it says on the tin'? Our perceptions of this capability, and any experience we have with them, will influence subsequent decisions and behaviors, reinforced by evidence of an expanding or declining physical presence and positive or negative communication between actors or decision makers who can influence take-up and diffusion.

The example of the roof pod, and more particularly the quotation by the property manager above, highlight another issue that is central to Peter Allen's work, the role of knowledge and ignorance in generating or constraining adaptive capability (Allen, 2001). Knowledge of who might manufacture the pod did not come from the direct network of contacts but through a secondary contact consistent with the strength of Granovettor's seminal weak ties (Granovettor, 1973); the tacit knowledge used to measure the roof space using a plumb line was not shared but formed the basis for trust in the competency of the installer. The recent move towards the customized mass production of the pod has raised a fundamental problem about the accurate measurement of slightly differing roof space and led to experimentation with state of the art point cloud data derived from specialist laser scanning. The core function of both approaches was to measure accurately the underlying drivers however they differed between the initial need for credibility and the subsequent need to mass produce. Knowledge has been a continual theme throughout the retrofit story e.g., the assimilation and transfer of knowledge about air tightness and carbon reduction among the property management and on site work force and the translation of that knowledge into modified site worker, and subsequently householder, behavior. This knowledge conversion (Nonaka & Takeuchi, 1995) may have been formal and explicit as in the provision of householder manuals and or tacit and informal through shared on site experience. There is insufficient space in this short paper to adequately cover the transfer of knowledge at the same structural level and between different hierarchical levels or to explore the fundamental importance of that transfer for new behavior, modified interpretive frameworks and thereby new knowledge requirements. It is however this ongoing dynamic, and the challenges generated by it, that underpins the quote from Peter Allen (2010) that we have used at the start of this chapter.

Conclusions: Models, meaning and narratives

This paper has emerged out of the authors' sympathy with Peter Allen's approach and our experience of working with him. In particular we recognize the significance of micro-diversity as the basis of complex multi-scalar interactions and their representation in models for exploring (un)sustainable futures. However we question whether such micro-diversity is, can, or should be incorporated fully into this modeling activity and moving on from this can the subtlety and nuance of complex micro level causation be reproduced by such models rather than accepting their undoubted potential for generating multiple futures with diverse, and potentially new attractors? Our case study demonstrates some of the features of Peter's approach to research. We see the complexity of multiple perspectives and interactions alongside the porosity of whatever boundaries are drawn for the purposes of study—while we draw the boundary at the house, who can predict how the knowledge gained by all during the process will affect paths to sustainability as it is taken to other contexts? However, we can analyse and describe the processes undertaken and make some qualified statements about what might render the course of action followed in the case study unsustainable.

Following Uprichard and Byrne (2006), we observe that:

1. The nuance mentioned above relates to multiple non linear interaction across and within levels. Indeed the levels themselves are artificial boundary conditions, albeit on occasions ones with tangible perimeters (administrative boundaries, acceptable suppliers etc.). Models, of necessity, must incorporate a limited number of clearly defined levels and their permeability is difficult to reproduce;
2. The ability to integrate the measurable with other data that convey meaning is problematic and can only be effectively modeled through abstractions and the creation of quantifiable representations;
3. We need to understand and follow the reflexive nature of individual and collective action and question whether non linear, multi-scalar models can do this.

In light of the above we argue strongly in favour of such non-linear modeling approaches for exploring (un)sustainable futures however we also argue that the diversity of individual and collective experiences, and the narratives associated with them, should be used on both the input and the output side of such a modeling endeavour. In defining a model, narratives and qualitative understanding should be used to help refine the core attributes within it (e.g., the subtleties of culture on behavior may by necessity be only implicitly represented by specific parameters that emerge from that behavior e.g., income, carbon generation). Secondly, those narratives and the insight from them should be used to help explore and explain the scenarios generated by the models and thirdly we need to more effectively involve stakeholders, local level actors, at the specification and interpretation stages of contextually grounded models. This leads to an additional, and we consider fascinating area of questions; do the same conditions exist for non human actors and how can the multidimensional nature of their influence, and how they are influenced, be represented?

We agree that there is a need to change our thinking from 'being' to becoming' and feel that this is key to gaining a handle on what might prove to be unsustainable in the future and identifying the knowledge required to generate the necessary adaptive capability to avoid those futures. We also recognize that Peter Allen sees local knowledge and local diversity as the driver of structural change, which of course then impacts upon those local contexts. Where we are less clear is exactly how that local diversity or those local contexts and narratives fit into futures models.

References

Allen, P. and Sanglier, M. (1981). "Urban evolution, self-organization and decision-making", *Environment and Planning A*, ISSN 0308-518X, 21: 167-83.

Allen P. (1989). "Modelling innovation and change" in S. van der Leeuw and R. Torrance (eds.), *What's New: a Closer look at the Process of Innovation*, ISBN 0044451431.

Allen P. (1999). "Policy relevant modelling in the Argolid: from sociological investigation to crop choice model," in M. Lemon (ed.), *Exploring Environmental Change Using an Integrative Method*, ISBN 9056991930.

Allen, P. (2001). "Knowledge, learning and ignorance," *Emergence*, ISSN 1521-3250, 2(4): 78-103.

Allen, P. (2007). "The complexity of change and adaptation," *Emergence: Complexity & Organization*, ISSN 1521-3250, 9(3): vi-vii.

Allen, P. (2010). What is the science of complexity? in A. Tait and K. Richardson (eds.), *Complexity and Knowledge Management*, ISBN 1607523558.

Cilliers, P. (1998). *Complexity and Postmodernism: Understanding Complex Systems*, ISBN 0415152879.

Eisenhardt, K., Furr, N. and Bingham, C. (2010). "Microfoundations of performance: Balancing efficiency and flexibility in dynamic environments," *Organization Science*, ISSN 1047-7039, 21(6): 1263-1273.

Flyvbjerg, B. (2001). *Making Social Science Matter: Why Social Inquiry Fails and How It Can Succeed Again*, ISBN 0521772680.

Geels, F. (2002). "Technological transitions as evolutionary reconfiguration processes: a multi-level perspective and a case study," *Research Policy*, ISSN 0048-7333, 31: 1257-1274.

Granovetter, M. (1973). "The strength of weak ties," *American Journal of Sociology*, ISSN 0002-9602, 78: 1360-1380.

Gregory, A. (2006). "The state we are in: Insights from autopoeisis and complexity theory," *Management Decision*, ISSN 0025-1747, 44(7): 962-972.

Gren, M. and Zierhofer, W. (2003). "The unity of difference: A critical appraisal of Niklas Luhmann's theory of social systems in the context of corporeality and spatiality," *Environment and Planning A*, ISSN 0308-518X, 35: 615-630.

Holling, C.S. (2001). "Understanding the complexity of economic, ecological, and social systems," *Ecosystems*, ISSN 1432-9840, 4: 390-405.

Jeffrey, P. and Seaton, R.A.F. (2004). "A conceptual model of receptivity applied to the design and deployment of water policy mechanisms," *Environmental Sciences*, ISSN 1001-0742, 1(3): 277-300.

Law, J. (1992). "Notes on the theory of Actor-Network: Ordering strategy and hetereogeneity," *Systems Practice*, ISSN 1094-429X, 5(4): 379-393.

Law, J. (2004). *After Method: Mess in Social Science Research*, ISBN 0415341752.

Lemon, M. Jeffrey, P. and Seaton, R. (1998). "Deconstructing the orange: The evolution of an agricultural milieu in Southern Greece," *Int. J. Sustainable Development*, ISSN 1741-5268, 1(1): 9-23.

Lemon, M. (ed) (1999). *Exploring Environmental Change Using an Integrative Method*, ISBN 9056991930.

Leydesdorff, L. (2000). "Luhmann, Habermas and the Theory of Communication," *Systems Research and Behavioral Science*, ISSN 1092-7026, 17(3): 273-288.

Newell, S. and Swan, J. (2000). "Trust and inter-organizational networking," *Human Relations*, ISSN 0018-7267, 53(10): 1287-1328.

Nonaka, I. & Takeuchi, H. (1995). *The Knowledge Creating Company*, ISBN 0195092694.

Madden, P. (2011). "Systems shift," http://sd.defra.gov.uk/2011/06/systems-shift/.

Reason, P. and Bradbury, H. (eds.) (2001). *Handbook of Action Research: Participative inquiry and Practice*, ISBN 0761966455.

Richardson, K., Cilliers, P. and Lissack, M. (2001). "Complexity science: A grey science for the 'stuff' in between," *1st International Conference on Systems Thinking in Management*, Geelong, Australia, 8-10 November pp 532-537.

Rittel, H. and Webber, M. (1973). "Dilemmas in a general theory of planning," *Policy Sciences*, ISSN 0032-2687, 4: 155-159.

Stone, M. and Barlow, Z. (eds.) (2005). *Ecological Literacy: Educating Our Children for a Sustainable World*, ISBN 1578051533.

TSB (2009). Retrofit for the future: Competition to cut carbon emissions in social housing.

Uprichard, E. and Byrne, D. (2006). "Representing complex places: a narrative approach," *Environment and Planning A*, ISSN 0308-518X, 38: 665-676.

Mark Lemon is a social scientist whose research over the past twenty years has covered a range of policy relevant issues relating to the human – technical interface particularly as it affects the natural environment and sustainable development. He is interested in the characteristics of integrative research and the development of trans-disciplinary, cross-cutting, skills and has published extensively in this area. He is currently pursuing research into the understanding of sustainable urban design. He has supervised over twenty Doctoral and MPhil students on a wide range of topics linked to understanding socio-technical aspects of the environment. He is currently leader for the Resource Use and Pollution module on the Climate Change and Sustainable Development Masters in IESD. He has over one hundred publications including over thirty in peer reviewed journals.

Paul Jeffrey, his research interests encompass the development of sustainable water use arrangements and the relationships between human (e.g., socio-cultural, psychological, behavioural, economic), natural (e.g., water quality, environmental) and technological (engineering, technology & infrastructure design) dimensions of water management. He has contributed to over 100 journal & conference publications in fields as diverse as water resources management, science & society, technology assessment, social justice, and complex systems. He holds a first class honours degree in Science & Society from Middlesex University as well as an MSc in Energy and Environment and a doctorate in Technology Policy from Cranfield. As a post-doc researcher he spent three years studying sustainable development issues at the Hebrew University of Jerusalem, returning to Cranfield in 1996.

J. Richard Snape, MEng, MA(Cantab.), is a research student at the Institute of Energy and Sustainable Development, De Montfort University. Following a first degree in Electronic Engineering, Richard worked in the railway industry before returning to academia. His research interests include sustainability, computational social simulation, agent based modelling and complexity.

Chapter Δ

Scaling in city systems

Michael Batty

We develop the idea that many distributions characterizing the attributes of human systems in general and city systems in particular are scaling. The classic example is the city size distribution, first demonstrated as scaling by Zipf (1949) for the evolution of US cities from 1790 to 1930, and popularized as the rank size distribution. Here we argue that the regularity posed by such distributions at single cross sections in time which appears unvarying over time, reveals a hidden micro volatility with respect to the elements or objects that compose such distributions. To illustrate these ideas, we introduce the notion of the rank clock which shows the degree to which a particular city or object changes its position over time. We demonstrate this first for population and income in metropolitan statistical areas (MSAs) in the US from 1969 to 2008 and then examine changes in population size for the US Census from 1790 to 2000. This leads us to consider distributions other than cities—skyscrapers in New York City, volumes of trips passing through stations defining public rail transport in London, and finally the distribution of firms by revenue from the Fortune 500. To link the various mnemonics we introduce to one another, we conclude with our example of the Rank Clock Visualizer which suggest directions for future research.

Regularities in city systems

A considerable body of research into city systems developed over the last 50 years and longer has drawn on classical analogies with Newtonian mechanics based on concepts of gravity and potential. Ideas about force, mass and potential have been instrumental in suggesting how individual populations associated with different locations ranging from small areas defining locations in space such as residential and commercial centres to entire cities and countries, interact with one another (Batty, 2013). The kinds of patterns that have been simulated tend to be those that are observable at a cross section of time and thus the focus has been on how the elements comprising such patterns scale with size. In particular power laws have been widely used to illustrate how the frequency of distinct locations and cities are organized with respect to their size measured directly or through their ranks, and how movements or interactions between such locations scale with the distance at which they are spaced. In general, the spatial landscape that has resulted and which is present to a greater or lesser degree everywhere in the contemporary and historical world consists of smaller numbers of larger and denser agglomerations which exert an effect on their smaller neighbors spaced in such a way that smaller clusters are nearer to one another than larger clusters which are further away. The classic landscape of the city system is thus defined by a hierarchy of ever larger hinterlands associated with ever larger cities. These scale in such a way that there are fewer larger cities and many smaller ones, usually following a power law which can be represented in its traditional form as a probability distribution of city sizes or in its cumulative form as a rank frequency.

This approach to cities is loosely referred to as social physics in that the regularities that are revealed are described by power laws of various kinds. The theories that have been developed are largely phenomenological in nature but much of this spatial logic is consistent with theories of the urban economy. These explain the way activities inside cities as well as comprising whole cities tend to be organized around dense clusters of highly accessible locations which fall off in density as smaller clusters develop at further distances from those that comprise the poles of the urban system. Central place theory after Christaller (1933) is consistent with this structure of cities organized in economic terms following the way land uses organise themselves around central markets. This is the theory originally due to von Thunen (1826), subsequently widely developed as the basis for urban economics from Alonso (1964) onwards.

This was more or less how we articulated the world of cities and city systems around 40 years ago. It was a world which is many senses was timeless in that the concern for pattern at a cross section in time was paramount while the notion that cities were systems in equilibrium and could be explained as such was the dominant viewpoint. In one sense, this is not surprising in that superficially the built environment and the urban morphology of cities changes much more slowly than their economic and social structure and as much of our interest was in this superficial appearance—as indeed was much of the focus of urban planning, there was an implication that one could explain a lot using the notion that cities were in equilibrium. Moreover this implied that if one were to change the urban system through planning, then it would adjust quickly enough to a new urban equilibrium. Cities were thus imagined to be well behaved in terms of their temporal dynamics. In fact, this is also quite surprising, not least because there had been very substantial economic and social changes during the first half of the 20th century although despite this and despite that fact that new technologies were forcing cities to spread out and get ever larger, cities still appeared similar in structure to what had been characteristic of their form since classical times and before. Only now at the beginning of the 21st century are we seeing some massive changes in their structure as we begin to grapple with new kinds of dynamics posed as much by the digital world as the way their economy and demography is changing.

It was into this milieu that Peter Allen and his colleagues from the Free University in Brussels threw down the gauntlet that set the world of urban modelling on course to end its focus on thinking of the city as being in equilibrium. In fact, several commentators and researchers were uncomfortable with the notion that temporal dynamics was at best implicit in both the theory and models that were being developed. These kinds of social physics and urban economics were then being used to fashion a series of simulation models to be used predictively to fashion new kinds of city plans and to test the impact of various urban policies designed to solve problems of poverty, transport congestion, and economic development, all of which were intrinsically dynamic in nature. Peter's approach in fact was much broader and deeper than simply proposing that static models be made dynamic by extending them to deal with temporal increments. It was based on the idea that change in urban systems was composed of qualitative innovations, that change was surprising and often novel, hence unexpected and that such radical shifts should be part and parcel of the machinery of urban prediction. In fact what Peter did and this is best articulated in his book Cities and Regions as Self-Organizing Systems: Models of Complexity (Allen, 1997), is to embed the notion that a random event could trigger such a change in the trajectory of the urban system at any point and that if certain criteria were triggered, a point of bifurcation—a switch in the direction of the trajectory of the system—would be encountered. In fact this notion cast doubt on the fact that models of cities even in their extended quasi dynamic mode, could generate such surprising changes, changes in fact that appear to happen in cities across many scales and dimensions as we will show here for entire systems of cities and related phenomena.

Although Peter Allen, Alan Wilson (1981), Peter Nijkamp and Aura Reggiani (1992) amongst others pushed the field towards dynamics, parallel developments in the complexity sciences tended to change the focus yet again. Onto the agenda came a concern for systems driven from the bottom up. Cities in this context were intrinsically dynamic composed of agents acting at the individual level from which the kinds of patterns that social physics grappled with emerged. The notion that many millions of small scale decisions would generate pattern and order in the spatial as well as other domains of the city system was highly consistent with earlier ideas about social physics. Emergent patterns generated from the bottom up manifested strong regularities implied in the sorts of power laws—order across size and scale marking the earlier theories of how city systems were spatially structured—that appeared to persist at the macro level for many cities and city systems over long time periods.

Developments in complexity theory which treat human and spatial systems as intrinsically non-linear, reveal a dynamics that is generated from the bottom up, implying a slight paradox in that patterns that emerge from the mix, often do not reveal any easily understood aggregate dynamics although being consistently associated with such a dynamics. In short, observing spatial patterns that are generated by micro processes reveals considerable self-similarity at different scales which generate self-similar forms and which imply a kind of stability that is oddly inconsistent with dynamic systems. In fact it appears that although some of these patterns are fractal in the large, they are not so in the small. Although they reveal massive regularity in the aggregate, the elements reveal massive heterogeneity and change in their relative positions and sizes. In short, fractal hierarchies that remain comparatively stable with the respect to how cities are organized and how they scale relative to one another achieve this macro regularity while at the same time admitting rather basic volatility between the elements that make up such patterns.

Let us state this paradox in starker terms: if we rank a set of cities by their population size that describes some sort of integrated city system—that is they are spatially complete in that they occupy a territory, a nation, a continent—the distribution tends to follow a power law which has strong regularity from time period to time period. In fact this regularity is unerringly strong so much so that Paul Krugman (1996) was prompted to say: "The size distribution of cities in the United States is startlingly well described by a simple power law: the number of cities whose population exceeds P is proportional to $1/P$. This simple regularity is puzzling; even more puzzling is the fact that it has apparently remained true for at least the past century." However if we were to examine the size distributions and the cities that compose them at any two times, we would find that cities moved quite quickly in terms of their ranks (and of course their sizes). Taking the two distributions of cities in the US in 1890 and 1990 which relates to Krugman's observation, in 1890 New York City was number 1 as it was in 1990. But Houston was not in the list of the top 100 cities in 1890, yet it had reached number 4 by 1990. In terms of the top 50 cities in the world at the time of the fall of Constantinople in 1453, only 6 remain today. This micro-volatility in the face of macro-stability implicit in the power law is puzzling to say the least in that we do not have good theories of why city systems can maintain their aggregate stability while at the same time shuffling the objects that make up this stability in such a way that the overall scaling appears almost static. It clearly relates to competition between the objects in some way that suggests that the relative size of any object is always constrained by some upper resource limit that remains largely undefined.

Dynamics was rarely considered in city systems prior to the development of non-linear thinking which resulted in ideas about catastrophe and bifurcations and it was Peter Allen and his group that alerted us not only to its importance but the mysterious properties which could be activated by chance events. In fact, power laws can easily be explained in part by the random growth of a set of objects and although such explanations are almost nihilistic, they do provide a starting point. In this essay, we will not develop the kinds of dynamics that Peter is associated with apart from involving randomness as a default explanation but we will explore the paradox of micro-volatility in the face of macro-stability that we have just illustrated and we will illustrate this for variety of competitive situations in cities where locations compete for pride of place and where only a handful can ever be significant. In fact, the kinds of bifurcations and innovations that characterise the dynamics that Peter is associated with might be used to explain particular moves up and down this hierarchy although our focus will not be on this kind of explanation per se. All we wish to do here is to pose the problem and illustrate its perplexing nature.

In what follows, we will first state the nature of the power law and examine its properties. Power laws exist everywhere in human systems and we will explore two varieties—first the laws that govern the frequency of sizes—in particular city sizes, but also the sizes of high buildings, hubs in transport networks, firms, and so on—and then second the laws that associate two or more size distributions with one another that can also show regularity and nonlinearity as for example in allometry. Our foray will take us first into city systems where we will illustrate the dynamics of power laws associated with city sizes and then we will illustrate a rather different dynamics associated with comparing the attributes of two size distributions associated with the same set of cities, in this particular case, income and population size which we find for the US urban system to be related by a positive allometric law. We will then explore three rather different systems. First we will explore high buildings that are distinguished by the fact that newer buildings tend to be higher than old while a building rarely declines in height due to the fact that high buildings tend to be demolished if they are changed at all. This poses a rather different dynamics that produces rather different patterns through time. Second, we examine the change in scaling associated with hubs in a network whose sizes are based on the number of travellers moving through these locations at different times of the working day. Last but not least, we explore the evolution of firm sizes with respect to the revenues they generate from the Fortune 500. These are in fact much more volatile than cities and transport hubs perhaps because they are not constrained physically to the same degree.

In all of these dynamics, we will resort to illustrating our dynamics in terms of changes in ranks (at least for the single size distributions) and we will exploit the idea of the rank clock, how ranks change over time where time is displayed as a clock which is organized not as the 24 hour clock per se but one which is calibrated to the time periods over which the dynamics is considered (Batty, 2006). Only in the case of examining travel volumes at transport hubs does the clock converge on our standard conception of the 24 hour clock over the working day. Having introduced these ideas, we will speculate a little on the morphology of the city system, not in terms of its shape or size but in terms of what these various scaling laws and rank clocks say about the overall dynamics, arguing that the real puzzle is to relate different dynamics to one another, in short moving on to examining how one perspective on the dynamics of city systems relates to another.

Power laws explained

As Krugman (1996) noted, the size distribution of cities in the United States follows the simplest power law where the size of a city P varies in inverse proportion to its rank r as $P \sim r^{-1}$. Expressed in terms of frequencies as a probability distribution, the frequency of the occurrence of a city of size P varies as $f(P) \sim P^{-2}$ which is the derivative of the previous rank size expression. This kind of manipulation is quite simple but it is worth noting that much confusion arises with power laws and rank size because different authors use one or other of these equations and the discussion as to the actual value of the power can become obtuse. In fact the idea that the power of the rank exactly equals 1 or the power of the frequency 2 is an ideal type although it does appear to be the consequence of a system developing competitively but randomly to a steady state (Gabaix, 1999). A more generic form however is to assume that the power varies which in probability form would be $f(P) \sim P^{-\alpha}$ and in rank size form is $P \sim r^{-1(\alpha-1)}$.

This power law has been applied to many size distributions. In its probability form, it is often contrasted with the Gaussian (or normal) distribution which is symmetric about its mean and is bell-shaped with two very thin tails covering the smallest and largest objects in the size distribution. In fact power laws essentially have long tails or fat tails that are skewed either right or left or both but in this context are usually skewed to the right where the long tail contains the largest objects for which there is no bound. Again there is confusion over fat, thin, long and heavy tails in the literature but here we will cut through all of this with a simple example taken from the population distribution of US cities by Metropolitan Statistical Area (MSA) for the year 2005. In Figure 1(a), we show the rank size distribution which illustrates the fact that the distribution follows a power law and because such a law is linear in its parameters when converted to logarithmic form, it is clear that the 366 cities that comprise these MSAs can indeed be approximated by a straight line as we show in Figure 1(b). In fact the power of the rank size is quite close to the idealized value of 1 at 1.0817 which translates to a value for α of 1.9245, again quite close to the ideal value of 2.

When we examine the relationship in Figures 1(a) and (b), there are worrying features of the distributions that suggest that the linear model might not be quite so good as the statistics imply. Although the proportion of the variance explained is 97.7%, the long tail in terms of Figure 1(a), which is the upper left of the graph in Figure 1(b), containing the biggest cities, slightly deviates from the line and appears to produce cities lesser in size than they should be while for the smallest values of the distribution, the right tail in Figure 1(b) seems to fall off faster than might be expected. Were we to consider a much larger number of cities—for only those with $P > 50000$ are in this set, then we would find that the distribution was more like a lognormal than a power function. We will return to these issues a little later but for now, this suffices to show that the argument is strong, notwithstanding these small deviations.

The second kind of power law that appears in city systems involves relationships not between the sizes of the objects that comprise a single distribution but between the comparative sizes of attributes of the same objects that comprise two (or more) distributions). Here we will use income and population but many different attributes of the same set of cities can be compared with respect to the nature of their interrelationships. Total income Y will rise as population rises but because of agglomeration economies—that is, the bigger the city, the more efficient its local economy is able to become—it appears that income increases at a faster rate than population and this can be represented in a scaling law of the form $Y \sim P^\beta$ where β is referred to as the allometric coefficient. This kind of scaling is that which lies at the core of allometry which is essentially the relationship between size and shape. In this context, if the relationship is more than linear—super-linear—where $\beta > 1$, then this implies that there is qualitative change in the attribute of the object with respect to its size and this is called positive allometry. If $\beta < 1$, then this too implies qualitative change which is sublinear and negative allometry, while if $\beta = 1$, then this is called isometry. The relationship to shape is implicit here but there are direct links to urban morphology with respect to the fractal geometry of cities (Batty & Longley, 1994; Batty, 2013).

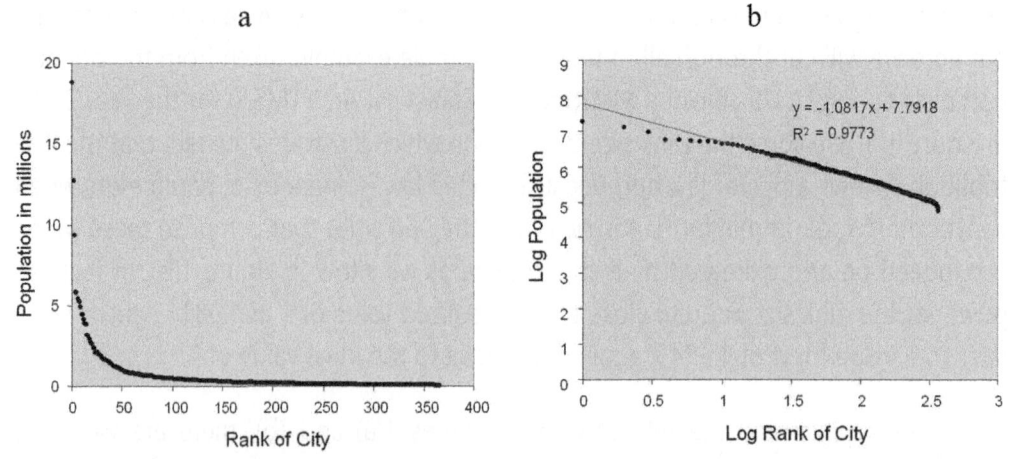

Figure 1 *The city size distribution for 366 US metropolitan statistical areas: a) the rank size distribution b) the logarithmic linear fit*

If we examine this relationship for our set of 366 US cities based on MSAs that we ranked in Figure 1, we find that the relationship is superlinear. We show this in Figure 2 where the value of the coefficient β = 1.0787 and the proportion of the variance explained is 98.4%. This implies that if we double the size of a city, the income will not simply double in proportion but 11% more than the simple proportion; that is if the income were 100,000, it would more than double to 211,000. In fact it is likely that allometric relationships of this kind operate within limited ranges for in this simple case, if a city were 100 persons in size, and it were to increase to 10 million, increasing by a factor of 100,000, then its income is unlikely to increase by 147% more than it would if this were simply in proportion to population size. The best demonstrations of superlinearly in recent years come from the work of the Santa Fe group who have shown for the same MSA data in 2002 that the allometric coefficient is 1.12 (Bettencourt *et al.*, 2007). There is some debate about this as in all such relationships because the objects in question—cities here—are not well-defined and there is considerable discussion as to the way they are bounded physically or economically as well as in terms of the measures of income and population used (Arcaute *et al.*, 2013).

In fact, we can relate these two different scaling laws for if one attribute varies regularly with another, and we have two types of scaling law, we can use one to derive the parameters of the other. For example, we have shown in Figure 1 that the rank size relationship for population is $P \sim r^{-1.0817}$ and if we generate a similar rank size relationship for income which we now estimate as $Y \sim r^{-1.1772}$ (where the proportion explained is also very high at 97.8%), we can estimate the allometric relationship from $Y \sim P^{-1.0787} \sim (r^{-1.0817})^\beta \sim r^{-1.1772}$. From this it is clear that we can estimate the allometric coefficient

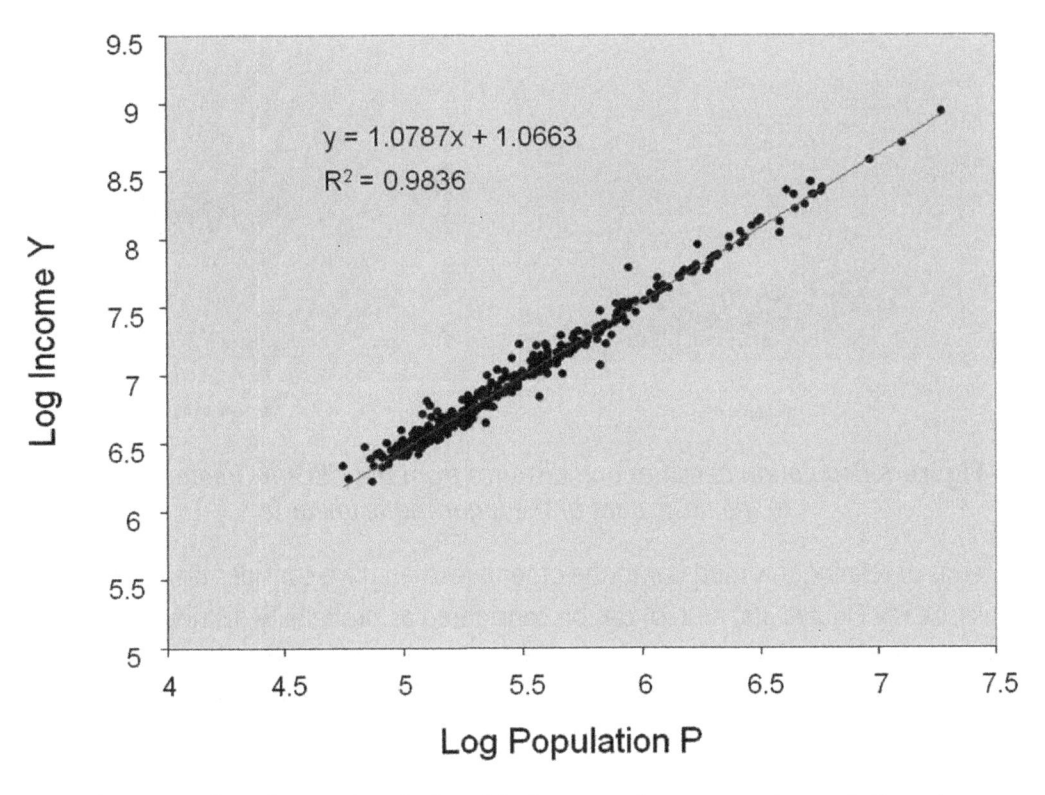

Figure 2 *The allometric relationship between income and population size*

$\beta \sim 1.1772/1.0817 = 1.0913$. This is close to the value 1.0787 which we have estimated directly.

The third scaling relationship that has become intrinsic to city systems relates to the arrangement of objects which compose a city with respect to the way its population is distributed across space. Probably the best way to consider this is to take the distribution of zones of population at different distances from the central business district (CBD) which tends to still be the dominant structuring force in historic cities of the industrial era and before. Many studies have shown that the density of population which we define as $\rho(d)_i$ where i is the distance from the CBD declines as a power law (or some similar inverse function) in such a way that $\rho(d)_i \sim d_i^{-\eta}$ where η is the power. In fact there are strong links in all of this to classical physics where the power in Newton's second law of motion is based on the inverse square law. We show this for 633 population zones in Greater London using population density at 2001; the distance decay against density is shown in Figure 3(a) and its fit using the logarithmic form is shown in Figure 3(b). The power is not 2 but 0.5307, which implies the slope is much lesser than that associated with the inverse square law but the proportion of variance explained is quite low at 22.8%.

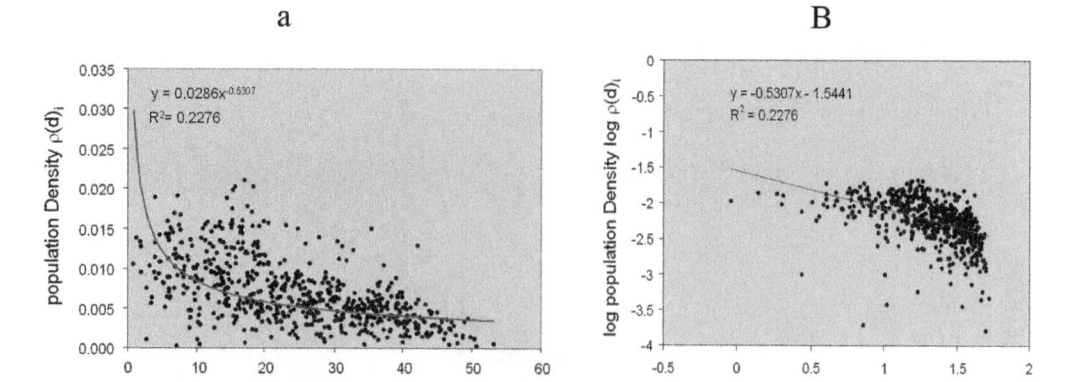

Figure 3 *Population densities and distance from the CBD in Greater London: a) the basic data b) the logarithmic linear fit*

In terms of relating this third law to the other two, then if we consider the distances as ranks, clearly Figures 3(a) and (b) can be considered as probability distributions where the frequency of population with respect to distance is the key explanatory focus of this perspective on the city system. In fact the distances are not ranks but we could quite easily envisage these distances as being sequential integers with the populations being aggregated or disaggregated to meet this constraint. If we use the distances as ranks, then the relationship becomes $\rho(d)_i \sim d_i^{-0.3099}$ but with an even lesser amount of the variance (18.8%) explained. In fact there are 7 major outliers which imply much lower densities than might be expected and these are probably due to the fact that entire zones might almost be empty of population—airports and large parks are cases in point, and if we were to exclude these from the original model now leaving 626 observations, the model is much improved to $\rho(d)_i \sim d_i^{-0.5729}$ with some 30.1% of variance explained. However there is much more noise in these explanations than we find in the domain of systems of entire cities and thus we will not take this third relationship any further in this essay as the spatial variation is too noisy to enable us to examine temporal change. Therefore we will focus on systems of cities in the first instance, then moving to other distributions such as skyscraper heights, volumes of hubs in transport networks, and firm sizes, amongst many other possible relationships whose form changes with respect to size and scale.

The dynamics of power laws:
Macro-regularities, micro-volatilities

If we examine how these rank size and allometric relationships change through time, there is quite remarkable regularity in that the power and allometric coefficients vary very little, notwithstanding there might appear some drift in their values. For example for the more recent MSA data than that we used above, we have distributions of population and income for 366 US cities defined on a standard areal base from 1969 until 2008 giving 40 time slices for which we can compute the rank size $1/(\alpha - 1)$ and allometric η parameters. Note that the income data is defined now as wages and this gives slightly higher values of the allometric coefficient than that defined for the 2005 distribution above. We show these values in the graph in Figure 4 where the mean value of η over 40 years is 1.108 with a very narrow standard deviation of 0.009. The mean for the rank size coefficient is 1.078 with a slightly higher standard deviation of 0.013. The fit of these relationships is also very high with the mean coefficient of variation R^2 for the allometric relation as 0.970 and for the population rank size relations as 0.969. From this, it is quite apparent that there is very little variability in these macro-relations. For the rank size rule, this confirms Krugman's (1996) observation of this regularity from a somewhat different data set for US cities over the last century, and indeed confirms Zipf's (1949) original demonstration of this in his analysis of US city data from 1790 to 1930. In terms of the income-population data, the regularity seems to imply much the same as the rank size data which we will explore below.

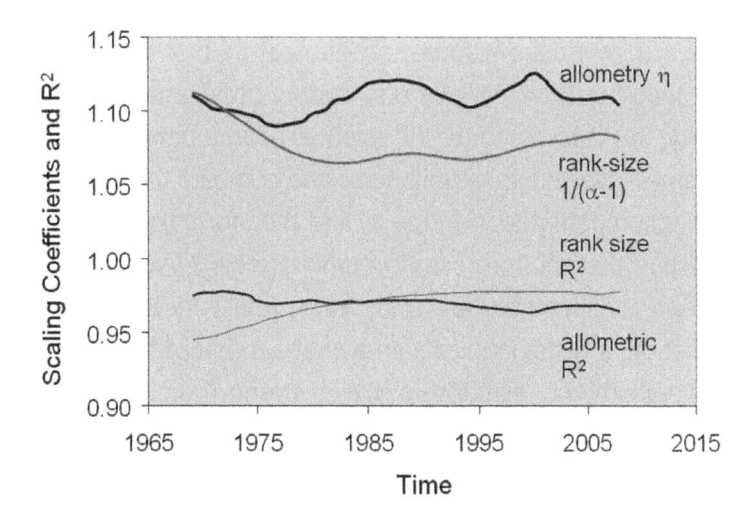

Figure 4 *Stability of the allometric and rank size coefficients*

Let us first examine shifts in the rank size distribution using the population data from 1969 to 2008, that is over the 40 year period. One simple measure of the rank shift for a set of objects that does not differ between two periods is to simply compute the sum of their absolute differences. Defining the rank at any time as $r_i(t)$, then the mean absolute shift can be computed as $\Psi(t + \tau: t) = \Sigma_i |r_i(t + \tau) - r_i(t)|/n$ where τ is the time interval and n is the number of cities. This measures the mean shift in ranks over all cities and gives a measure of the volatility of the distribution. For 100 cities, if every rank were to move 1 place, then the number of feasible shifts would be 99 and the mean would be a shift approximately 1. The mean maximum number of shifts would be no more than 50 which would mean that the shift in ranks for feasible combinations would be no more than 50% of the maximum rank. Given our knowledge of competitive systems, it is likely that the shift in ranks will increase over time and we will indicate this is the case in the examples that we explore in the next section. For the 40 year period from 1979 to 2008, there is an average rank-shift of 35 which is about 10% of the 366 set of MSAs. This is not as great as we have seen for other systems (Batty, 2006) for the US urban system over the last half century has become quite mature and the dramatic changes which were associated with the rise of great cities during the westward expansion mainly occurred before 1969.

To examine the volatility of two size distributions, or rather one distribution with respect to two attributes such as population and income for cities which lie at the basis of allometric scaling, we need a slightly more sophisticated measure. First let us compare the shift in ranked incomes of the cities between 1969 and 2008 with that for population. This mean shift is 41, quite close to the value of 35 which is the mean shift for the population of cities but a little greater and this probably reflects much of what we might intuit about how cities and their relative prosperities have changed over the last half century. What we need is some sort of measure that shows not only how the ranks shift over time but how the joint distribution of income and population shifts. One way of doing this is to compute the relative income per capita $y_i(t) = Y_i(t) / P_i(t)$ for each of the times involved, then rank these and compare the different ranks across time. Thus we first compute $y_i(t)$ and $y_i(t + \tau)$ and then form $\Psi(t + \tau: t)$ for these ranks. If we do this we find that the mean shift in ranks is now 77 which is about twice the shift for population and income respectively. In fact these measures pick up more and more of the variation, at least implicitly and what is perhaps so surprising is the fact that although the allometric relationship is very strong (from Figure 3), there is substantial shift—something in the order of 20 percent of the income per capita associated with the 366 MSAs. We have not developed this analysis for the MSAs in terms of a close analysis over all time periods, simply assuming that the differences in rank are likely to increase in time. But to develop these ideas further, we must examine more specific trajectories through time and to this end, in the next section we will develop various mnemonics which enable us to examine in more detail the way cities change through time in terms of their income and population.

The concept of the rank-clock

Many size distributions with a heavy tail that contain the largest objects (cities) in the set appear to be quite stable across time and if all one knew was that the distributions were composed of the same objects, then the most likely hypothesis would be that the distribution of the objects would be similar across time. We know this is not the case from the analysis of MSAs in the previous section and this is made even starker when we examine the distribution of objects over longer time periods. If we plot the top 100 cities defined from the start date of the US Census in 1790 until the year 2000, the rank size distribution appears extremely stable as we show in Figure 5(a) but if we then plot the shift in ranks there is considerable movement of cities in terms of the size and rank up and down the hierarchy. In Figure 5(a) we display one measure of this shift by plotting the year 2000 city sizes according to the 1940 ranks and one can see that the smaller sizes tend to shift more than the larger. We can enhance this by noting the cities that are in the top 100 ranks over the 210 year period from 1790, can be displayed individually by plotting their ranks and coloring them according to spectrum that begins with red and transitions through to yellow, then green to blue as cities appear in the ranking (using the typical heat map convention). So the first city at the top rank in 1790 is colored red and the last city to appear in the ranking over the next 210 years is colored blue with the transition evenly spaced according to the heat map color spectrum. We show this for what we call the rank space which is the size versus rank graph—the so-called Zipf (1949) plot—in Figure 5(b) but this is a particularly messy form for many more than a few objects that comprise the distribution. What this plot does show however is that there are several very distinct trajectories defining the space: for example cities that shoot into the space from outside the top 100 towards the top and vice versa, cities that remain at the same rank defined by vertical lines on the plot, cities that oscillate up and down in terms of rank and so on.

A much better mnemonic is to make time explicit and to suppress size for after all, rank is a synonym for size and if the focus is simply on relative position then rank and time are somewhat more illustrative of volatility in the Zipf plot than rank and size (Batty, 2006). What we do is to plot time as a regular clock around its circumference defining the beginning of the time in question—in this case 1790 at the noon-mid-

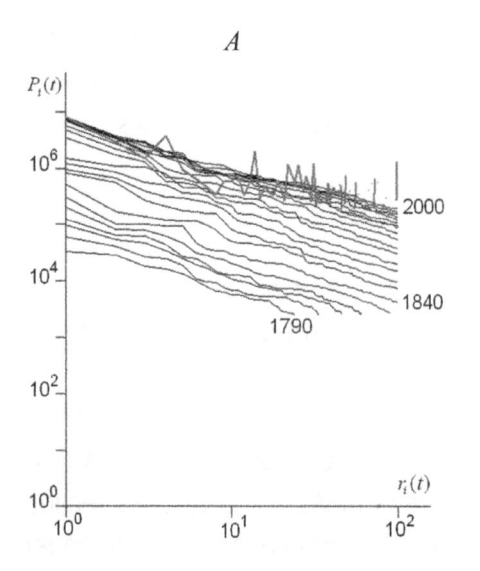

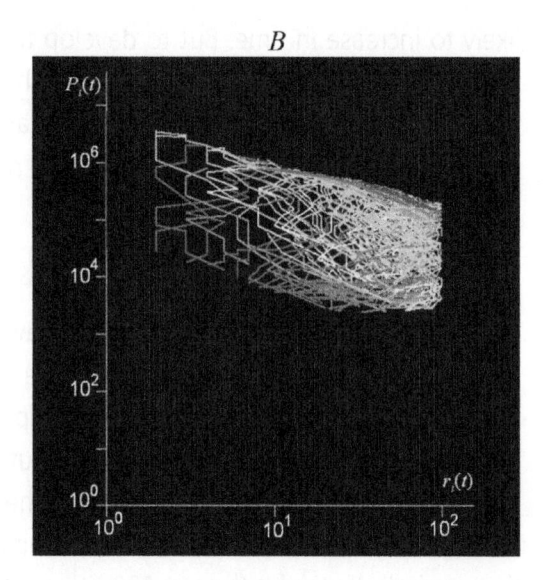

Figure 5 *The rank size distribution and rank shift (a), and changes to individual cities (b) in the rank space from 1790 to 2000*

night position with the years running in the clockwise direction around the clock until the hand reaches back to noon-midnight at the end of the time period in question, in this case at the year 2000. We can then plot the rank of the city as a radial from the centre of the clock at the appropriate time where we can organise the radial from rank 1 at the centre to rank 100 (or whatever is the upper limits of rank) on the circumference, or the other way around, using a linear or logarithmic scale. Here we will use the simplest linear scale with the highest rank at the centre of the clock and the lowest on the circumference. For our 210 year city size distribution taken for the cities defined in the US Census, we show some typical trajectories which compose part of the rank clock in Figure 6 where it is clear that different cities are associated with quite different trajectories.

What is fascinating about this particular clock is that this is the temporal signature of the development of the US urban system. New York City is the anchor of the clock being number 1 in rank ever since the Census began in 1790. In some respects, the city is the fulcrum of the entire American system. The opening up of the mid-west, California and the south is also marked out with first Chicago (around 1840), Los Angeles (1890), Houston (1900), then Phoenix (1950) flying into the clock from outside the top 100. Several colonial towns established in the 18th century and before such as Charleston (SC) lose rank and fall out of the top 100 while some colonial settlements in the vicinity of Washington DC and the northern part of the south lose rank and then begin to stabilise as sprawl makes an impact post world war 2. Rust-belt cities such as Buffalo (NY) lose rank systematically from the early 20th century onwards. There

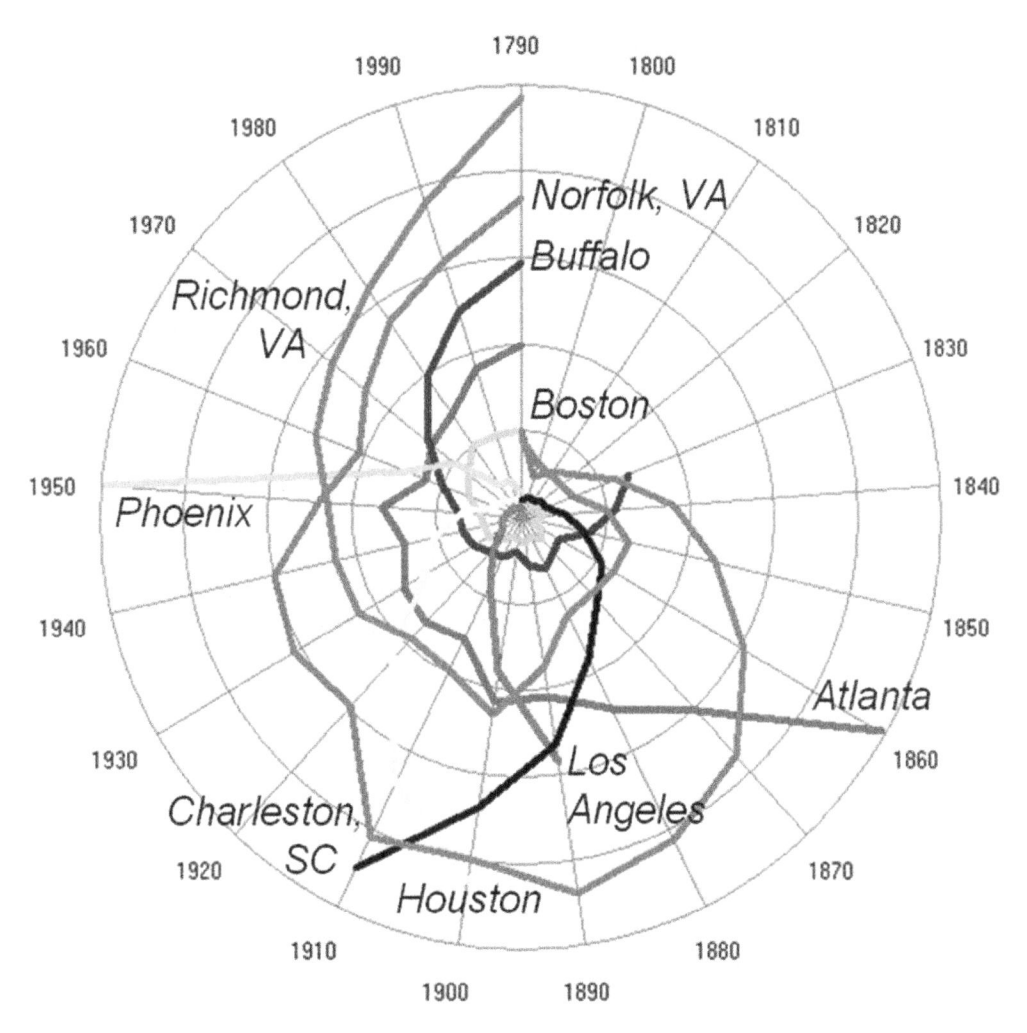

Figure 6 *The rank clock and typical trajectories for US cities from 1790 to 2000*

are few cities that enter and leave the top 100 in any significant way but some such as Atlanta zoom in only to lose rank as they stabilise although to an extent, boundary changes and suburban sprawl complicates the picture. We have not attempted any classification of different trajectories so far but the prospect exists for such analysis in future work.

We also illustrate the complete rank clock for all cities that are in the top 100 from 1790 to 2000 in Figure 7. Many cities enter and leave this exclusive set. Before 1840, there were less than 100 cities catalogued in the US Census, and thus the rank clock in Figure 7(a) shows this build up. We have not normalized any of these cities for boundary changes, so our analysis is inevitably crude. Moreover, although after 1840, there are only 100 cities ranked at each time period, in fact over the 210 year period there are 266 cities that are part of the top 100, and this is itself a measure of the volatility of the set, with there being 2.6 times the number of cities appearing in the top 100 over this period. If these cities entered and left the top 100 uniformly, this would mean that on average, about 8 cities would enter and leave the top 100 each time period, about 8% per decade.

What Figure 7(a) clearly illustrates impressionistically through the collage of various trajectories (by their color) is the substantial volatility of the US system over two centuries. Of the cities in the top 100 in 1840, only 20 cities remained in the top 100 by 2000. This again is consistent with the rate of change being about 8% per decade. To generate a better measure of this shift, we can approximate a half life for how long a city remains in the top 100 for every decade. For any time, we can plot the number of cities that are in the top 100 at this time for every decade prior to and after that time slice in question and the resulting curves show the rate at which cities join and leave the top 100. These plots are illustrated in Figure 7(b) where the vertical axis shows the cities in the top 100 (or less than this before 1840 in all those cities listed) before, at, and after each year in question. The half life for each time is the number of years it takes for 50% of the cities to join the top 100 and to leave. This half life will be different on the upswing than the downswing as Figure 7(b) reveals but we can compute it as an average for both sides of the curve. It is also clear that the average rate of decay appears to be getting greater over time, and this is consistent with the fact that historically the first cities have a higher probability of lasting longer than later cities as indicated by the increasing rate of decay for later decades of urban growth. The overall average half life is about 45 years which means that before the top 100 cities

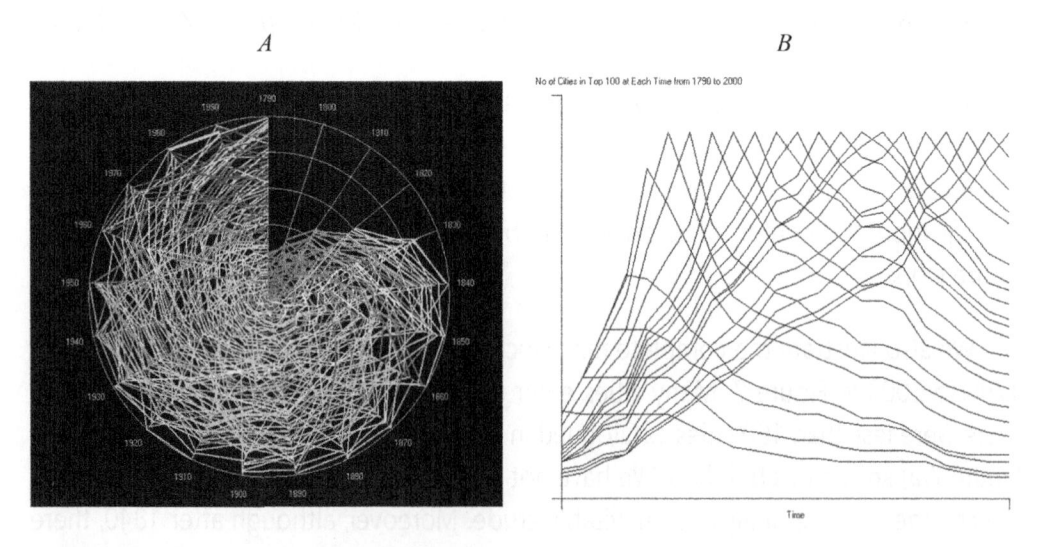

Figure 7 *The US rank clock (a) and number of cities defining the half lives (b) of cities at each time*

are noted at any time, 50% of them will be there 45 years before this date and remain 45 years after. In fact over the 210 year period, the quarter life is much longer at about 150 years but in fact these estimates are complicated by the fact that the time series is limited and the number of cities is less than 100 for the first 50 years of the analysis. Nevertheless, all these estimates give some sense of the considerable micro-volatility that exists as cities move up and down the hierarchy.

To complete this analysis of city systems, we will return to the first data sets that we examined earlier for population and income associated with the 366 MSAs from 1979 to 2008. The great advantage of this data base is that the MSA boundaries are stable and the set of cities is complete, thus implying that the growth dynamics associated with changes in size and rank is clearer than that of the 100 top ranked cities from the US Population Census data. In Figure 8, we show the rank size distributions associated with the population and income measures for the 366 MSAs. These are at a much finer temporal interval than those we examined from the Population Census and thus any measures of shift need to be normalized if comparisons are to be made. In Figure 8(a) and 8(c), we show the rank size distribution for population and income. These are extremely close to one another as we illustrated previously but when we examine their individual clocks in Figures 8(b) and 8(d), these show more volatility. Nothing equivalent to the volatility for the city size distributions from the Census is seen in these distributions but the time period is much shorter and the urban system is much more mature. In fact the population clock implies that the core cities remain as core but that there are several smaller cities that rise up the hierarchy while a few established cities drift down more gradually. Of course this clock implies changes in rank not size so although Phoenix which is ranked 35 in 1969 increases its rank to 12 in 2008, it more than doubles in size. In fact Houston increases its population almost three times while its rank goes from 14 to 6. In the mid range of the hierarchy, Las Vegas increases its population from about 267,000 to 1.87 million, some 7 times, and its rank from 114 to 30, some 3.8 times. These are substantial shifts given the fact that the system is mature, with a growth rate in metropolitan populations of only around 0.5% per annum.

When we examine the income distribution, there is more volatility with one very obvious shift in income due to oil being discovered in Fairbanks, Alaska in the late 1960s. There was a boom in pipeline and related infrastructure construction in 1975-1977 leading to a big increase in income and then the local economy collapsed back to its former trajectory. This is clearly seen in the income rank clock in Figure 8(d). The last distribution relates income to population as income per capita and the rank size and its clock are shown in Figures 8(e) and 8(f). These is considerable mixing implied

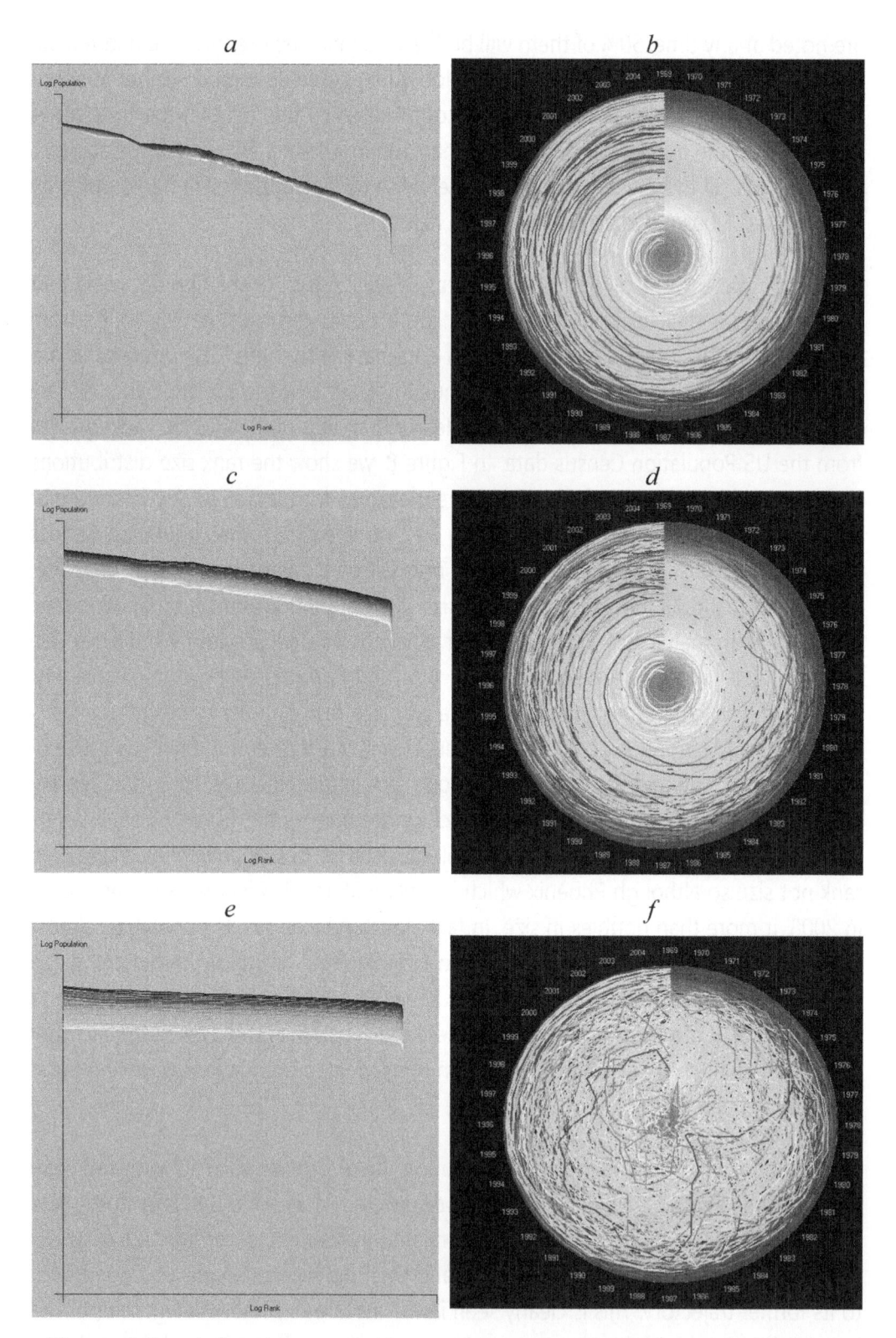

Figure 8 *Population, income and income per capita: Respective rank size (a,c,e) and rank clocks (b,d,f) for 366 MSAs from 1969 to 2008*

by this clock despite the impressively smooth macro distributions with respect to their form over time. In fact the mixing is so considerable in an impressionistic way that it is hard intuitively to reconcile such substantial variations in the income per capita of different cities with the very strong positive allometry displayed earlier in Figures 2 and 4. We really cannot explain this. Quite clearly as we form composite distributions by taking ratios of more basic data, then the noise from one combines with the other and differences—variances—can become magnified. There is in fact little experience of such mixing and we lack real intuition as to what are its consequences but it is enough to cast doubt on many of the more stable and regular relationships that we often begin with such as power laws and this suggests that our knowledge of this whole area is primitive and subject to much more profound research than anything that we have done so far. This remains unfinished business but it is clearly the way forward.

Related distributions: High buildings, transport hubs, firm sizes

There are many systems where competition in space and time determine how their constituent elements grow or are manufactured in size. If we disaggregate populations and examine the internal distribution of such clusters in cities, we have already seen that these intra-urban elements follow power laws, albeit with considerably more noise associated with their spatial and temporal arrangements than entire cities. If we now divert our attention from actual amounts of such activities to the physical environment which accommodates them—from people to the buildings in which they reside or work—we also find that the sizes of these buildings follow power laws. In fact if we examine high buildings, which are usually defined as being greater than 6 stories, certainly greater than 10 (which require elevators for their operation), then their distribution can also be shown to follow the rank-size rule. There is a major difference between high buildings and population sizes in that buildings do not grow or decline—at least in the same way as populations; they are manufactured and rarely are stories taken off them through partial demolition and rarely are they added to. There are exceptions of course but in our analysis here we will exclude these occasional cases. Buildings do get demolished but in our analysis we have excluded these too for we will illustrate these ideas only on extant skyscrapers—buildings greater than 40 metres—in New York City from the year 1909 until 2010 also measuring their height in metres. There are 516 buildings in this set but we will only ever plot the top 100. In fact the rank clock is quite different from that for cities. As the century progresses, a building which is number 1 in rank does not stay there for long. Skyscrapers have got successively higher as building technologies and materials have

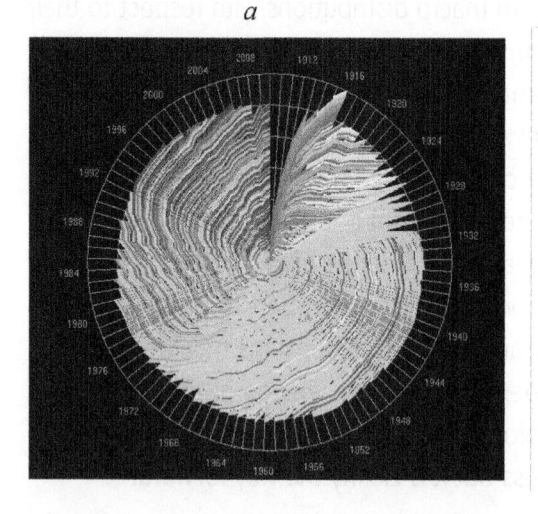

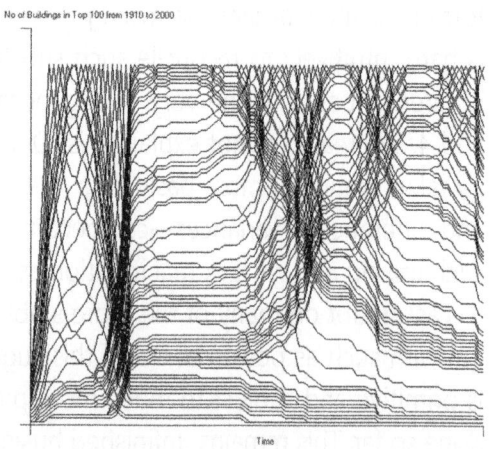

Figure 9 *New York City skyscraper heights: 1910 to 2000: a) Rank clock and b) Number of buildings defining the half lives at each time*

progressed and thus the rank clock is marked by a continuing downward spiral of earlier high buildings, many of them leaving the top 100 during the 101 years that the clock portrays. We show the clock in Figure 9(a) and the downwards spiral provides the dominant morphology of these dynamics. In fact at the start of the clock in 1909, there is rapid growth to 119 high buildings 'in the top 100' because there are several ties for height but the rest of the clock is contained with the envelope of 100.

The other feature of this dynamics is that the clock shows quite distinctly the waves of skyscraper building that have dominated New York. At the start of the period in the early 20th century before the first world war there was a great wave of such building. Then again after the war in the mid 1920s to early 1930s, the boom, which preceded the great recession, was a time of massive investment in high buildings. In fact the highest buildings in the city which still dominate the skyline like the Empire State and Chrysler buildings were constructed then and if you look at the core of the clock—the top 10 ranks—you will see that these are dominated by buildings (colored green) which were constructed in the 1930s. In fact earlier buildings are overwritten at the level of resolution used in the clock and some of the early high buildings such as the Woolworth building only reappear on the clock once the 1930s wave of building subsides. There are waves in the 1960s and 1980s and then more recently in the 2000s but buildings in general have not been much taller then those built in earlier times. It is elsewhere in the world where the highest buildings have been built recently, in the Middle East and in China.

An even more graphic demonstration of these dynamics is given by the plot of half lives for the 101 years that comprise these competitive processes which are firmly linked to boom and bust. In Figure 9(b), as in Figure 7(b), we show these half lives where it is clear that the dynamics produces clusters that relate to specific 'economic events'. Remember that the half live for the set of one hundred buildings that exist at a given point in time, is the number of years between the time in question and the time when only half the number of buildings at this time remain in the system. We can of course compute half lives for buildings that are entering the system as we indicated earlier and these ultimately compose the 100 in question. This is often different in time span from those that are leaving the system, being knocked out by higher buildings being constructed where the progression is often slower. The times of rapid building are clearly picked out by the half life plots in Figure 9(b) where the 1910s, 1930s, 1950s, 1960s and 1980s and 2000s are periods of very rapid growth in high buildings with half lives on the upswing much shorter than the downswing which are more subdued. In fact this figure shows how hard it is to produce an average half life for the entire series. In fact for periods of rapid growth (boom), the upswing half life appears to be about 4 years whereas the subsequent downswing is about 25 years. Also the downswing half life seems to be shortening whereas the upswing is less variable. Overall we estimate that the average upswing half life to be about 10 years and the downswing 20 years but the volatility and the dominance of booms and busts complicates the picture.

Our second example refers to the size of hubs in spatial networks. It is very clear that the evolution of networks is governed by competitive forces that enable a limited number of hubs to gain more than proportionate numbers of links. Translated into volumes of activities flowing on such links into hubs, this gives rise to distributions that are similar to power laws. This was first demonstrated by Barabasi and Albert (1999) but it pertains to many developments in network science pioneered during the last 20 years. There is some debate as to whether or not these distributions are scaling for their derivations using laws of proportionate effect tend to generate log-normal distributions but as most of these distributions are modelled with respect to their heavy tails, then power laws can form a good approximation to these. Moreover there are likely to be more constraints on the form of these distributions due to the fact that spatial networks are constrained in space and cannot generate the numbers of links that are generated in unconstrained network structures. In short, planar graphs which dominate spatial networks do not manifest scaling in their pure form (Barthelemy, 2011).

Our example constitutes the hubs that define the rail stations on the London underground and overground where the volumes of travellers entering and exiting these hubs define their size on a typical weekday in November 2010. The dynamics of these hubs relates to the fact that during a typical day, all the hubs only operate for 20 hours for the system is closed from 1-20am to 5-20am each day. The dynamics is also dominated by the morning peak and the evening peak hours and the volumes reflect this. We have organized the data which is available on a second by second basis into bins of 20 minutes each, of which there are 72 defining the 24 hour day. 12 of these are empty as there are no trains running. There are a total of 6.2m entries and 5.4m exits (the difference is due to open barriers where the RFID (Oyster) card from which the data is taken, is not used) and we will aggregate these entries and exits to form the volumes for each of the 666 hubs that define the system.

We show the 60 different rank size distributions in Figure 10(a) where it is clear that the differences pertain to different volumes at different times of the day. The dominant cluster of trips is during the two peaks that are clearly evident in Figure 10(a) but the shape of these distributions is quite similar at each 20 minute interval as we show in Figure 10(b) where they are collapsed onto one another. We achieve this by taking the distributions from their mean size and normalizing by their variance so that they are comparable. It is also clear from Figures 10(a) and (b) that the distributions are not scaling but are much closer to lognormal. We show the clock in Figure 10(c) from which it is clear there is enormous variability in the way hubs move up and down in the hierarchy of ranks during the day. What we see from this is that some hubs that have low volumes early in the day pick up in the morning peak and then collapse back in the middle of the day to rise again in the evening peak. These are inner suburban hubs whereas those in the central business and shopping districts tend to remain the biggest during the whole day and close down last. To really explore the meaning of these dynamics, it is necessary to know the actual spatial configuration of rail lines and hubs and to this end, the *Rank Size Visualizer* that we introduce towards the end of this essay indicates how one can make progress in generating much more satisfactory explanations of these spatial network dynamics.

Our last example throws away the spatial dimension and concentrates entirely on time. Just as city size and income distributions follow power laws, so do firm sizes (and related characteristics) and there is a long tradition of work in this area stretching back to Simon (1957) and Gibrat (1931). The Fortune 500 database records the top 500 firms in the US at each year, listing their size in terms of revenues and profits. We have actually taken the top 100 firms from this set and examined their distributions over the 40 year period from 1955 to 1994 (when the index was recalibrated) but even

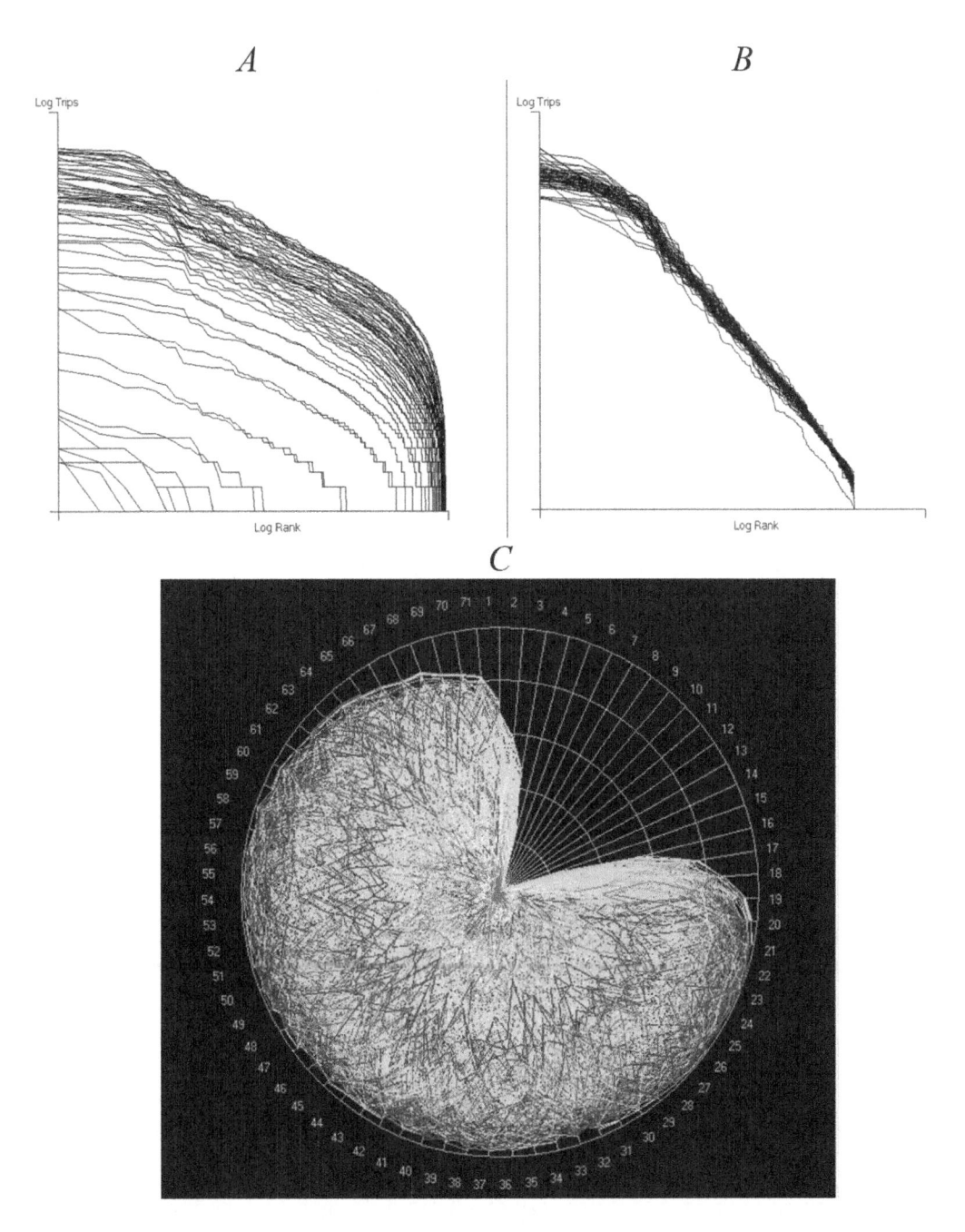

Figure 10 *Sizes of London station trip volumes (November 2010) a) Rank size b) Collapsed rank size, and c) Rank clock*

over this period there are only 39 firms that were present in 1955 that still exist in the set in 1994 (Batty, 2010). In fact 20 years on (today in 2014), most of these are gone. In Figure 11(a), we show some of the key revenue trajectories that define typical firms. This 40 year period marked the beginnings of de-industrialization in America and also the rise of the computer industry. It does not quite cover the rise of financial services

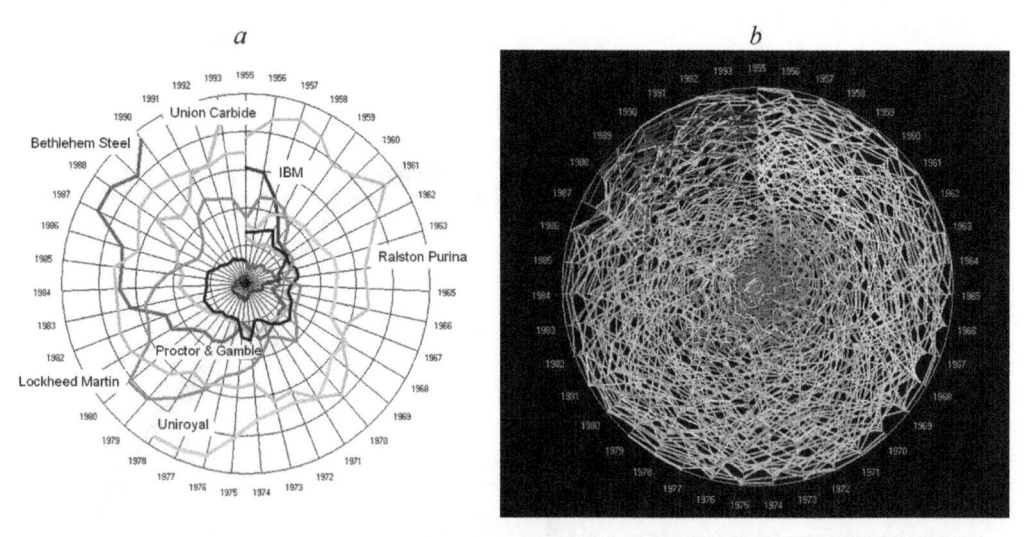

Figure 11 *Rank clocks for firm sizes 1955 to 1994, for the top 100 from the Fortune 500: a) Typical individual firm trajectories and b) All trajectories*

that have characterized the last 20 years but Figure 11(a) does indicate how iron and steel firms declined, IT firms (that is IBM in fact) became established, as well as the mixed fortunes of aerospace companies which were dominated by the ebb and flow of funding for the military-industrial complex.

The rank clock in Figure 11(b) shows very considerable volatility. Although we have not shown the regularity of the rank size distributions (which in fact show the same dramatic regularity we have seen throughout the examples in this essay), there is massive mixing within the clock with much more stable half lives on the upswing and downswing for all firms where these are both about 25 years, In fact as with cities, these half lives are decreasing in numbers of years as the economy has picked up momentum and globalized over the last half century. The biggest issue however with economic systems such as those composed of firms, is that the units of definition change much more frequently than cities. Cities of course can amalgamate with one another, even acquire adjacent less powerful territories but their essence usually remains the same or at least changes very much more slowly than such firm amalgamations, Firms are dominated by a series of mergers and acquisitions and are prone to reinventing themselves, making different products and capturing different customer bases but often keeping the same organizational structure and name. In fact this is dramatically true for software companies and the IBM that is pictured in Figure 11(a) which rises from relative obscurity to the top 5 by the mid 1990s, has completed reinvented itself since then, moving away from hardware to software and thence to software services and increasingly to consulting on matters such as urban and health care technologies that are very different from their original core business. This explains the increased volatility of firms compared to cities and other inanimate systems.

Next steps: The rank clock visualizer

There are many different ways of visualizing the kind of space-time dynamics that we have focused upon here. Besides the rank space which is directly related to the rank size, the rank clock exploits the analysis by examining rank and time suppressing size (assuming rank is related directly to size). But there are other measures and even within the rank space and rank clock one can reorient the visualization from inside to out, and use various transformations of the axes from linear to logarithmic. We can of course produce size clocks where size instead of rank is displayed but in all these mnemonics, what we really require is the hot-linking of one display to another. Moreover as space tends to be implicit in these visualizations, linking the various objects and their variation in time to their spatial location is important, particularly as we can animate these visualizations in such a way that the user can explore many more dimensions simultaneously than those that are possible on the printed page.

We have developed a visualizer* for these kinds of scaling distributions that enable the user to link the objects in the clock to their spatial location. The user can generate a rank clock for many different distributions which so far are confined to mainly to city size and building height distributions but do include the Fortune 500 data from 1955 to 2010 (with the re-normalizations shown in 1994-5) and the distribution of US baby names. The London tube hubs are in the set as are several distributions of UK, US and Japanese populations by small area/cities. Users can plot the rank clock with the highest rank at the centre or the edge, choose any time periods from the maximum available for each data set and identify specific objects by name on the clock and map. The link to the spatial distributions uses either Open Street Map or Google Earth and the user can point to either a city or object on the map or globe or on the rank clock and its equivalent in clock, map or globe will show up. There is no animation of the clocks so far in this interface but this will be done in time for there are many easy extensions like this.

In the data set, we have world city sizes from the Population Division of the UN Department of Economic and Social Affairs from 1950 to 2010 (some 576 in all, greater than 1 million population each). We show the rank clock of these in Figure 12 from

---

*. This is the so-called Rank Clock Visualizer developed by Oliver O'Brien of CASA, UCL and it is available at http://casa.oobrien.com/rankclocks/

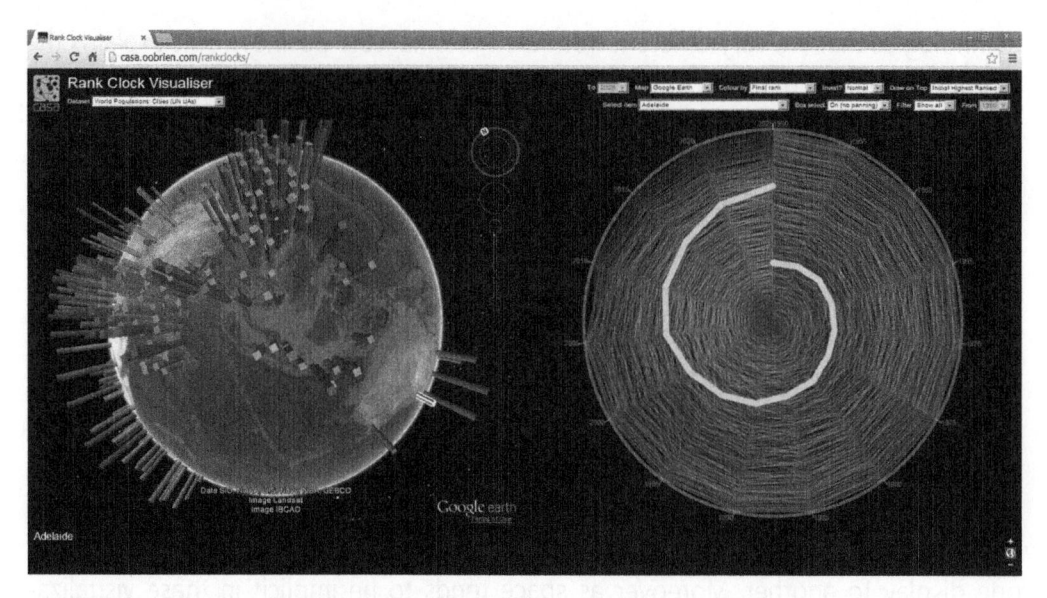

Figure 12 *Rank clocks of UN urban areas 1950 to 2010 from the rank clock visualizer showing the trajectory for Adelaide, Australia*

the Rank Clock Visualizer where we have picked out Adelaide in South Australia, a good example of a 1 million population city that is stable in population but declining in rank. The Google Earth display alongside lets the user visualize all these cities and their sizes and a user can click on a city and see where its trajectory lies on the clock or identify the name of the city from a drop down list and activate its trace on the clock and on the map or globe. It is not yet possible to zoom into the clock to identify a hot link to a trajectory because the level of resolution is too fine from the static display but all these extensions are possible and will be explored in future work. The world cities distribution and its rank clock are shown in Figure 12.

We have not so far explored the possibility that the morphology of the rank clock itself provides a shorthand for the kinds of dynamics that characterise the system of interest. We have noted that the individual trajectories might be classified but the shape of the clock also varies and we can see from those illustrated here how different they might be. For example a clock where there is no change in rank for any of its objects would be a perfectly circular spectrum from a central red point to a circumference colored blue. Where an object always enters at number 1 and then falls by one rank at each subsequent time, the clock would appear as symmetric set of outward spirals, not unlike the New York City skyscraper clock. If the objects entered at number 1 and then went to number 100 at the next time, then this would be a set of radials with the colors moving around the clock symmetrically from red to blue. Other kinds of regular geometry can be constructed that bound the space of possible morphologies but what we need is to tie the dynamics more consistently to this geometry. In fact a purely regular dynamics is not very useful as it is quite unrealistic in city systems whereas the boom-bust structure of economic dynamics is much more likely. Changes in social taste are also likely to be reflected in urban dynamics and our quest must be to begin to identify different dynamics that are associated with different geometries of clock so that a deeper, more structured picture of the way this world of cities works can be generated.

To conclude, it is worth saying a little more about the notion of scaling in city systems. Our argument began by suggesting that city size distributions were scaling, following power laws in their heavy tails although always predicated on the basis that their underlying distribution is more likely to be lognormal. Power laws are thus a good approximation to the heavy tails but no more than this. And they simply represent our starting point. The rank clock is a good device if we can assume that population is related to rank by a simple logarithmic transformation as is consistent with a power law but once we have the idea of the clock, the fact that it relates rank to population can be conveniently forgotten. The clock has its own integrity in that it displays a kind of dynamics that can be explored more generally and if a rank clock and a size clock are defined, one related to the other, then all kinds of novel animations and exploration suggest themselves. It is in this spirit that the Rank Clock Visualizer has been developed.

In future research, we need much better statistics that pertain to the different kinds of dynamics and their variation over time and space. The idea of the half life needs to be put on a more consistent footing and defined more rigorously. But in the perspective that Peter Allen has brought to the study of space and time in city and regional systems, we need to explore the extent to which the kinds of dynamics that define bifurcations, tipping and turning points, and even catastrophes relate to scaling. Little has been done to date but there are strong hints in the notion of fractals, self-similarity and hierarchy that need to be exploited in linking this kind of aggregate spatial analysis to the dynamics of spatial modelling that follows the tradition of Allen and others. I believe after 25 years or more of consistent but slow development in the field of spatial dynamics, there are now many fertile ideas that will see this field of ours explode intellectually and in terms of applications during the next 25 years.

References

Allen, P.M. (1997). *Cities and Regions as Self-Organizing Systems: Models of Complexity*, ISBN: 9056990713.

Alonso, W. (1964). *Location and Land Use: Toward a General Theory of Land Rent*, ISBN 0674537009.

Arcaute, E., Hatna, E., Ferguson, P., Youn, H., Johansson, A., and Batty, M. (2013). *City Boundaries and the Universality of Scaling Laws*, http://arxiv.org/abs/1301.1674

Barabási, A-L., and Alberts, R. (1999). "Emergence of scaling in random networks", *Science*, ISSN 0036-8075, 286(5439): 509-512.

Barthelemy, M. (2011). "Spatial networks," *Physics Reports*, ISSN 0370-1573, 499: 1-101.

Batty, M. (2006). "Rank clocks," *Nature*, ISSN 0028-0836, 444: 592-596.

Batty, M. (2013). *The New Science of Cities*, ISBN 0262019523.

Batty, M. and Longley, P. (1994). *Fractal Cities: A Geometry of Form and Function*, ISBN 0124555705.

Batty, M., (2010). "Visualizing space-time dynamics in scaling systems," *Complexity*, ISSN 1099-0526, 16(2): 51-63.

Bettencourt, L.M.A., Lobo, J., Helbing, D., Kuhnert, C, and West, G.B. (2007). "Growth, innovation, scaling, and the pace of life in cities," *PNAS*, ISSN 0027-8424, 104: 7301-7306.

Christaller, W. (1933, 1966). *Central Places in Southern Germany*, ISBN 0131226304.

Gabaix, X. (1999). "Zipf's Law for cities: An explanation," *The Quarterly Journal of Economics*, ISSN 0033-5533, 114: 739-767.

Gibrat, R. (1931). *Les Inegalites Economiques*, Librairie du Recueil, Sirey, Paris, France.

Krugman, P. (1996). "Confronting the mystery of urban hierarchy," *Journal of the Japanese and International Economies*, ISSN 0889-1583, 10: 399-418.

Nijkamp, P. and Reggiani, A. (1992). *Interaction, Evolution and Chaos in Space*, ISBN 364277511X (2012).

Simon, H.A. (1957). *Models of Man: Social and Rational*, John Wiley and Sons, Inc., New York.

Thunen, J.H. von (1826, 1966). *Von Thünen's Isolated State*, P. Hall (ed.) and Carla M. Wartenberg (trans.), Pergamon Press, Oxford, UK.

Wilson, A.G. (1981). *Catastrophe Theory and Bifurcation: Applications to Urban and Regional Geography*, ISBN 0415687969 (2012).

Zipf, G.K. (1949, 1965). *Human Behavior and the Principle of Least Effort*, ISBN 161427312X (2012).

Michael Batty is Bartlett Professor at University College London where he chairs the Centre for Advanced Spatial Analysis (CASA). His research involves the development of computer models of cities and regions, and he has published numerous books and articles in this area, such as *Cities and Complexity* (MIT Press, 2005) which received the Alonso Prize of the Regional Science Association, and *The New Science of Cities* (MIT Press, 2013, and www.complexcity.info). He is editor of the journal *Environment and Planning B: Planning and Design*. In 2013, he received the Lauréat Prix International de Géographie Vautrin Lud. He was made a Fellow of the British Academy in 2001, received the CBE award in the Queen's Birthday Honours in 2004 and elected a Fellow of the Royal Society in 2009

Chapter E

Robustness in complex systems: The role of information 'barriers' in network functionality

Kurt A. Richardson

In this supposed 'information age' a high premium is put on the widespread availability of information. Access to as much information as possible is often cited as key to the making of effective decisions. Whilst it would be foolish to deny the central role that information and its flow has in effective decision making processes, this chapter explores the equally important role of 'barriers' to information flows in the robustness of complex systems. The analysis demonstrates that (for simple Boolean networks at least) a complex system's ability to filter out, i.e., block, certain information flows is essential if it is not to be beholden to every external signal. The reduction of information is as important as the availability of information.

Introduction

In the Information Age the importance of having unfettered access to information is regarded as essential—almost a 'right'—in an open society. It is perhaps obvious that acting with the benefit of (appropriate) information to hand results in 'better' actions (i.e., actions that are more likely to achieve desired ends), than acting without information (although incidents of 'information overload' and 'paralysis by (over) analysis' are common). From a complex systems perspective there are a variety of questions/issues concerning information, and its near cousin knowledge, that can be usefully explored. For instance, what is the relationship between information and knowledge? What is the relationship between information, derived knowledge, and objective reality? What information is necessary within a particular context in order to make the 'best' choice? How can we distinguish between relevant information and irrelevant information in a given context? Is information regarding the current/past state of a system sufficient to understand its future? Complexity thinking offers new mental apparatus and tools to consider these questions, often leading to understanding that deviates significantly from (but not necessarily exclusive of) the prevailing wisdom of the mechanistic/reductionistic paradigms. These questions may seem rather philosophical in nature, but with a deeper appreciation of the nature of information, and the role it plays in complex systems and networks, we can begin to design more effective and efficient systems to facilitate (rather than merely manage) information creation, maintenance, and diffusion.

The science of networks has experienced somewhat of a renaissance in recent years with the discovery of particular network topologies, or architectures, in natural and human systems. These topologies include both small-world (Watts & Strogatz, 1998) and scale-free networks (Barabási & Albert, 1999). Barabási (2007) has shown that the World Wide Web has a scale-free architecture which essentially means that it has relatively few high-connected nodes (i.e., nodes containing many inputs and outputs) and relatively many low-connected nodes (i.e., nodes containing few inputs and outputs). However, there are significant limitations to network representations of real world complex systems. Barabási (2007) himself says that "The advances [in network theory] represent only the tip of the iceberg. Networks represent the architecture of complexity. But to fully understand complex systems, we need to move beyond this architecture and uncover the laws that govern the underlying dynamical processes..." Boolean network modeling is a relatively simple method to facilitate this move beyond mere architecture. Although the networks considered herein have random topologies (rather than scale-free or small-world) the modeling of such simple dynamical networks allows researchers to explore, albeit in a limited fashion, the emergent underlying dynamical processes of real systems such as the Internet.

In this chapter, I would like to examine one particular aspect of complex dynamical networks: How barriers to information, and its flow, are essential in the maintenance of a coherent functioning organization. My analysis will be necessarily limited. A very specific type of complex system will be employed to explore the problem, namely, Boolean networks. And, a rather narrow type of information will be utilized: a form that can be 'recognized' by such networks. Despite these and other limitations, the resulting analysis has applicability to more realistic networks, such as human organizations, and useful lessons can be gleaned from such an approach. We shall see that such an approach may offer the possibility of complementing mechanistic approaches to the design of information systems—in which the most important characteristics are engineered—with approaches that allow certain characteristics to emerge from the interaction of the information system's components. The chapter begins with an introduction to Boolean networks and certain properties that are relevant to the analysis herein.

Boolean networks: Their structure and their dynamics

Given the vast number of papers already written on both the topology and dynamics of such Boolean networks there is no need to go into too much detail here. The interested reader is encouraged to look at Kauffman [4] for his application of Boolean networks to the problem of modeling genetic regulatory networks. A short online tutorial is offered by Lucas (2006), but the basics are provided herein.

A Boolean network, which is a particularly simple information system, is described by a set of binary gates $S_i=\{0,1\}$ interacting with each other via certain logical rules and evolving discretely in time. The binary gate represents the on/off state of the 'atoms' (or, 'agents') that the network is presumed to represent. So, in a genetic network, for example, the gate represents the state of a particular gene. The logical interaction between these gates would, in this case, represent the interaction between different genes. The state of a gate at the next instant of time (t+1) is determined by the k inputs or connectivity $S_{j_i^m}$ (m=1,2,...,k) at time t and the transition function f_i associated with site i:

$$S_i(t+1) = f_i\left(S_{j_i^1}(t), S_{j_i^2}(t), ..., S_{j_i^k}(t)\right)$$

There are 2^k possible combinations of k inputs, each combination producing an output of 0 or 1. Therefore, we have 2^{2^k} possible transition functions, or rules, for k inputs. For example, if we consider a simple network in which each node contains 2 inputs, there are $2^2=4$ possible combinations of input—00, 01, 10, 11—and each node can respond to this set of input combinations in one of $2^{2^2}=16$ different ways. Figure 1 illustrates a simple example. The figure shows not only an example network (a), but also its space-time evolution (b) and its phase space attractors (c) which will be discussed next. Figure 1d shows the sixteen different Boolean functions that can be created from two inputs.

The state, or configuration, space for such a network contains 2^N unique states, where N is the size of the network (so for the simple example shown, state space contains 64 states). Because state space is limited in size, as the system is stepped forward it will eventually revisit a state it has visited previously. Combine this with the fact that from any state the next state is unique (although multiple states may lead to the same state, only one state follows from any state—a characteristic known as irreversibility), then any Boolean network will eventually follow a cycle in state space. As a result, state space (or phase space) is connected in a non-trivial way, often contain-

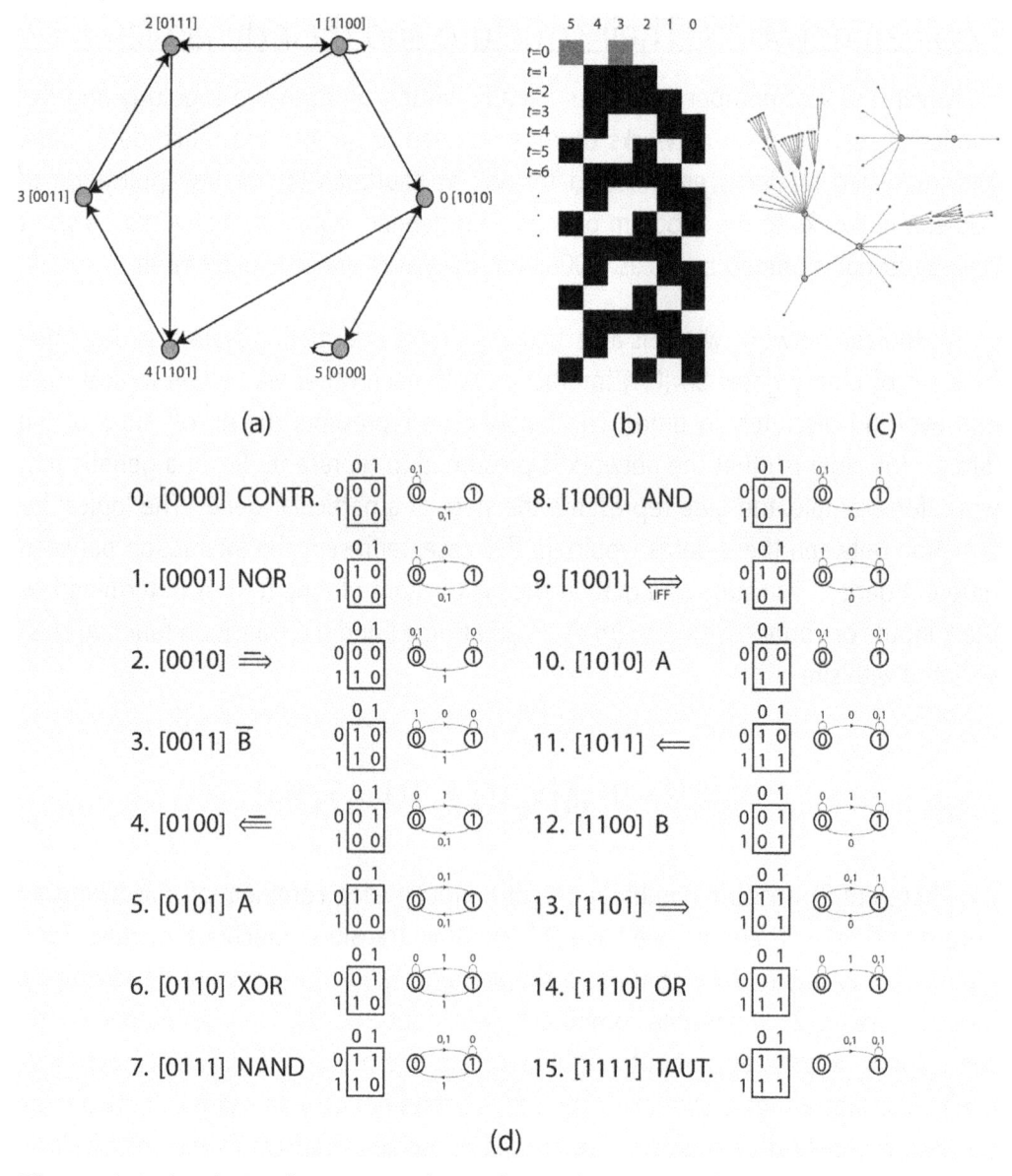

Figure 1 *A simple Boolean network (a) shows the set-up of a particular network with N=6. Each node is numbered 0 thru 5. The binary sequences in between the square brackets represent the Boolean function for each node. Note that nodes 2 and 5 are connected to themselves. (b) shows the space-time diagram for the evolving network. Each column shows the evolution of a particular nodes, and each row shows the overall network state at a particular time-step. In this case the network was seeded at t=0 with a random sequence. After an initial settling down period, the sequence converges of a period-3 cycle. (c) shows the two state space attractors that characterize this particular network: a period-3 (p3) and a period-2 (p2). (d) shows all the sixteen Boolean logic functions that can be constructed from two inputs depicted in the standard way with both look-up tables and machine state diagrams. In this study the two rules CONTR. (contradiction) and TAUT. (tautology) are excluded.*

ing multiple attractors each surrounded by fibrous branches of states, known as transients. Figure 1c shows the state space attractors, and their associated basins, for the particular Boolean network shown in Figure 1a. The transition functions for each node were chosen randomly, but did not include the constant rules 0 (0000), or 15 (1111) as these force the node to be input-independent—nodes with constant transition functions that do not change from their initial state.

Figure 1c shows the two attractors for this particular example: one period-3 attractor basin containing 56 states, and a single period-2 attractor basin containing the remaining 8 states (a total of 2^5=64 states). In this example only five states actually lie on attractors and the remaining 59 lie on the transient branches that lead to these four cycle states. When a Boolean network is simulated, it is usually seeded with an initial random configuration. As the network is stepped forward one time-step at a time, there is an initial settling down period (sometimes referred to as the relaxation time) before the system converges onto a particular attractor. There are often many different routes to a particular attractor, and these are represented by the branching structures (or, transient branches) shown in Figure 1c. This relaxation period might be compared to the settling down period in human systems that occurs when a new strategy is implemented and people take time to find their place in the new strategy before the overall system stabilizes in one particular direction or another. Of course, this relatively stable period may only be temporary as adjustments to strategy are made in response to changing requirements often triggered by particular environmental conditions.

Figure 2a is another way of visualizing how state space is connected. Each shade of grey represents one of the two attractor basins that together characterize state space. The figure comprises a grid with state 000000 at the uppermost left position and state 111111 at the lowermost right position. It shows, for example, that the 8 states that converge on the period-2 attractor are distributed in four clusters, each containing two states. This particular representation is called a destination map, or phase space portrait, and we can see that even for small Boolean networks, the connectivity of state space is non-trivial. Figure 2b is a destination map for an N=15, k=2 Boolean network (whose state space is characterized by four attractors: 2p2, 2p3), to illustrate how quickly the complexity of state space connectivity, or topology, increases as network size, N, increases. The boundaries between different shades are known as separatrices, and crossing a separatrix is equivalent to a move from one attractor basin into another (adjacent) attractor basin, i.e., if the system is pushed across a separatrix then there will be a qualitative change in dynamic behavior—this is also known as bifurcation.

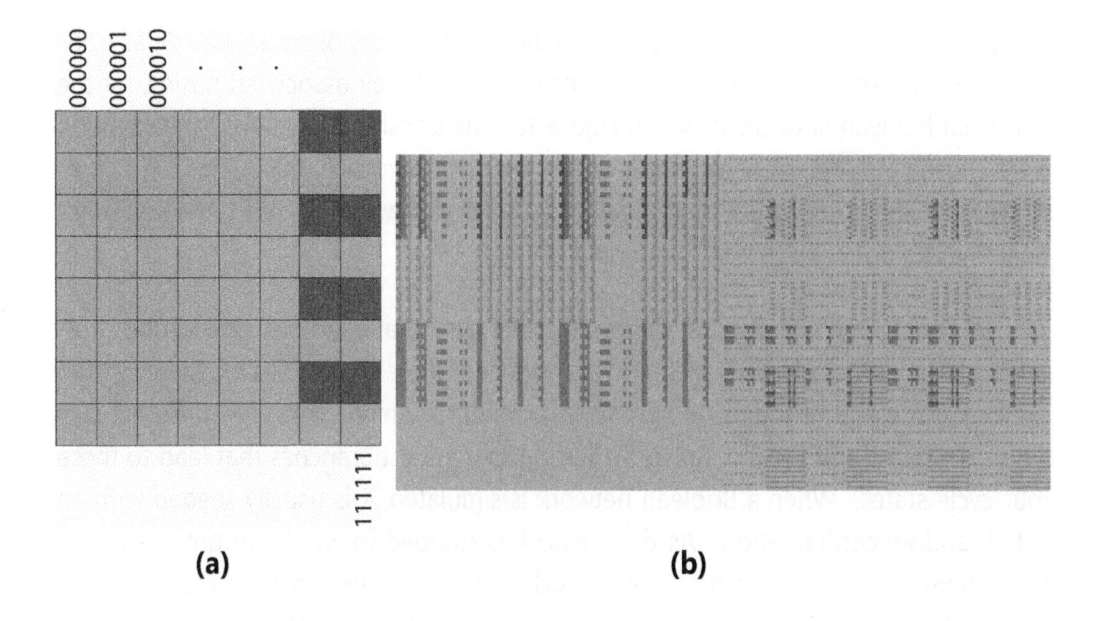

Figure 2 *The destination maps for (a) the network depicted in Figure 1a (N=6, k=2), and (b) a particular N=15, k=3 network (the grid showing individual states is not included for clarity).*

It is all well and good to have techniques to visualize the dynamics that results from the Boolean network structure (and rules). But how are a network's structure and its dynamics related? It is in the consideration of this question that we can begin to explore the role of information flows in such systems.

The potentially complex dynamics that arises from the apparently simple underlying structure is the result of a number of interacting structural feedback loops. In the Boolean network modeled to create the images in Figure 1, there are seven interacting structural feedback loops: two period-1, two period-2, two period-3, and one period-5, or 2P1, 2P2, 2P3, 1P5 for short. The P5 structural loop in this case is: Node 0 → 4 → 3 → 2 → 1 → 0. It is the flow (and transformation) of information around such structural loops, and the interactions between different loops (or, inter-loop connectivity) that results in the two state space attractors in Figure 1c; 1p1, 1p3 for short (here I use 'P' for structural, or architectural, loops and 'p' for state, or phase space loops, or cycles). The number of structural P-loops increases exponentially (on average) as N increases, for example, whereas the number of state space p-loops increases (on average) in proportion to N (not $\sqrt{N}$ as Kauffman, 1993, reported, which was found to be the result of sampling bias by Bilke & Sjunnesson, 2002).

For networks containing only a few feedback loops it is possible to develop an algebra that can relate P-space to p-space, i.e., it is possible to construct some very simple and robust rules that relate a network's architecture to its function much in the same way as it is rather trivial (when compared to complex nonlinear dynamical systems) to build serial computers with particular functionality. However, as the number of interacting p-loops increases this particular problem becomes intractable very quickly indeed, and the ability to bootstrap from P- to p-space becomes utterly impossible.

Sometimes the interaction of a network's P-loops will result in state space (p-space) collapsing to a single period-1 attractor in which every point in state space eventually reaches the same state. Sometimes, a single p-loop will result whose period is much larger than the size of the network—such attractors are called quasi-chaotic (and can, incidentally, be used as very efficient and effective random number generators). Most often multiple attractors of varying periods result, which are distributed throughout state space in non-trivial ways.

Before moving on to consider the robustness of Boolean networks, which will then allow us to consider the role information barriers play in network dynamics, it should be noted that state space, or phase space, can also be considered to be functional space. The different attractors that emerge from a network's dynamics represent the network's different functions. For example, in Kauffman's analogy with genetic regulatory networks, the network structure represents the genotype, whereas the different phase space attractors represent the resulting phenotype, each attractor representing a different cell type. Another example might be in team dynamics in which the modes of forming, storming, norming, and performing are each represented in state space by a different attractor. A further example might be in computer network operations. Here again, different operational modes, including both stable and unstable (e.g., a crash) modes can be considered as different state space attractors. An appreciation of a system's state space structure tells us about the different responses a system will have to a variety of external perturbations, i.e., it tells us which contextual archetypes the system 'sees' and is able to respond to.

Dynamical robustness

The dynamical robustness of networks is concerned with how stable a particular network configuration, or operational mode, is under the influence of small external signals[1]. Most information systems will exhibit a range of qualitatively different modes

---

1. Some complexity researchers distinguish between external and internal perturbations. This is a problematic dichotomy as it requires us to determine if a perturbating signal comes from within a system or from its environment. To make such a determination requires us also to have a complete definition

of operation. Only some of those modes will be desirable, and others (like system wide crashes) will be undesirable. An understanding of a system's dynamical robustness provides insights into which modes are most likely (i.e., which ones dominate state space), and how easily the system's behavior can be 'pushed'—deliberately or not—into a different mode. In Boolean networks we can assess this measure by disturbing an initial configuration (by flipping a single bit/input, i.e., reversing the state of one of the system's nodes) and observing which attractor basin the network then falls into (after some relaxation period). If it is the same attractor that follows from the unperturbed (system) state then the state is (qualitatively) stable when perturbed in this way. An average for a particular state (or, system configuration) is obtained by perturbing each bit in the system state, or each node, and dividing the number of times the same attractor is followed by the network size. For a totally unstable state the robustness score would be 0, and for a totally stable state the robustness score would be 1. The dynamical robustness of the entire network is simply the average robustness of every system state in state space. This measure provides additional information concerning how state space is connected in addition to knowing the number of cyclic attractors, their periods, and their weights (i.e., the volume of state space they occupy). By way of example, the average dynamical robustness of the network depicted in Figure 1a is 0.875 which means the network is (qualitatively) insensitive to 87.5% of all possible 1-bit perturbations.

---

of what the system itself is. For example, Boolean networks, as mentioned above, have been used to describe genetic regulatory networks. In the field of genetics, mutations in the nucleotide sequence of the genetic material of an organism can occur in response to environmental conditions (such as exposure to ultraviolet or ionizing radiation, chemical mutagens, and viruses), or can be induced by the organism itself (by cellular processes such as hypermutation). However, the concept of internal perturbations in the context of Boolean networks is meaningless. Such networks are 100% deterministic, and so the only way they deviate from a pre-determined trajectory is in response to external intervention. This raises the possibility that the relationship between real world perturbations and modeled perturbations is in itself a complex issue. It is quite reasonable that if we observe some real world systemic behavior that we are tempted to label as some kind of internal (to the system) perturbatory mechanism, then all we are really acknowledging is the incompleteness of the model we're using to understand that system. In this sense all perturbations can only be external; external to the model we are using to describe the system of interest, but not necessarily external to the real world system we're trying to describe with our model.

More on structure and dynamics: Walls of constancy, dynamics cores and modularization

As we have come to expect in complex systems research, there is always more to the story than what at first meets the eye. This is also the case for the relationship between network structure and dynamics. Although information flows (and is transformed) around the various structural networks, certain logical couplings 'emerge' that force particular nodes into one state or another, keeping them in that state for as long as the network is simulated. In other words, once the network is initiated/seeded and run forward, after some seemingly arbitrary transient period (the relaxation time) some nodes cease to allow information to pass. These 'fixed', or 'frozen' nodes effectively disengage all structural feedback loops that include them—although these structural loops still exist, they are no longer able to cycle information continuously around them. As such we can refer to them as non-conserving information loops; information that is placed onto such loops will flow around the loop until a frozen node is reached, after which the information flow is blocked. A number of such nodes can form walls of constancy through which no information can pass, and effectively divide the network up into non-interacting sub-networks, i.e., the network is modularized. This process is illustrated in Figure 3.

Nodes/loop types and structural attractors

The process is actually more complicated than what is depicted in Figure 3. When the occurrence of 'frozen' nodes is examined more closely we find that some nodes are not frozen in all the different operational modes (attractors). As such we can distinguish between four different types of information loop:

1. *Type 1*: Those that contain frozen nodes (that are frozen to the *same* state, 0 or 1) in all modes of operation (attractors);
2. *Type 2*: Those that contain the frozen nodes in all modes of operation, but the frozen state may *differ* from one mode of operation to another;
3. *Type 3*: Those that contain frozen nodes in some modes of operation, but may be active (i.e., contain no frozen nodes) in other modes—these act as non-conserving loops in some contexts, but as conserving loops in other contexts);
4. *Type 4*: Those that contain no frozen nodes in any modes of operation, i.e., information conserving loops.

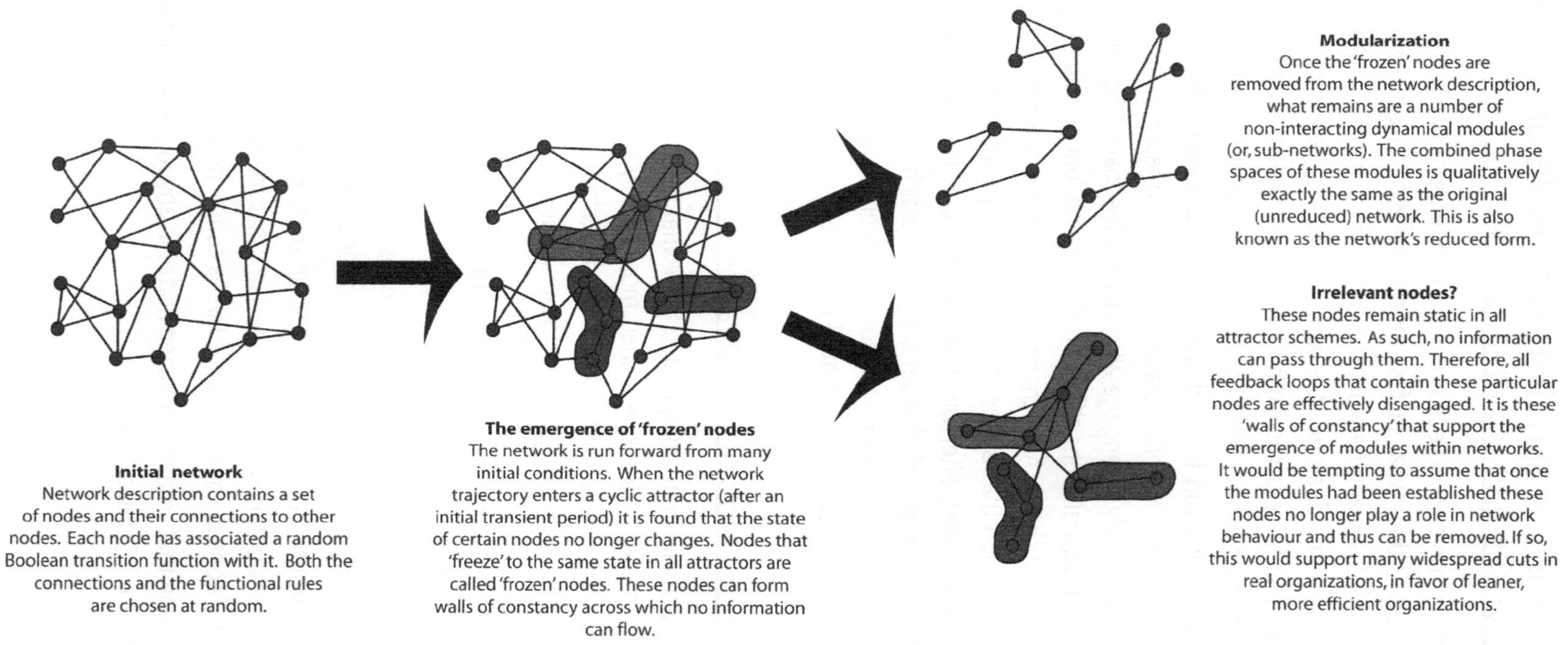

Initial network
Network description contains a set of nodes and their connections to other nodes. Each node has associated a random Boolean transition function with it. Both the connections and the functional rules are chosen at random.

The emergence of 'frozen' nodes
The network is run forward from many initial conditions. When the network trajectory enters a cyclic attractor (after an initial transient period) it is found that the state of certain nodes no longer changes. Nodes that 'freeze' to the same state in all attractors are called 'frozen' nodes. These nodes can form walls of constancy across which no information can flow.

Modularization
Once the 'frozen' nodes are removed from the network description, what remains are a number of non-interacting dynamical modules (or, sub-networks). The combined phase spaces of these modules is qualitatively exactly the same as the original (unreduced) network. This is also known as the network's reduced form.

Irrelevant nodes?
These nodes remain static in all attractor schemes. As such, no information can pass through them. Therefore, all feedback loops that contain these particular nodes are effectively disengaged. It is these 'walls of constancy' that support the emergence of modules within networks. It would be tempting to assume that once the modules had been established these nodes no longer play a role in network behaviour and thus can be removed. If so, this would support many widespread cuts in real organizations, in favor of leaner, more efficient organizations.

Figure 3 *The process of modularization in complex networks*

Given that it is the existence of frozen nodes, or not, that determines whether an information loop is Type 1, 2, 3 or 4, we can also label nodes as Type 1, 2, 3 or 4 also. Throughout the rest of this paper when we talk of frozen nodes, or non-conserving information loops, we are only concerned with Type 1 as these do not contribute to the gross characteristics of state space at all. But before moving on to consider the role of Type 1 nodes/loops at length I'd like to briefly mention the role of Types 2 and 3.

Assuming a particular network contains all types of nodes/loops, when the system is following different attractors, different sets of nodes/loops will be 'active'. As we shall see below, only the dynamic core (which comprises all types except Type 1 nodes/loops) contributes to the number and period of attractors in state space. However, the allocation of node/loop type is a global one in that we must consider the behavior of all nodes and loops in all modes of operation. If we consider the behavior of all nodes and loops in only one particular mode then each node becomes either frozen or not, and each loop becomes either conserving or non-conserving, i.e., Types 2 and 3 are effectively either Types 1 or 4 at the local attractor level. As a result there exists another level of modularization. The modularization process discussed above is the result of the emergence of only Type 1 nodes/loops and as such the modules identified in this way are the largest in size, but the fewest in number (because every operational mode—attractor type—is considered). However, for particular modes there may be additional nodes that are frozen (and therefore a greater number of non-conserving loops) and so certain modules may actually divide ('modularize') further. In this localized scenario, although the static structure of the network (i.e., the pre-simulated structure) is the same in all modes of operation, the emergent dynamic structure is (potentially) different for each state space structure (attractor). Peter Allen (2001) uses the term 'structural attractor', which is strongly related to the emergent dynamic structure discussed here. The relationship between structural attractors and state space attractors is complex (as one might expect). For example, the same structural attractors may account for different state space attractors (although in the case of Boolean networks I would expect those different state space attractors to be the same in most respects, such as period for example).

I will not explore the role of Type 2 and Type 3 nodes/loops further herein, but I suspect that the key to understanding the dynamics of (discrete) complex networks at both the global (state space wide) and local (for particular state space attractors) levels is through an understanding of how these different types form and interact.

The emergence of a dynamic core

The identification of 'frozen' nodes is non-trivial (and even 'non-possible') before the networks are simulated. Although the net effect is the same as associating a constant transition function (i.e., contradiction or tautology) with a particular node, the effect 'emerges' from the dynamic interaction of the structural feedback loops (and is often dependent on initial conditions). It is a rare case indeed that these interactions can be untangled and the emergent frozen nodes identified analytically beforehand.

As already mentioned above, these 'frozen' nodes (Type 1) do not contribute to the qualitative structure of state space, or network function. A Boolean network characterized by a 1p3, 1p2 state space, like the one considered above, will still be characterized by a 1p3, 1p2 state space after all the (Type 1) frozen nodes are removed. In this sense, they don't appear to contribute to the network's function—they do block the flow of information, but from this perspective have no functional role. Another way of saying this is that, if we are only concerned with maintaining the qualitative structure of state space, i.e., the gross functional characteristics of a particular network, then we need only concern ourselves with information conserving loops; those structural feedback loops that allow information to flow freely around the network. This feature is illustrated in Figure 4 where the state space properties of a particular Boolean network, and the same network after the non-conserving information loops have been removed (i.e., the network's 'reduced' form), are compared. Although the transient structure (and basin weight) is clearly quite different for the two networks, they both have the same qualitative phase space structure: 2p4—we might say that they are functionally equivalent.

The process employed to identify and remove the non-conserving loops is detailed in Richardson (2005a). As network size increases it becomes increasingly difficult to determine a network's state space structure. As such reduction techniques are not only essential in facilitating an accurate determination, but also in research that attempts to develop a thorough understanding of the relationship between network structure and network dynamics (as, already mentioned, only information conserving structural loops contribute to the network's gross functional characteristics). Reduction techniques basically sample state space and identify which nodes are frozen to the same state (0 or 1) in all attractors (see the definition of Type 1 nodes above). These nodes, and nodes that have no outgoing connections (and any connections associated with these nodes), are removed from the network description to form a 'reduced' version. Another way of saying this is that all non-conserving information loops are identified and deactivated. What remains after this reduction process is

B	A	O
0	0	0
0	1	1
1	0	1
1	1	0

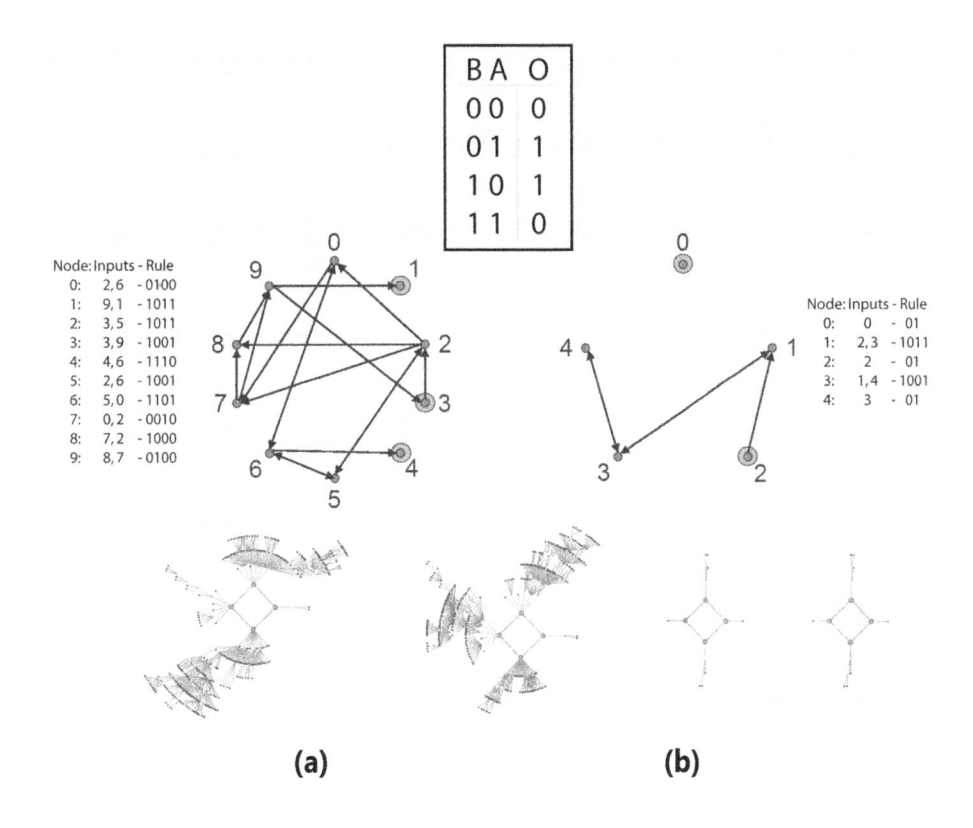

Node: Inputs - Rule
0: 2,6 - 0100
1: 9,1 - 1011
2: 3,5 - 1011
3: 3,9 - 1001
4: 4,6 - 1110
5: 2,6 - 1001
6: 5,0 - 1101
7: 0,2 - 0010
8: 7,2 - 1000
9: 8,7 - 0100

Node: Inputs - Rule
0: 0   - 01
1: 2,3 - 1011
2: 2   - 01
3: 1,4 - 1001
4: 3   - 01

(a)   (b)

Figure 4 *An example of (a) a Boolean network, and (b) its reduced form. The nodes which are made up of two discs feedback onto themselves. The connectivity and transition function lists at the side of each network representation are included for those readers familiar with Boolean networks. The graphics below each network representation show the attractor basins for each network. The state spaces of both networks contain two period-4 attractors, although it is clear that the basin sizes (i.e., the number of states they each contain) are quite different. The window located at the top-middle of the figure illustrates how different rules are applied (in the two-input case). The different combinations of the two-input, A and B, are mapped to outputs, O. In the example given (which is for rule [0110]) the 'rule table' shows that if both A and B or zero or one, then the next state (O) of the node that has A and B as inputs, will be 0. This is the essence of how Boolean networks are stepped forward in time.*

known as the network's dynamic core. A dynamic core contains only information conserving loops. The majority of (random) Boolean networks comprise a dynamic core (which may be modularized) plus additional nodes and connections that do not contribute to the asymptotic dynamics of the network (i.e., the qualitative structure of state space). A full description of a system's dynamic core is the smallest description of the system that still contains the essence of what that system's function is: all other structure is superfluous.

If we consider a clock: at the heart of most clocks found in the home there is some kind of time-keeping mechanism and another mechanism that represents the time kept to observers of that clock. Of course, these mechanisms are often surrounded by ornate cases and other sub-systems designed for aesthetic reasons, or to protect the main mechanisms from external influence. We could regard the central time-keeping mechanisms as this particular system's dynamic core, with the other frills and flourishes largely being irrelevant if our concern is with the time-keeping functionality of the clock. It should be noted though that in this example, the existence of modules, sub-systems, barriers, etc. are engineered into the system and that the dynamic core is trivial to identify (once a particular functional perspective is chosen). In the Boolean networks considered in this study, the existence of modules, sub-systems, barriers, etc. emerges through the nonlinear interaction of the various system components, and so details of their existence cannot (normally) be determined beforehand. It is likely that in real world networks such as the World Wide Web (WWW) some of the structure will be hard-wired (designed) into the network, and other structures (which may not be easily identified) emerge.

Another way of thinking about a complex network's dynamic core is in the process of modeling itself. Complex systems are incompressible (e.g., Cilliers, 1998; Richardson, 2005b). This statement simply asserts that the only complete model of a particular complex system is the system itself[2]. This is, however, quite impractical, especially when considering systems such as the WWW or a natural ecology. Instead of regarding our models as efforts to represent a system in its entirety we can say, using the language developed above, that the aim of any modeling process is to find an adequate representation of a system's dynamic core, i.e., those processes that are central to the functionality of the system.

---

2. Of course there is no such thing as a complete model. Models, by their very nature, are always incomplete. The importance of the concept of incompressibility is to highlight that simplifications of complex systems can often lead (but not necessarily so) to understanding that is qualitatively incomplete. Contrast this situation to that of complicated systems—i.e., systems that comprise many components that may also be related via nonlinear interactions but without the connectivity profile (topology) that would lead to the multiple nonlinear and interacting feedback loops that distinguish complex systems—for which simple representations (models) can always be constructed that are qualitatively complete.

The role of non-conserving (structural) information loops

As the gross state space characteristics of a Boolean network and its dynamic core (or, reduced form) are the same, i.e., (in this sense) they are functionally equivalent, it is tempting to conclude that (Type 1) non-conserving information loops—information barriers—are irrelevant. If this were the case then it might be used to support the widespread removal of such 'dead wood' from complex information systems, e.g., human organizations. The history of science is littered with examples of theories which once regarded such and such a phenomena as irrelevant, or 'waste', only to discover later on that it plays a very important role indeed. The growing realization that 'junk DNA' (DNA being probably the most intricate information system known to us, other than the brain perhaps) is not actually junk is one such example. What is often found is that a change of perspective leads to a changed assessment. Such a reframing leads to a different assessment of non-conserving information loops. Our limited concern, thus far herein, with maintaining a qualitatively equivalent state space structure in the belief that a functionally equivalent network is created, supports the assessment of non-conserving information loops as 'junk'. However, this assessment is wrong when explored from a different angle.

There are at least two roles that non-conserving information loops play in random Boolean networks:

1. The process of modularization, and;
2. The maximization of robustness.

Modularization in Boolean networks

We have already briefly discussed the process of modularization above. This process, which we might label as an example of horizontal emergence (Sulis, 2006), was first reported in Bastolla & Parisi (1998). It was argued that the spontaneous emergence of dynamically disconnected modules is key to understanding the complex (as opposed to ordered and quasi-chaotic) behavior of complex networks. So, the role of non-conserving information loops is to limit the network's dynamics so that it does not become overly complex, and eventually quasi-chaotic (which is essentially random in this scenario: when you have a network with a high-period attractor of say, 250-1, which is not hard to obtain, it scores very well indeed against all tests for randomness. One such example is the lagged Fibonacci random number generator, another is linear feedback shift registers (LFSRs) in microelectronics).

In Boolean networks the resulting modules are independent of each other, so the result of modularization, is a collection of completely separate subsystems. This independency is different from what we see in nature, but the attempt to understand natural complex systems as integrations of partially independent and interacting modules is arguably a dominant theme in the life sciences, cognitive science, and computer science (see, for example, Callebaut & Rasskin-Gutman, 2005). It is likely that some form of non-conserving, or perhaps 'limiting', information loop structure plays an important role in real world modularization processes. Another way of expressing this is: organization is the result of limiting information flow.

The concept of modularization (which is an emergent phenomenon) appears to be similar to Herbert Simon's concept of near decomposability (Simon, 1962/2006). In his seminal paper "The Architecture of Complexity" Simon developed his theory of nearly decomposable systems, "in which the interactions among the subsystems are weak, but not negligible" (p. 474). Simon goes on to say:

> *At least some kind of hierarchic systems can be approximated successfully as nearly decomposable systems. The main theoretical findings from the approach can be summed up in two propositions: (a) in a nearly decomposable system, the short run behavior of each of the component subsystems is approximately independent of the short-run behavior of the other components; (b) in the long run, the behavior of any one of the components depends in only an aggregate way on the behavior of the other components.* (p. 474)

The process of modularization described above is slightly different, but complementary. Near decomposability suggests that complex systems evolve such that weakly interacting subsystems emerge that, on sufficiently short timescales, can be considered as independent from each other (although it says little about the dynamic processes that lead to this state). This phenomenon is clearly observed in Boolean networks, except that in these particular complex systems the emergent subsystems (modules) are completely independent of each other—information does not flow from one subsystem to another. If there were no communications with the system's environment then we could simply consider these independent subsystems in isolation. However, we shall see in the following discussions on dynamical robustness that the 'padding' between these subsystems—the (Type 1) non-conserving information loops, or frozen nodes—plays an important role in the overall system's (the network of subsystems) dynamics in the face of external 'interference'. (It should be noted that, although the emergent modules discussed here are independent, in larger modules sub-modules may emerge that can indeed be viewed as weakly interacting subsystems. Near decomposability and modularization are related, but not in a trivial way.)

Dynamic robustness of complex networks

Dynamic robustness was defined above, in a slightly different way, as the stability of a network's qualitative behavior in the face of small external signals. In this section we will consider the dynamic robustness of an ensemble of random Boolean networks and their associated 'reduced' form to assess any difference between the two.

To do this comparison the following experiment was performed. 10^6 random Boolean networks with $N=15$ and $k=2$ (with random connections and random transition functions, excluding the two constant functions) were constructed. For each network its dynamic core was determined using the method detailed in Richardson (2005). The average dynamical robustness was calculated for both the (unreduced) networks and their associated dynamic cores. The data from this experiment is presented in Figure 5, which shows the relationship between the unreduced and reduced dynamic robustness for the 10^6 networks considered. The black points shows the average value of dynamic core (or, reduced) robustness for various values of unreduced robustness. On average, the dynamic robustness of the reduced networks is typically of the order of 20% less than their parent (unreduced) networks. Of course, the difference for particular networks is dependent on specific contextual factors, such as the number of non-conserving information loops in the (unreduced) networks (the extent of the dynamic core, in other words). This strongly suggests that the reduced networks are rather more sensitive to external signals than the unreduced networks. In some instances the robustness of the reduced network is actually zero meaning that any external signal whatsoever will result in qualitative change. This generally occurs only in networks which have small dynamic core sizes compared to the network size. What is also interesting, however, is that sometimes the reduced network is actually more robust than the unreduced network. This is a little surprising perhaps, but not when we take into account the complex connectivity of phase space for these networks. This effect is observed in cases when there is significant change in the relative attractor basin weights as a result of the reduction process and/or a relative increase in the orderliness of state space.

In complex systems research it is important to consider the process of averaging data (and adding error bars extending a certain number of standard deviations). The data in Figure 5a is clearly multi-modal and as such one has to take care in inter-

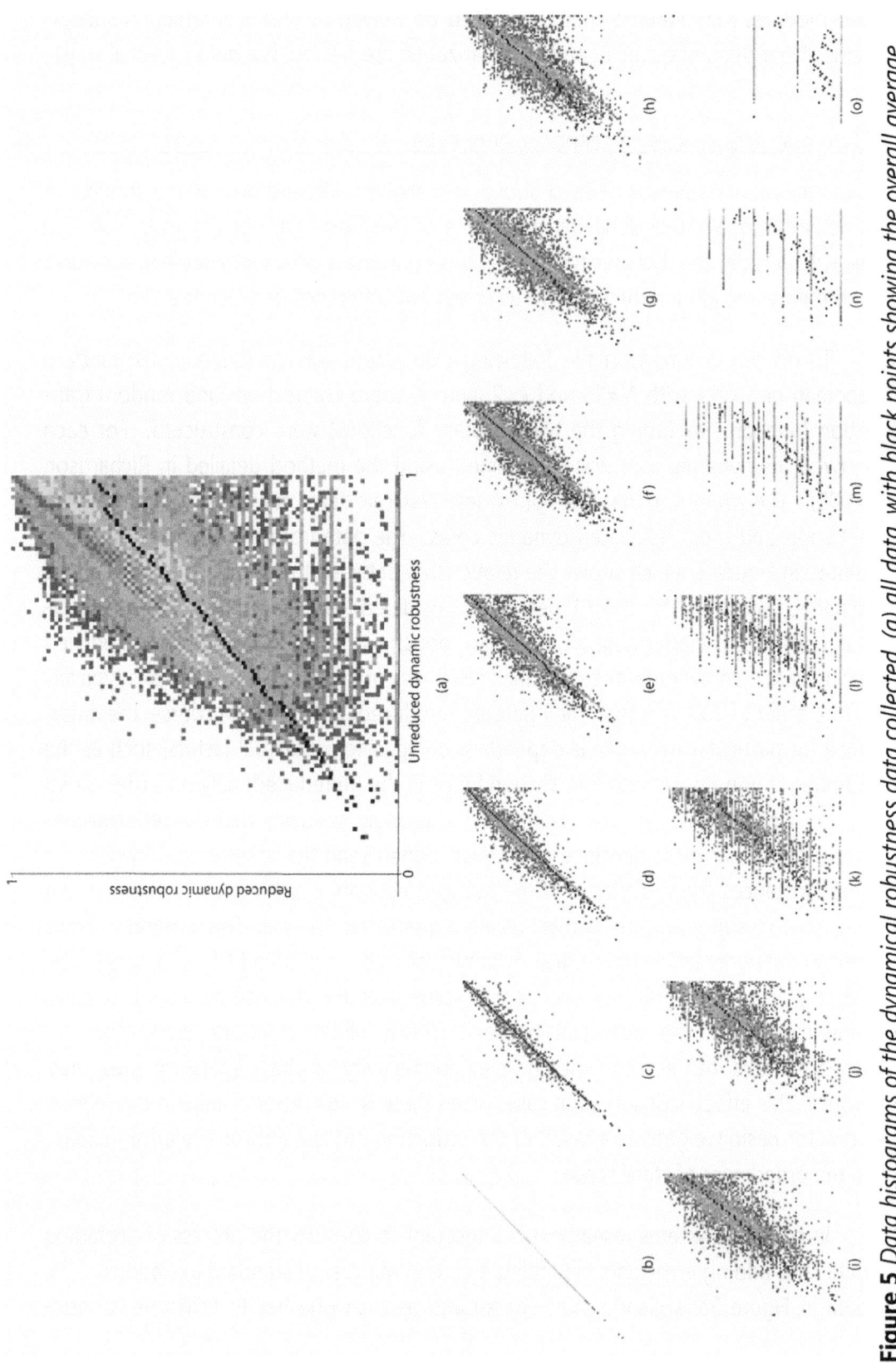

Figure 5 *Data histograms of the dynamical robustness data collected. (a) all data, with black points showing the overall average, (b) data associated with a dynamic core size of 15, (c) 14, (d) 13, (e) 12, (f) 11, (g) 10, (h) 9, (i) 8, (j) 7, (k) 6, (l) 5, and (m) 4. Data for dynamic core sizes of 3, 2, and 1 are not included.*

preting the physical meaning of the average. The different shades of grey in Figure 5 indicate the number of data points that fall into a particular data bin (the data is rounded down to 2 decimal places leading to 100 data bins). So the darkest grey regions contain many data points (> a^{10} where a is the relative frequency of the data, $a = 2.4$ in this case) and the lightest grey regions contain only one (or a^0) data point. If we had blindly calculated the average values and attempted to interpret its meaning we would have missed the importance of dynamic core size completely. Figures 5b—5m show only the data for particular sizes of dynamic core. This helps considerably in understanding the detailed structure of Figure 5a. The various diagonal 'modal peeks' relate to networks with different dynamic core sizes, and the different horizontal structures correlate with networks containing smaller dynamic cores which can have only limited values of dynamic robustness, i.e., as the size of the dynamic core decreases the data appears more discrete and as the core size increases the data appears more continuous.

Further analysis was performed to confirm the relationship between network structure, dynamic core structure and, state space structure. This included comparing the number of structural feedback loops in the overall network to the number of (information conserving only) loops in network's dynamic core. Figure 6 shows the data for all 10^6 ($N=15$, $k=2$) networks studied with all dynamic core sizes superimposed on each other. The data indicates that *on average* the dynamic core of a network has between 30% and 60% fewer structural feedback loops; all of them information conserving loops. On average only ~42% of all structural feedback loops contribute to the (global) functionality of a particular network. It should of course be noted though that this proportion is strongly dependent upon dynamic core size. In the next section we shall consider the implications of this in terms of state space characteristics and dynamic robustness.

State space compression and robustness

In Boolean networks, each additional node doubles the size of state space. In fact, in any discrete system (or any system that can be approximated as discrete), when an additional node is added, the size of state space is increased by a factor equal to the total number of states that the additional node can adopt. So in a human system, like a team for example, the addition of an extra member will increase state space by a factor equal to the number of different responses that the additional member exhibits when operating in that particular system. This needn't result in the appearance of new state space structures (attractors), but the volume of state space is greatly increased. So even if 'frozen' nodes contribute nothing to the qualitative structure of

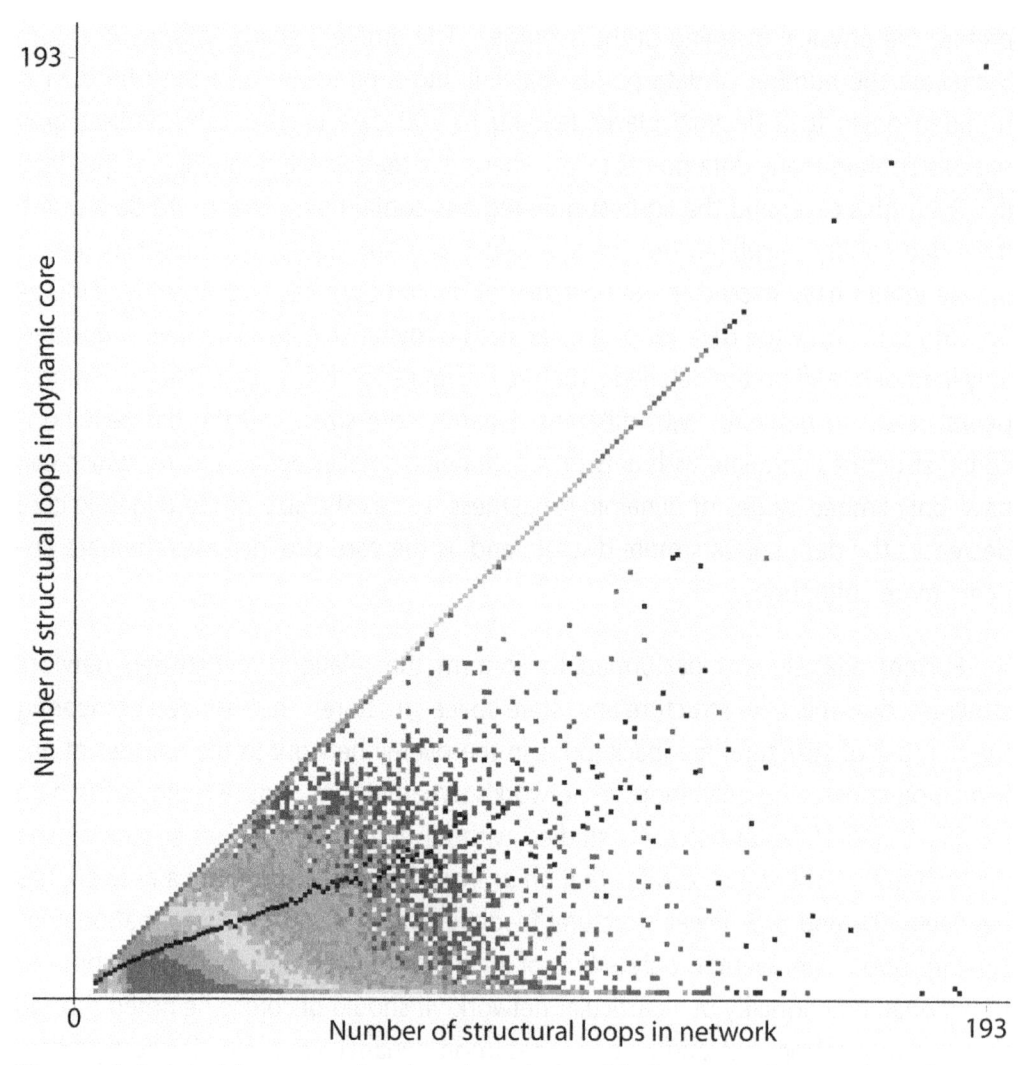

Figure 6 *A data histogram showing the relationship between the number of structural feedback loops in the unreduced networks, and the number of active structural loops in their dynamic cores. The black points represent the average number of structural loops in dynamic core.*

state space (which I shall now refer to as *first order functionality*), they at least increase the size of phase space. As an example, the phase space of an $N=20$ network is 1024 times larger than an $N=10$ network. Thus, node removal significantly reduces the size of phase space. As such, the chances that a small external signal will inadvertently target a sensitive area of state space, i.e., an area close to a separatrix, therefore pushing the network toward a different attractor, are significantly increased: leading to a kind of *qualitative chaos*. This explains why we see the robustness tending to decrease when non-conserving information loops are removed: the (emergent) buffer between the system and the outside world—the system's environment—has been removed.

Self-organization requires a container (self-*contained*-organization). Non-conserving information loops function as a kind of container. So it seems that, although non-conserving information loops do not contribute to the long term behavior of a particular network, these same loops play a central role as far as the dynamical stability is concerned. Any management team tempted to remove 80% of their organization in the hope of still achieving 80% of their yearly profits (which is sometimes how the 80:20 principle in general systems theory is (mis)interpreted in practice) would find that they had created an organization that had no protection whatsoever to even the smallest disturbances from its environment—it would literally be impossible to have a stable business.

It should be noted that the non-conserving information loops do not act as impenetrable barriers to external signals (information). These loops simply limit the penetration of the signals into the system. For example, in the case of a modularized network, the products of incoming signals may, depending on network connectivity, still be fed from a *non-conserving* information loop onto information *conserving* loops for a particular network module. Once the signals have penetrated a particular module, they cannot cross-over into other modules (as the only inter-modular connections are via non-conserving loops). As such it would seem that non-conserving loops may play an *information distribution* role.

It should also be noted that, even though a particular signal may not cause the system to jump into a *different* attractor basin, or bifurcate, it will still push the system into a different state on the *same* basin. The affect of signals that end-up on non-conserving information loops is certainly not nothing. So, although the term information 'barriers' is used herein, these barriers are semi-permeable.

Balancing response 'strategies' and system robustness

Figure 7 shows a data histogram for the number of state space attractors versus (un-reduced) dynamic robustness. The plot shows that the robustness decreases rapidly initially as the number of state space attractors increases. Remembering that the number of state space attractors can also be regarded as the number of contextual archetypes that a system 'sees', we see that versatility (used here to refer to the number of qualitatively different contexts a system can respond to, or is sensitive to) comes at the cost of reduced dynamic robustness assuming the same resources are available. Considered in this way robustness and versatility are two sides of the same coin. We would like for our systems to be able to respond to a wide variety of environmental contexts with minimal effort, but this also means that our systems might also be at

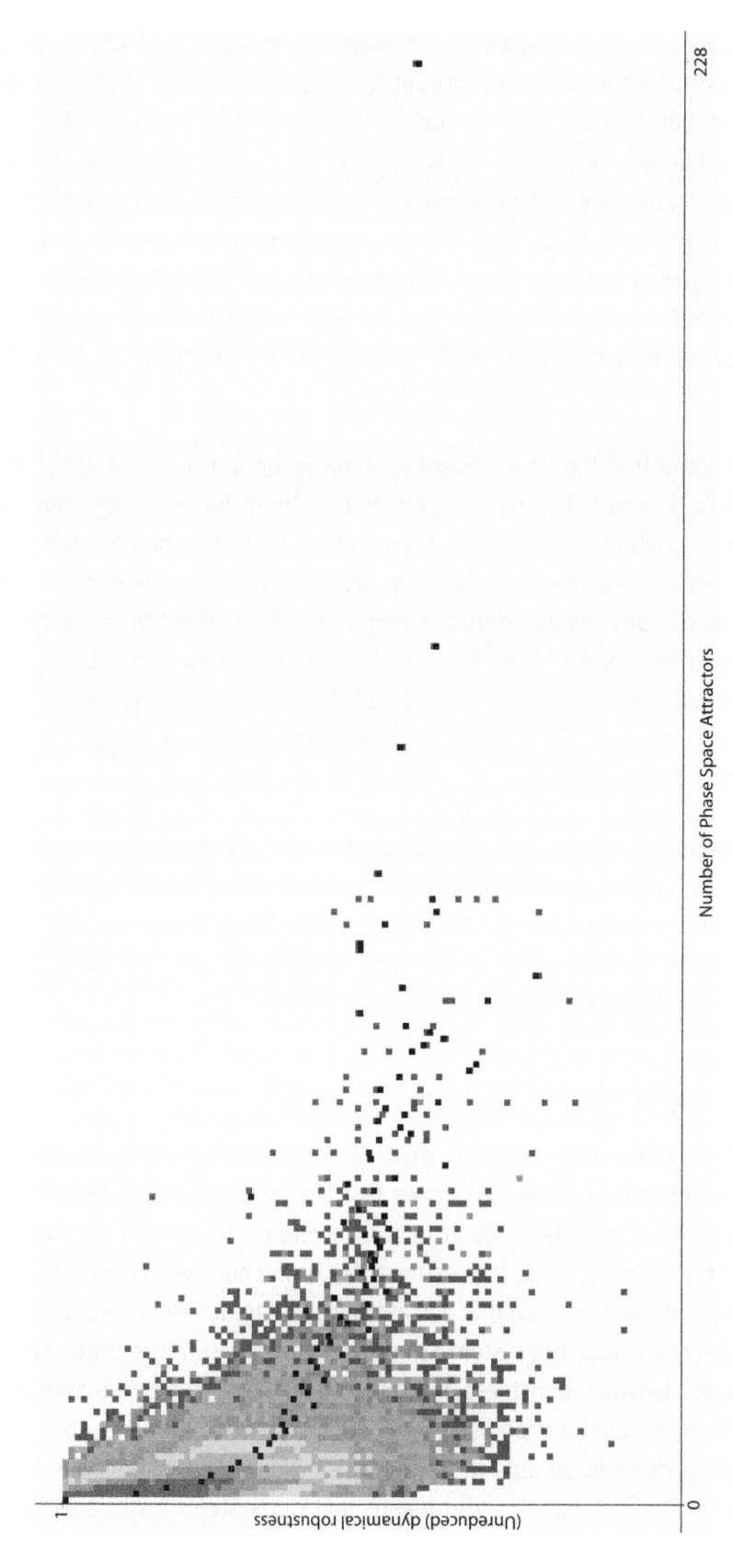

Figure 7 A data histogram showing the relationship between the number of state space attractors (which is also related to network flexibility/versatility) and the (unreduced) dynamical robustness. The black points represent the average dynamic robustness for increasing numbers of phase space attractors (the greyscaling is scaled using a^n, where $a = 2.4$ and n is an integer between 0 and 10 inclusive).

the mercy of any environmental change. A system with only one state space attractor doesn't 'see' its environment at all, as it has only one response in all contexts, whereas a system with many phase space attractors 'sees' too much—there is a price to pay for being too flexible.

This is only the case though for fixed resources, i.e., given the same resources a system with few state space attractors (modes of operation) will be more robust than a system with a greater number. This is because, on average, the system with more state space attractors will have a larger dynamic core and so the buffering afforded by non-conserving loops will be less pronounced. It is a trivial undertaking to increase phase space by adding nodes that have inputs but no outputs, or 'leaf' nodes (which is equivalent to adding connected nodes that have a constant transition function)—we might refer to this as a first-order strategy. This certainly increases the size of phase space rapidly, and it is a trivial matter to calculate the effect such additions will have on the network's dynamical robustness, i.e., the increased dynamical robustness will not be an emergent property of the network. Increasing the robustness of a network (without changing the number, period and weight of phase space attractor basins, i.e., functionality) by adding connected nodes with non-constant transition functions is a much harder proposition. This is because of the new structural feedback loops created, and the great difficulty in determining the emergent 'logical couplings' that result in 'frozen' nodes which turn structural information loops (that were initially conserving) into non-conserving loops. It is not clear at this point in the research whether there is any preference between these two strategies to increase dynamical robustness, although the emergent option is orders of magnitude harder to implement than the other, and will likely change functionality.

One key difference between the straightforward first-order strategy and the more problematic (second-order) emergent strategy is that the first-order enhanced network would not be quantitatively sensitive to perturbations on the extra (buffer) 'leaf' nodes, i.e., not only would such perturbations not lead to qualitative change (a crossing of a separatrix, or a bifurcation), but the position on the attractor cycle that the system was on when the node was perturbed (i.e., cycle phase) would also not be affected—there would be no effect on dynamics whatsoever. This is because the perturbation signal (incoming information) would not penetrate the system further than the 'leaf' node, as by definition it is not connected via any structural (non-conserving or conserving) feedback loops: they really are impenetrable barriers to information. If the emergent strategy was successfully employed, the incoming signals would penetrate and quite possibly (at most) change the cycle phase, i.e., quantitative (but, again,

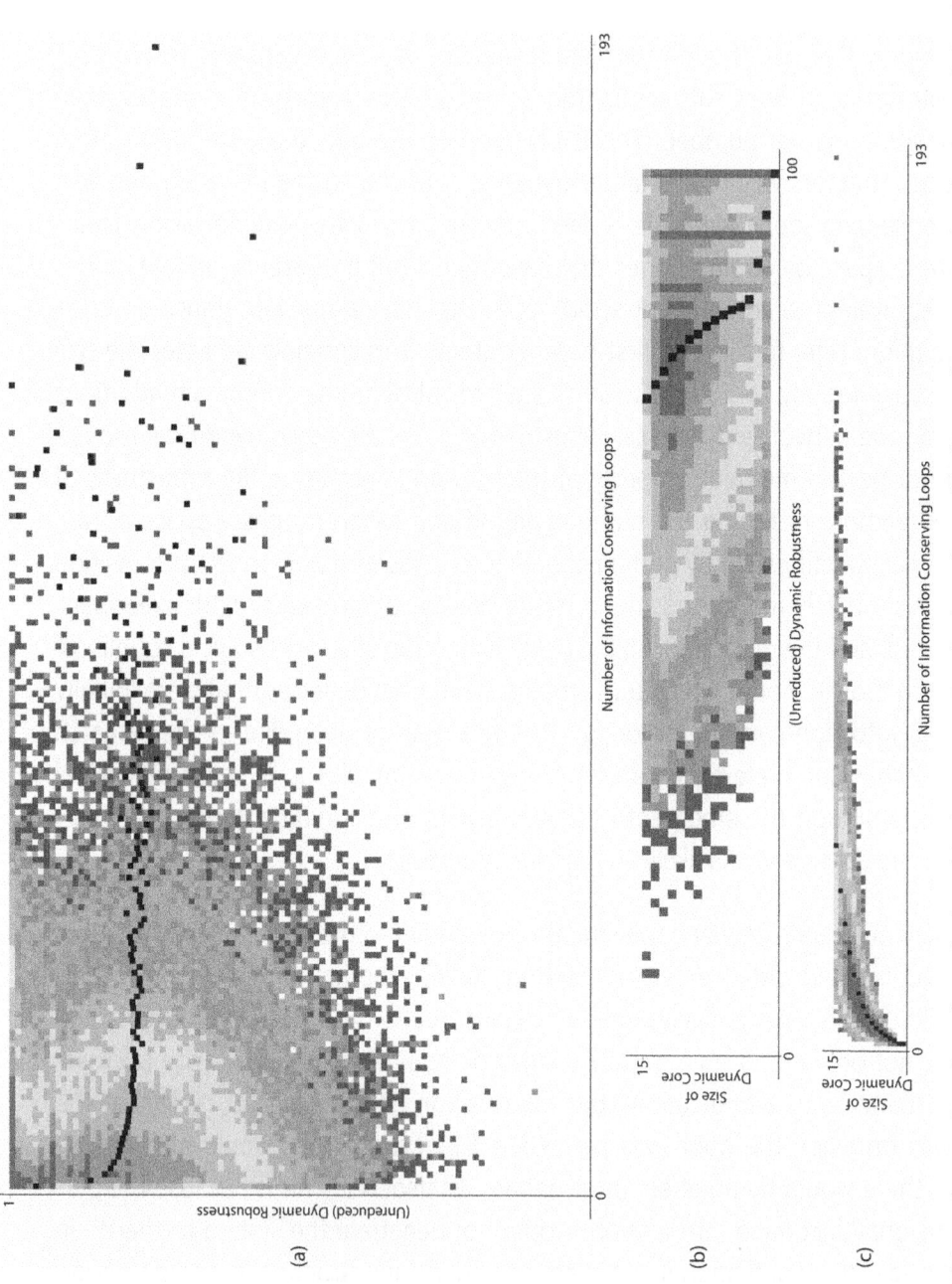

Figure 8 *Data histograms showing the relationship between: (a) the number of information conserving loops and the (unreduced) dynamic robustness (black points represent the average dynamic robustness); (b) the (unreduced) dynamic robustness and the size of the dynamic core (black points represent the average dynamic robustness for given dynamic core size); and; (c) the number of information conserving loops and the size of the dynamic core (black points represent the average number of information conserving loops for given dynamic core size).*

not qualitative) change would occur. Either one of these response traits may or may not be desirable depending on response requirements.

To complete the data analysis Figure 8 shows three data histograms that show: (a) how the (unreduced) dynamic robustness varies with the number of information conserving loops; (b) how the size of the dynamic core varies with (unreduced) dynamic robustness, and; (c) how the size of the dynamic core varies with the number of information conserving loops. Figure 8a shows that on average a network's dynamical robustness initially tends to decrease rapidly with the number of information conserving loops, but then quickly approaches saturation and stabilizes. Figure 8b shows this relationship from the difference perspective of dynamic core size rather than the number of information conserving loops connecting the elements of the dynamic core. This shows the average dynamic robustness decreasing steeply (and approximately linearly) for core sizes 0 to 3. The dynamic robustness continues to drop for core sizes > 3, but transitions to a very much slower rate (a rate which does increase slightly with increasing core size). Finally, Figure 8c shows that the average number of information conserving loops grows at an increasing rate with dynamic core size. Together these data show a preference for smaller dynamic core sizes (and hence lower numbers of conserving information loops, and therefore a higher proportion of non-conserving information loops). However, smaller dynamic core sizes lead to fewer phase space attractors and hence lower versatility as defined above. This, again, highlights the trade-off between flexibility and stability.

Discussion

From the analysis presented herein it is clear that non-conserving information loops—information barriers—play an important role in a network's global dynamics: they are not 'expendable'. They protect a network from both quantitative and qualitative (quasi) chaos. Quantitative chaos is resisted by the emergent creation of modules, or structural attractors—through the process of modularization—which directly reduces the chances of state space being dominated by the very long period attractors associated with quasi-randomness in Boolean networks. Whereas qualitative chaos—the rapid 'jumping' from one attractor basin to another in response to small external signals—is resisted by the expansion of state space, which reduces the possibility of external signals pushing a system across a state space separatrix, i.e., into another mode of operation.

Much more effort will be needed to fully understand the design and operating implications of this research for real world information systems. In one sense, Boolean networks are at the heart of all technology based systems—computers are themselves very complicated Boolean networks. However, these sorts of (complicated) Boolean networks do not usually have architectures that contain the many nonlinear interacting feedback loops that the (complex) Boolean networks discussed herein have. Complicated Boolean networks, such as computers, generally process information serially, whereas complex Boolean networks can process information in parallel. In fact, there is ongoing research that considers complex Boolean networks as powerful parallel computers (see for example, Mitchell, 1998). It may well be that the sort of research presented here has little impact on the design of engineered systems such as computers and technology-driven information nodes as these are generally hard-wired not to exhibit emergent behavior. It is in the area of soft-wired systems, perhaps, that this research may offer some insight. For example, although the individual components of the Internet are hard-wired, the way in which its global architecture is developing is through emergent processes. Recent advances in network theory provide powerful tools to understand the static structure (architecture) of complex networks such as the WWW. But, equally powerful dynamical tools will be needed to understand the dynamic structure of the WWW which is likely quite different. There is clearly someway to go to fully understand what abstract concepts such as non-conserving information loops, or 'frozen' nodes, refer to in real world information systems, but such research is beginning to provide us with the language and tools to facilitate our understanding of the structural and state space dynamics of a wide range of complex information systems like the Internet, the WWW, or even large decision networks to mention just a few.

Another useful avenue of further research may come by considering the 'activity' of each node. In this paper, we could have defined 'frozen' nodes as those with an activity of zero, i.e., their state does not change with time. The remaining nodes will have different activity levels, e.g., a node that changes state on every time step (a period-2 node) could be defined to have activity 1. The dynamic core structure of a particular network (containing non-interacting modules) is found by removing all nodes with activity 0. However, different structures will be revealed if the 'activity threshold for removal' is varied from 0 to 1. In this way, researchers would find further modularization, although modules (sub-structure) found this way would not be completely independent (the weakly interacting modules of Simon's 'nearly decomposable systems' perhaps); systems of interacting sub-systems (with varying degrees of interaction strength) within the same system would be found, which would provide further insight into the relationship between emergent structure and function. Furthermore, it is possible that such an 'activity threshold' analysis would be easier to apply to real world systems than the analysis performed herein which is only concerned with a zero activity threshold.

As mentioned at the beginning of this paper, there are certain limitations to the use of Boolean networks as surrogates for real world organizations. To close this paper, I'd like to briefly consider some obvious limitations, strategies and implications for overcoming them.

Perhaps the most obvious shortcoming is the simplicity of the transition functions themselves. However, although *Boolean* networks only allow each node to adopt two different states, there is nothing fundamentally different about extending the analysis to include an arbitrary number of different states, and an arbitrary degree of transition function complexity. The important step is the *discretization* of state space. As such we wouldn't expect the fundamental results presented herein to change.

A trickier shortcoming to address is the fact that in these discrete systems each node has only one identity, and so there is only one response to any particular signal. In human systems, there is ample evidence to suggest that human decision-makers adopt one of several different identities depending on certain contextual factors. There are (at least) two ways to address this shortcoming. The easy way is to assume that the modeled structure is valid only for periods shorter than the average time between 'identity shifts'. This would be an interesting parameter to enumerate! Another, more sophisticated approach would be to represent each node as a sub-network itself for which the different sub-network attractors would represent different identities[3]. Again, as long as state space is discrete, we would still expect to see the emergence of non-conserving information loops and a similar role for them.

This strategy of increasing the complexity of transition functions and reframing what we consider a 'node' to be (either a single node or a connected sub-network) can potentially even be employed to make the (seemingly non-adaptive) network *appear* to be adaptive, with changing connectivity at some levels, new entities emerging, and evolving identities. The frame of reference from which the network is considered enables these features to be 'seen' or not. Again, the important element is discretization of state space. I explore this particular issue of the relationship between non-adaptivity and adaptivity in discrete systems at much greater detail in Richardson (2005). What would become a challenge as we arbitrarily increased the complexity of such networks (other than the considerable challenge of actually constructing and simu-

---

3. The sub-system will still only respond to a specific signal in one way, but it will be able to 'resolve' a broader range of signals.

lating them) is not that the fundamental importance of non-conserving information loops would diminish, but that the *physical interpretation* of them would change as we 'stood further back' from the network. If we take this to the extreme, assume that a network model of fundamental matter is correct—an exquisitely complex network of simply connected simple entities (supernodes?)—then what role do non-conserving information loops (at that fundamental level) play at the human level? And what is the equivalent to non-conserving loops (at the fundamental level) at other levels of 'reality'? I certainly won't try to answer these questions here (partly because I believe they have so many different answers).

My reason for ending the paper on such a philosophical note is to simply warn against the temptation to dismiss the results of a particular analysis just because the model appears to be too simplistic. There is a deeply profound relationship between simplicity and complexity that scientists are only just beginning to understand. That being said, it might be that very simple models are more suitably employed as metaphors rather than analogies. If we chose to use the Boolean network model as a direct analogy (which I certainly think is possible at a much higher level of compositional complexity) we might find that the *exact* idea of a 'non-conserving information loop', for example, is so abstract that it would make a useful physical interpretation incredibly difficult to extract. If however, we chose to regard the model as a *useful*[4] metaphor then we are free to exploit the idea of a 'non-conserving information loop' (for example) without feeling obliged to stay honest to its exact (in the Boolean framework) definition; it becomes a useful tool for thinking rather than a tool to replace our useful thinking.

To conclude, I would suggest that 'barriers' (both impenetrable and semi-permeable) to information flow play a central role in the functioning of all complex information systems. However, the implications (and meaning) of this for real world systems is open to many different interpretations. At the very least it suggests that 'barriers'

---

4. I believe that the results of studying any simple nonlinear feedback model can be applied usefully as a (good) metaphor because there is a sense in which 'laws' uncovered for these systems are universal and valid for even the most complex systems. In short I am suggesting that even the most simple systems models (which here means at least containing many interacting nonlinear feedback loops) are good analogies (although the interpretation step is problematic) for real world complex systems. The difficulty, however, is that 'laws' for complex systems do not carry the same weight as 'laws' for simple systems: the difference in implied wording is subtle but the meaning is very different. A law for a simple system can and should be read as "it is always the case that...", whereas a law for a complex system should be read as "it tends to be the case that." The main implication of this difference in wording is that for a complex system a 'law' is only close to being absolutely true for large ensembles of systems (in the same way that statistics only works for large samples), whereas for a specific system contextual factors can conspire to 'break' the 'law', making the utilization of such 'laws' problematic.

to information flow should be taken as seriously as 'supports' to information flow (although, paradoxically, a good 'supporter' is inherently a good 'barrier'). This may seem obvious given the dual role of all types of boundary—i.e., to keep in *and* keep out, or to enable *and* disable—but this change in focus offers, perhaps, an interesting counterpoint to the emerging cultural perspective that suggests that having as much information as possible can only be a good thing.

References

Allen, P. M. (2001). "What is complexity science? Knowledge of the limits to knowledge," *Emergence*, ISSN 1521-3250, 3(1): 24-42.

Barabási, A.-L. (2001). "The physics of the Web," *Physics World*, ISSN 0953-8585, http://www.barabasilab.com/pubs/CCNR-ALB_Publications/200107-00_PhysicsWorld-PhysoftheWeb/200107-00_PhysicsWorld-PhysoftheWeb.pdf.

Barabási, A.-L. and Albert, E. (1999). "Emergence of scaling in random networks," *Science*, ISSN 0036-8075, 286(15 Oct), 509-512.

Bastolla, U. and Parisi, G. (1998). "The modular structure of Kauffman networks," *Physica D*, ISSN 0167-2789, 115: 219-233.

Bilke, S. and Sjunnesson, F. (2002). "Stability of the Kauffman model," *Phys. Rev. E*, ISSN 1539-3755, 65, 016129, http://pre.aps.org/abstract/PRE/v65/i1/e016129.

Callebaut, W. and Rasskin-Gutman, D. (2005). *Modularity: Understanding the Development and Evolution of Natural Complex Systems*, ISBN 9780262513265.

Cilliers, P. (1998). *Complexity and Postmodernism: Understanding Complex Systems*, ISBN 9780415152877.

Kauffman, S.A. (1993). *The Origins of Order: Self-Organization and Selection in Evolution*, ISBN 9780195079517.

Lucas, C. (2006). "Boolean networks: Dynamic organisms," http://www.calresco.org/boolean.htm.

Mitchell, M. (1998). "Computation in cellular automata: A selected review," in T. Gramss, S. Bornholdt, M. Gross, M. Mitchell, & T. Pellizzari (eds.), *Nonstandard Computation: Molecular Computation, Cellular Automata, Evolutionary Algorithms, Quantum Computers*, ISBN 9783527294275, pp. 95-140. Note that cellular automata are essentially Boolean networks with an ordered architecture rather than a random one.

Richardson, K.A. (2005). "Simplifying Boolean networks," *Advances in Complex Systems*, ISSN 0219-5259, 8(4): 365-381.

Richardson, K.A. (2005). "The hegemony of the physical sciences: An exploration in complexity thinking," *Futures*, ISSN 0016-3287, 37: 615-653.

Simon, H. (1962/2006). "The architecture of complexity," *Proceedings of the American Philosophical Society*, ISSN 0003-049X, 106(6): 467-482. Reprinted in K.A. Richardson, D. Snowden, P.M. Allen and J.A. Goldstein (eds.) (2006), *Emergence: Complexity &*

Organization 2005 Annual, Volume 7, ISBN 9780976681434, pp. 499-510.

Sulis, W.H. (2006). "Archetypal dynamical systems and semantic frames in vertical and horizontal emergence," in K.A. Richardson, J.A. Goldstein, P.M. Allen, and D. Snowden (eds.), *Emergence: Complexity & Organization Annual Volume 6*, ISBN 9780976681458, pp. 353-374.

Watts, D.J. and Strogatz, S.H. (1998). "Collective dynamics of 'small-world' networks," *Nature*, ISSN 0028-0836, 393(4), 440-442.

Kurt A. Richardson is the owner Emergent Publications, a publishing house founded in 2003 that specializes in complexity-related publications. He provides a variety of technical support (software, hardware and systems) to the likes of NASA, Lockheed Martin, General Dynamics, etc., through his consulting company, Exploratory Solutions. He has a BSc(hons) in Physics (1992), MSc in Astronautics and Space Engineering (1993) and a PhD in Applied Physics (1996). Kurt's current research interests include the philosophical implications of assuming that everything we observe is the result of underlying complex processes, the relationship between structure and function, analytical frameworks for intervention design, and robust methods of reducing complexity, which have resulted in the publication of 35+ articles and more than 10 books. He is the Managing/Production Editor for the international journal *Emergence: Complexity & Organization* and is on the review board for the journals *Systemic Practice and Action Research*, and *Systems Research and Behavioral Science*, and has performed adhoc reviewing for *Journal of Artificial Societies and Social Simulation* and *Nonlinear Dynamics, Psychology and the Life Sciences*. Lastly, Kurt is a qualified spacecraft systems engineer and was a senior engineer for NASA's Gamma-Ray Large Area Space Telescope (GLAST, now Fermi) for which he received a NASA Achievement Award.

Chapter Z

Noise and information: Terrorism, communication and evolution

Robert Artigiani

When "Modern" reductionist science was applied to humans and their institutions, the results ranged from irrelevant to threatening. Fortunately, "science" was not frozen in its Modern form, and by the later twentieth century it had changed enough to describe nature as an evolutionary process in which complex, integrated wholes emerge that expand environments by transforming noise into information. This nature could be mapped by computer simulations, and pioneers like Peter Allen thought the new paradigm might successfully model societies. This essay develops one such model and applies it as a tool for understanding terrorism. Its conclusions demonstrate the value of Allen's findings to policy making.

Introduction

Terrorists are hard to think about. Their brutal use of suicide bombers, in particular, appears too irrational to comprehend (Andriolo, 2002). But studies regularly find suicide terrorists to be "normal" people, often from broadly middle class backgrounds with better-than-average educations (Atran, 2003; McCauley, 2009; Pape, 2005; Post et al. 2003; Post, 2006). In other words, classical scientific techniques indicate that incomprehensible terrorists are indistinguishable from everybody else. Equating terrorists with normal people is not likely to improve our understanding of them. Borrowing from Prigogine's dissipative structures theory (Prigogine, 1980), this paper will construct a model to think with which explains how the behavior of apparently "normal" people interacting in stressed contexts is radicalized (Hutchinson, 2007). In the process it will show how religion can unintentionally serve as a vehicle for social evolution.

The contexts that most profoundly affect human choices and actions are the social systems upon which survival depends. Tens of thousands of years ago, crude and simple systems had only limited effects on behavior. Groups of people who marked themselves as members of collectives that hunted, gathered, bred, and fought together formed when population growth desiccated local environments so badly supplies of small, easily caught animals and rapidly ripening, easily gathered fruits were exhausted (Stiner *et al.*, 1999; Kuhn *et al.*, 2001). Humans were not equipped by nature to hunt large, dangerous animals alone, and protecting slow growing plants required cooperative efforts. By correlating their behavior as members of systems, therefore, people solved problems they could not solve for themselves.

With agriculture more extensively correlated efforts produced surpluses that allowed populations to continue growing. People then had no choice but to preserve their survival systems by making societies reliable locaters, producers, defenders, and processors of resources. The behavioral innovation most successful at producing systems able to find, exploit, and protect ever more resources was specialization. Specialization meant that individuals restricted the range of their activities and refined the skills they exercised. Thus, some farmed, some soldiered, some made tools and weapons, while others managed resources and policed performances. Although specialization made social systems more effective, it also made biological survival dependent on preserving societies.

Social systems offered protection from natural selection. However, to preserve systemic order actions had to be regularized and predictability took precedence over biological needs and desires. To be members of social systems individuals needed to know how to act so they could "fit in." Unable to rely on their genes for guidance, members of social systems needed new scripts to describe the specialized tasks they now performed. Early societies guided individuals ritualistically. More complex ones used centralized authorities to correlate behaviors. But time-consuming rituals inhibited adaptation, while neither rulers nor subjects could keep track of each other and remember endless lists of specific injunctions. Moreover, as circumstances changed, rules carved in stone turned into liabilities. Specialists needed more flexible rules to guide themselves, and maps for fitting in had to be distributed among all members of complex social systems.

To be simultaneously present in the brains of many people maps of how societies worked were written in symbolic terms, and new kinds of symbols emerged that mapped how actions affected the stability and survivability of the systems on which people depended. Symbols of behavioral effects mapped the meaning of actions. Maps of meaning enabled people to choose between actions by anticipating their systemic effects and could massively distribute responsibility for processing flows. If people chose incorrectly and threatened systemic stability, they were punished. Conversely, actions that improved societal well-being were rewarded. Experiencing rewards and punishments—observing the consequences of their own actions (Gouldner & Perterson, 1963)—specialists acquired self-consciousness.

Located in the brains of their members, guides for fitting into societies are "cognitive maps" (Tolman, 1948). Analogs of DNA, collectively shared cognitive maps describe societies and prescribe the behaviors necessary to preserve them. Representing human environments, obliging people to control their instincts, and guiding specialized behaviors, cognitive maps are written in the language of values, ethics, and morals (VEMs). VEMs replicate social systems by scripting social roles. Social roles are equivalents of niches in ecosystems and, by mapping the meaning of actions, VEMs motivate self-conscious individuals to choose in socially selected ways. VEM scripts, of course, are as subject to selection as DNA molecules. So for a society to succeed its VEMs must script social roles that match environmental realities.

Schematically, societies can be diagramed as three-tiered systems, with each level following its own internal rules. At ground level, elementary forces operate thermodynamically. At the second level, people act by processing sensations. Finally, at the symbolic level, information develops logically. Independently, each level could "run away." But in social systems the levels coevolve, with each acting as the others' selective environments. Sounds associated with phenomena become languages, languages become symbolic, and symbols guide behaviors selected by collectively shaped environments. Whimsical actions then become regularized into social roles that are morally sanctioned by shared VEMs and rewarded with material resources. If rewards match expectations, the validity of VEMs is demonstrated, societies are stable, people feel good about themselves, and roles are replicated in successive generations. Interactions between the three levels are diagramed in Figure 1.

While this model fits social system of all sorts, no two societies are completely alike, for they self-organize in different circumstances under the influence of distinct personalities. We can expect populations of human beings facing similar stresses to respond in similar ways. Every society is a unique solution to an N-body problem. They cannot be understood in terms of their separate dimensions, which are nonintegrable (Prigogine, 1996). Consequently policy-makers must recognize that there are always elements about human behavior that elude description. Human behavior is only understandable from the perspective of the social systems in which actions occur. Moreover, variations between societies make their VEMs and roles incommensurable. Many will invent agriculture, and, to process the increased energy agriculture releases, successful social systems become specialized, hierarchical, and statist. But whether emer-

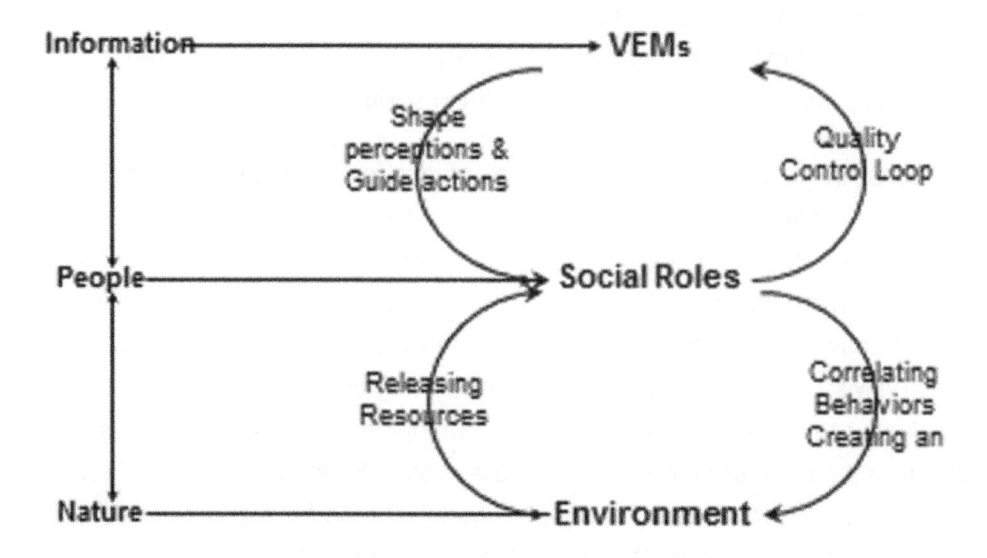

Figure 1 *Interaction and coevolution between levels*

gent complex societies are organized like Egypt or Assyria depends on their leaders, their local conditions, and their information storage systems. In any case, when social systems compete, success or failure will measure the effectiveness of their VEMs.

The results of inter-societal competition need not be profound. Societies regularly conquer one another, but in general conquest merely replaces one oligarchy with another. The lives of most people go on without many noticeable differences. Farmers continue to farm, artisans to manufacture, soldiers to fight, and priests to "manage." Following conquests, the people playing privileged social roles usually change, and new oligarchies often revise the heavenly hierarchies to map changes in relative power. Laws could change, as well. But, unless there are fundamental changes in technology or social organization, what conquered people do will bear more than a family resemblance to what their ancestors did.

The emergence of "Modern" society, however, has led to changes about as radical as those associated with the rise of civilization. Organized into nation-states, structured by new urban classes, operating powerful technologies, and mapped by science, Modern societies have spread all over the globe. And wherever Modern societies have intruded they have quickly established military dominance, introduced economic exploitation, revolutionized sociologies, and undermined traditional VEMs. A schematic model for Modern systems is approximated in Figure 2.

Modern societies

Traditional societies aim for stability, while Modern societies must be adaptive. Traditional societies focus on stability because they self-organize in preindustrial attractors. Once choices about herds, crops, and irrigation are made the goal is to replicate them generation after generation. Small surpluses discourage experimentation, so traditional societies "lock-in" to behaviors they find effective. Imitating ancestral behaviors, collective actions affect the environment in ways that release "normal" resource flows and legitimate inherited VEMs. Fearing breakdown through deviations from established norms, ritual and religion bind people to traditional actions by proclaiming inherited roles and relationships sacred. When ritual, religion, and performance function properly, role-players identify with the morally sanctioned whole so effectively that—barring external perturbations—the self is lost in the experience of solidarity.

Modern societies, by contrast, organize in more powerful flows. Modern industrial and communication technologies access new types of energy at increasingly greater rates. Probed and altered by powerful technologies, the environments in which Modern societies are embedded are more dynamic than were the environments of Traditional societies. Dynamic environments shift frequently, and, given their potential for

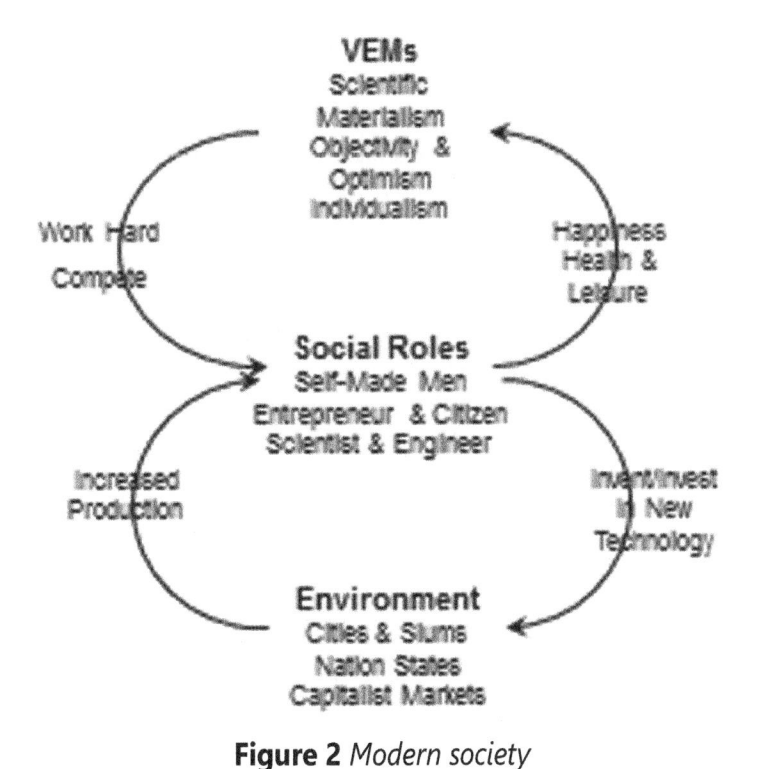

Figure 2 *Modern society*

disruptions, Modern societies must anticipate changes by reading their environments in fine detail. Social systems read environments through their members, and the more individualized their members, the more detailed readings they can make. Meanwhile, adjusting Modern societies to altered environments redefines roles and alters relationships in dramatic ways. To carry out system-saving adaptations members of Modern societies must periodically refashion identities and transform relationships.

To adapt, Modern societies bind members loosely, and to map flexible bonds, Modern societies substitute secular, pragmatic VEMs for moralizing religious symbols. In environments where new resources are suddenly available and great wealth is there for the taking, secularized VEMs pragmatically legitimizing individual successes are socially selected. Overtly functional, Modern VEMs work in all sorts of circumstances because individuals can decouple from relationships whenever their utility is lost. Treating VEMs quantitatively and making them rational gives Modernity advantages that led Europeans to suppose their approach to organizing societies was the best possible. Figure 3 attempts to illustrate that Modern and Traditional societies are incommensurable because they communicate messages about different realities using mutually indecipherable codes.

Competing with new kinds of systems, Traditional societies have had to choose between victimization and Modernization. Modernizing has its allures, for it opens avenues of advancement. Modernizing can make nations stronger, populations healthier, and free individuals wealthier. But the process of Modernizing has been brutal. In Europe and America, Modernization occurred amidst commitments to open markets that unleashed technological and economic forces communities were incapable of controlling or directing. Failure to develop adequate safeguards allowed resources to be wasted and the ecosphere to be imperiled. Nations trying to catch up to advanced states, like Russia and China, have used totalitarian ideologies to legitimate forcefully modernizing top-down. In places like Japan, where tradition was more actively involved in modernization, domestic stability was purchased at the expense of self-defeating foreign aggressions. Some states did manage Modernization more smoothly and established states generally use surpluses to build more humane societies. But new sources of wealth often have been used to gloss over violations of human rights, and injustices within and between nations remain with potentially corrupting long-term consequences.

Contrasting Social Systems

Traditional	Modern
Typical until ca. 1750	Emerge after 1750
Power = f(Land + People)	Power = f(Energy + Organization)
Expansive – conquer and enslave aliens	Intensive – motivate domestic workers
Reflect local traditions	Adapt to global markets
Organic	Atomistic
Relationships matter	Achievements matter
Family – Patron/Client	Business – Investor/Customer
Loyalty	Honesty
Coercion/Shame	Discipline/Reward
Solidarity	Adaptability
Regional & caste identities	National identities
Tribal bonds	Legal rules
Custom dominates	Calculation dictates
Concrete & unique	Abstract & universal
Position/Privilege	Mobility/Rights
Value order – cultivate stability	Value opportunity – relish fluidity
Look to the past – preservation	Look to the future – improvement
Roles = f(birth)	Roles = f(money)
Blood	Enterprise
Community comes first	Individuals come first
Religions dominate	Markets dominate
Goal = Dignity =	Goal = Wealth =
f(how well roles are played)	f(how competitive player is)
Spiritualistic	Materialistic
Fatalism – Endure nature's assaults	Optimism – Exploit environments
Cult of the *beau geste*,	Cult of the rational
Sacrifice esteemed as noble	Success explained by character

Figure 3 *Characteristics distinguishing traditional from modern societies*

Modern individuals

Modernized individuals saw themselves as missionaries who would properly profit by globalizing their societal form. The rest of the world, however, has often been disappointingly unwilling and/or ill-prepared to modernize, and for good reasons. The transition to Modernity is perilous, and Modernity's secular VEMs are anathema to traditionalists. Mapping the world scientifically rather than religiously, for instance, Modern society looks impious. Valuing worldly rewards makes Modernity look sinful, while rationality makes it look ruthless if not cowardly. Liberating women, of course, makes Modernity look weak. Finally, idolizing individuals makes Modernity look selfish to traditionalists (Margalit & Buruma, 2002).

Traditional judgments reflect societies adapted to rugged conditions, where unity must be instantaneous, unreflective, and enduring. Thus, every member of a clan, family, or tribe can be held responsible for what other members do. In such circumstances, members who break solidarity find themselves alone (Salzman, 2008). Since the environments in which Traditional societies operate are typically unforgiving, isolation is tantamount to death. Hence, to traditional peoples Modernity isolates and threatens as individuals are elevated above society and calculation supersedes commitment. Nor is this judgment unfounded, for the bloodlessly functional and transient relationships that allow Modern societies to adapt, alienate individuals required to alter their defining social roles and relationships.

Nevertheless, distrusting Modernity does not halt its march, and as globalization occurs it inevitably alters traditional boundary conditions and resource flows. Such environmental shifts destabilize societies sufficiently to render inherited roles dysfunctional and reduce individual commitments to family and tribe. Welcome or not, globalizing Modernity creates mismatches with inherited VEMs and roles that allow members to flirt with new principles as they seek new roles. Once Traditional societies are destabilized, established elites lose status and authority. Thus, resistance to modernization is often led by clerics who defend traditional roles using religious VEMs that map impoverished and marginalized societies no longer able to solve their members' problems. Then the undeniable lure of Modern wealth and opportunities excites some to run the risk of transitioning to Modernity by pursuing its roles and adopting its VEMs.

Transitioning to Modernity can be unbearably stressful. As solidarity dissolves, individuals leave familiar places behind, abandon loved ones, and lose essential protections. Venturing into new territories with only American pop-culture to guide them, people know too little about the Modern environment and the systems into which they aspire to fit. They rarely know what resources are available and which niches are most rewarding. Nor have they acquired the skills necessary to fill desirable social roles. Because significant skills cannot be acquired simply by mimicking celebrities, transitioning usually means getting advanced educations in foreign languages that focus on science and economics rather than religion and morality. With education the spiritual dimension of traditional life fades, and religion survives mostly as hollow rituals that do not jeopardize developing "Modern" identities. Despite difficulties, some individuals successfully adjust, no doubt because they have clear goals and the challenges of achieving them are sufficiently demanding to sustain their psychological stability.

But the spiritual inadequacies of Modern life may surface among the children of successful Modernizers. Their reversion to Traditional ways suggests that, absent the immediate struggle to succeed, Modernity fails to provide recent assimilators with the support needed to deal with relentless self-awareness and instability. The narrow materialism and the precarious individualism of Modern society leave cultural immigrants feeling empty and purposeless, especially if opportunities for full membership and further advancement are blocked (Roy, 1995). To restore their sense of belonging and purpose second generation cultural immigrants might commit themselves to destroying the world their parents worked so long to join—as did the youths who attacked the London Transit (Malek, 2006; Kirby, 2007).

Even more terrifying is the position of those who sacrificed traditional roles and relationships to fit into the Modern world only to find themselves repulsed. Leaders of the "9/11" attacks fall into this category. Several had left home, travelled to Europe, and entered college engineering programs. These actions constitute breaks with the past that leave transitioning individuals in a nether world. Ceasing to be members of their tribal groups they shed traditional identities. But when their transition stalls, they fail to acquire membership in Modern systems. Rejection by the world for which so much has been sacrificed must be a bitter pill indeed, while suffering "social death" leaves individuals without systems to solve problems they cannot solve for themselves. In either case, individuals puzzled and anxious about their worth may translate their negative sense of self into hostility to Modernity. If isolation and vulnerability turn individuals abandoned in an alien environment from yearning to hate, they might embrace a new Social Role, "The Terrorist." The role of terrorist can be a powerful attractor because its extreme characteristics reflect the anomic dread shared by both the rejected and the aimless heirs of successful assimilators. As Figure 4 suggests, the failure to fit in or be assimilated can produce negative self-images individuals overcome by defining themselves in opposition to Modernity.

Membership and identity

Our model suggests membership and identity entail aligning the three tiers of belief, behavior, and circumstance, as Figure 5 indicates.

While aligning levels may appear easy, the search is confusing and the results fortuitous. Finding ways to fit in is punctuated by false-starts, mistakes, guilt, and heightened self-awareness. Fundamentalism, one response to the trauma of transition, plays a critical part in the journey from alienation to terrorism and might actually support modernization. Fundamentalists depict themselves as religious traditionalists (Lawrence, 1995), which hardly seems an appropriate strategy for locating new roles and VEMs let alone modernizing. However, traditional religion is the only language people in transition have available. Luther and Calvin made it the vehicle for unintentionally modernizing Europe when they legitimized bourgeois roles in emerging nationalist and capitalist environments. When Islamic Fundamentalists call the alienated back to religion, they may also be doing something other than reviving the ancient faith that restores morality and saves souls. Indeed, Fundamentalism could be acting as a catalyst, exciting actions in which it plays no part. If so, it facilitates the behavioral and cognitive maneuvers by which the alienated fit into viable social systems that, ultimately, no more reflect Fundamentalist positions than undiluted Western concepts of Modernity do.

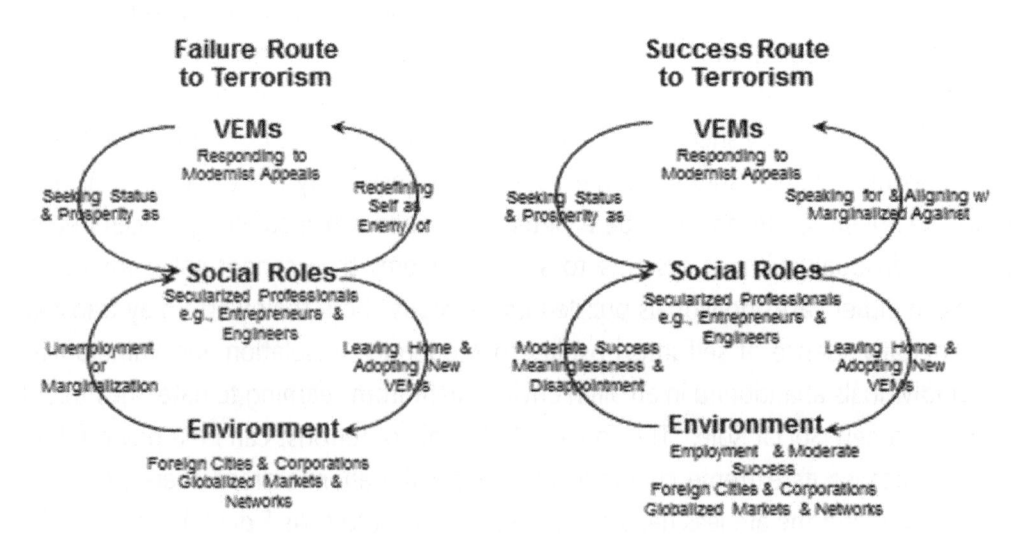

Figure 4 *Breeding terrorists*

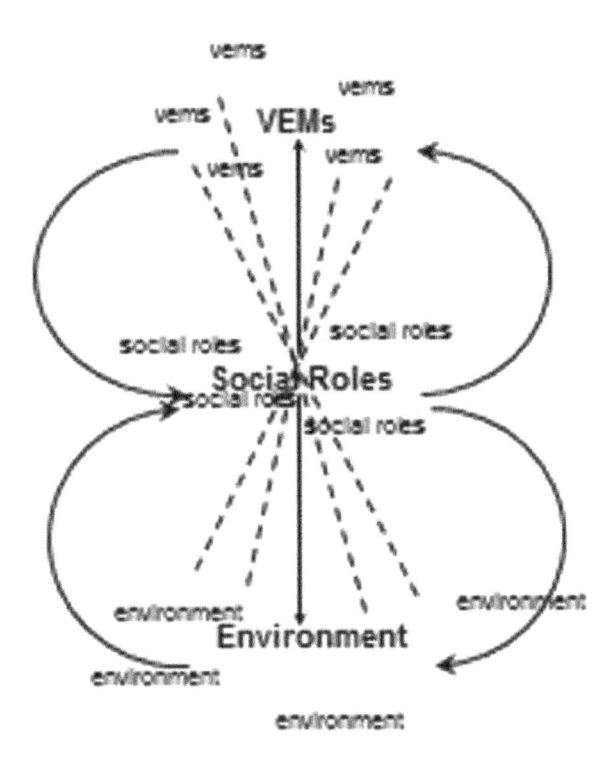

Figure 5 *Alignment and empowerment*

But new roles cannot be legitimated nor new niches mapped by remaining loyal to established churches, which sanction the restrictions and positions Traditional society imposes. Inherited VEMs condemning assertiveness expose individuals whose aspirations have awakened and who are advancing themselves to terrible pangs of conscience. Fundamentalism resolves such mental conflicts by shifting blame from individuals to societies. Denouncing accepted religious concepts that moralize inherited roles and relationships as corrupt innovations, Fundamentalism strips Islam of the symbols linking it to Traditional societies. Proclaiming their intention to purify the faith, the single-minded allegiance to religious fundamentals liberates the frustrated from the status-quo. Thereafter, alienated individuals caught in the throes of transition can explore a vast range of possibilities. Since all things are permitted to those who believe in a God "without qualities," a dis-accreted God purged of attributes committing people to traditional norms can legitimate every possible behavior. Fact-free VEMs decoupled from coevolutionary constraints "run away." Worshipping a God uninhibited by human and natural circumstances, Fundamentalism can enable terrorists to act ruthlessly.

Traditional in name only (Ismail, 2003), Fundamentalists seeking matches with altered environments invest scriptural VEMs with new meanings. The spiritual struggle every Muslim is taught to wage within himself, for instance, becomes physical jihad against Westerners in general. Similarly, by combining unbounded altruism with cosmic fatalism, the suicide Muslims are taught to reject becomes martyrdom. Equally important, Fundamentalists united in opposition to a common enemy find themselves once more members of a functioning community with meaningful identities (McCauley, 2009; Sageman, 2008; Nasiri, 2006). No longer failed assimilators or goal-less heirs of successful businessmen, terrorists gain membership in a system of dedication and sacrifice that gives them purpose and meaning: Suicide bombers will instantly go to Paradise through an act that leads others to a new society (Neuman, 2008; Routledge & Arndt, 2009; Strenski, 2003).

Fundamentalism catalyzes transitions for it excites and legitimates action even in circumstances where uncertainty typically paralyzes. Enjoined to re-establish their continental empire, Islamic Fundamentalists pursue a universal goal that can attract marginalized peoples from all parts of the world (Roy, 2004) with the widest range of motivations (Sageman, 2004). Becoming the agents who restore Islamic glory, Fundamentalists believe they will achieve prominent social roles in a new, stable, and rewarding ummah (Berman & Laitin, 2008). Aligning an abstract God with a universal goal, the terrorist role unites the rootless and rejected in a mystical community symbolized by "salvation." Symbolizing the meaning of membership in a self-organized society with a promising future, "salvation" maps ecstatic sensations. In this heightened state, suicide becomes sacrifice, a gift that will be reciprocated by remembering the bomber as a hero to educate successive generations to see the same world and respond to it in trustworthy ways. Far from being psychotic, the role of religious bomber appears perfectly logical to people whose aspirations have been frustrated

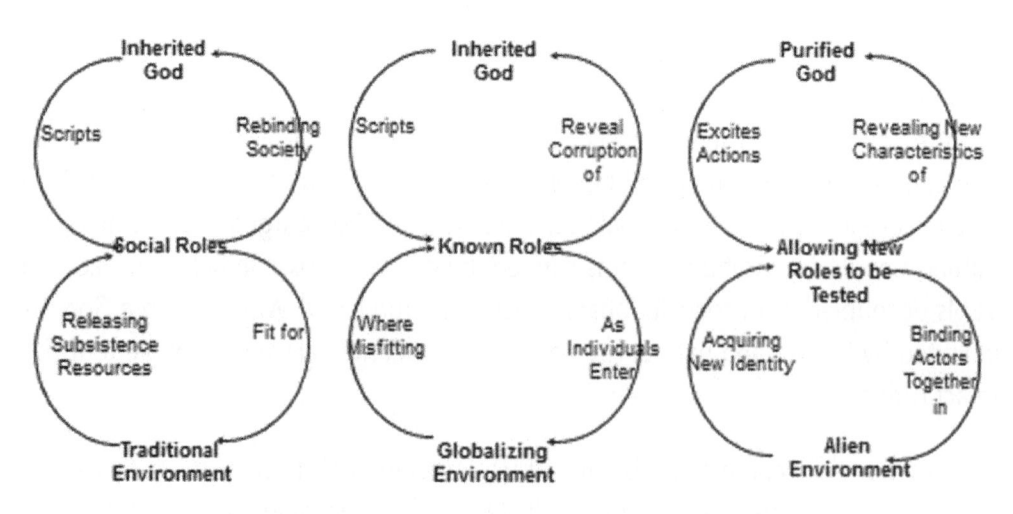

Figure 6 *Fundamentalism and social change*

for no acceptable reason. Thus Fundamentalism transforms the negative experience of failed transitions into positive feedbacks that can drive societies to new evolutionary forms (Maruyama, 1963).

Moreover, the violence needed to pursue universal ambitions by a God decoupled from existing institutions allows persons in transition to seize viable roles by adapting the environment to them. Thus, although the terrorist goal of reestablishing the Islamic empire seems unrealistic, its pursuit has practical consequences. Meanwhile, fundamentalism excites commitments and justifies sacrifices, while training camps reshape personalities. As traditional fighters since at least China's "Boxers" have recognized, reshaping personalities is essential to successful transitions. Playing Modern roles requires the sort of self-control and inner-directedness Max Weber described. And, as Weber said, it is religion—now embracing its "prophetic" mode—that facilitates recalibrating cognitive maps so that resources are more intensively accessed. Eventually, the intolerant and vicious ideologies of terrorists are likely to settle into the accommodating and judicious hypocrisies of normal life.

Discussion

The shift from bewildered and frustrated individual to convinced and committed terrorist is comparable to a "phase change" or "paradigm shift." Individuals are in one situation at one moment and in a radically different one soon after. Fundamentalists consider such transformations "revelations" (Packer, 2006). Complexity theorists call such moments "bifurcation points," moments of choice between possibilities. It is impossible to tell what event tips a process on to one or another branch of a bifurcation tree, for such microcosmic activities are probably idiosyncratic and random. No one can say that a particular financial loss, bigoted slur, official humiliation, or political event made confused and frustrated individuals decide God has called them to offer their lives as gifts to a new society (Coolsaet, 2005). That decision occurs deep in the black boxes that are individual human minds (Federal Research Division, 1999). But we can see how alternative outcomes compete for the allegiance of individuals in transition. On the one hand, there are the attractions of Modernity: prosperity, health, pleasure. On the other hand, there are its drawbacks: loneliness, meaninglessness, vulnerability. Compounding these alternatives are the legacies of Traditional societies, their naturalness and grace, their humiliation and weakness (Hamid, 2008).

Individuals react to these competing options in many ways. But their choices undoubtedly reflect their experiences and the people around them. Hostility or rejection by Moderns probably inclines people toward taking up arms against the

Modern. Terrorist groups now have recruiters to tilt the frustrated and resentful into violence, a choice made all the more attractive because it provides high-status identities, expectations of immortality, and promises of economic relief for surviving relatives. More welcoming attitudes by representatives of more flexible societies might incline perplexed and conflicted individuals to follow a more peaceful path.

But it is difficult for members of Modern societies to see the benefits of increased openness. Conflating criticisms of their societies with terrorism and deviations from their VEMs and roles as attacks on Modern civilization, leaders like President G.W. Bush saw messages from people in transition as "noise." Since noise threatens systemic stability, isolated members of Modern societies treat terrorism as a potentially mortal blow aimed at their very essence. Outsiders treated as bad and violent often react by exaggerating their foreignness and becoming terrorists (Elias & Scotson, 1995). Modern societies then respond in kind to terrorism, for fighting external enemies provides some sense of community and value for individuals defined only by their private possessions. This interaction harms both sides by interrupting the transition to Modernity. Treating societies as complex systems that survive by evolving shows this response is not only ineffectual but self-defeating.

Complex Modern societies adapt by turning noise into information. Noise in this case refers to environmental flows, phenomena, or situations systems do not have agents or processes for exploiting or containing. Since complex systems are far from equilibrium (Prigogine, 1996), they can be easily destabilized by external perturbations. And since complex systems are embedded in dynamic environments, perturbations are regular parts of their experience. To retain dynamic stability, therefore, Peter Allen regularly advised complex social systems to confront their futures as "clouds of possibilities" (Allen, 1994). He meant that complex systems should not permanently commit themselves to particular arrangements because, adapting to a surprise-filled environment, they must always be willing to change. As Professor Allen also realized, adapting depends on the internal resources available to systems. Homogeneous systems that exclude variations, therefore, turn out to be non-adaptive, even when their current structure happens to be just what the environment is looking for (Allen & Lesser, 1991).

But systems that preserve wide ranges of alternative behaviors, which are themselves clouds of possibilities, dramatically increase their chances of successfully adapting to unexpected future events (Allen, 2006). To locate threats and opportunities, Modern societies break up tightly-bonded tribal groups, individualizing members so that environments can be probed in a finer-combed manner. To preserve options,

Modern societies stop equating alternative behaviors with sin, relaxing their judgments to favor innovation. But, as Peter Allen recognized, successful systems favor conservative strategies, and in following this path Modern societies unintentionally create conditions favorable to terrorism. Committed to the VEMs and roles that favored globalization, Westerners have insisted globalized societies universally exactly mimic their version of Modernity. Consequently, they treat transitioning individuals who cling to Traditional behaviors and mannerisms with suspicion if not hostility. When modernizing is made an all or nothing experience individuals in transition feel they must either become agents of the rich and powerful who oppress their families and friends or enemies of Modernity. The choice reduces to being traitors or terrorists.

Peter Allen's more nuanced approach to social policy welcomes diversity as the way for Modern societies to preserve adaptive options (Allen et al., 2006). Diversity can provide Modern systems with the sensibilities to empathize with peoples in transition and even the ability to recognize the benefits of moderating their own cognitive maps. Finding ways to assimilate more of the populations from which terrorists emerge—turning noise into information—benefits both citizens of Modern systems and individuals transitioning from Traditional ones.

Of course, determining the exact policies by which to incorporate marginalized elements from transitioning societies is difficult. But computer simulations using complex systems models could help policy makers adjust the rates at which traditional environments and roles are modified. Properly tuned, the transition from Traditional to Modern systems could be made less threatening. If so, the incomprehensible noise of terrorism might become the information by which a more integrated global society reached new levels of complexity. Incorporating some traditional elements by assimilating modernizing people might reduce the internal stresses associated with Modernity. A calmer, happier, and more secure population could retain its dynamic stability in an even greater number of niches.

References

Abadie, Alberto (July 3, 2006), "Poverty, political freedom, and the roots of terrorism" Working Paper 10859 National Bureau of Economic Research http://www.newstatesman.com/200607030030

Allen, P.M. and Lesser, M. (1993) "Evolution: Cognition, ignorance and selection" in E. Laszlo, I. Masulli, R. Artigiani, and V. Csányi (eds.), The Evolution of Cognitive Maps, ISBN 2881245595

Allen, P.M., Strathern, M. and Baldwin, J. (2006). "Evolutionary drive: New understandings of change in socio-economic systems," *Emergence: Complexity & Organization*, ISSN 1521-3250, 8(2): 2-19.

Allen, P. M. (1994)"Evolutionary complex systems: Models of technology change," in L. Leydesdorff and P. van Besselaar (eds.), *Evolutionary Economics and Chaos Theory: New Directions in Technology Studies*, ISBN 9781855672024, pp 1-17.

Allen, P.M. and Lesser, M. (1991)"Evolutionary human systems: Learning, ignorance, and subjectivity," in P.P Saviotti and J.S. Metcalfe (eds.), Evolutionary Theories of Economic and Technological Change, ISBN 9783718650781, pp. 160-71.

Allen, P. M., Strathern, M. and Baldwin, J. (2006)"Evolution, diversity and organization," in E. Garnsey and J. McGlade (eds.), *Complexity and Coevolution: Continuity and Change in Socio-Economic System*, ISBN 97808236166664, pp 22-60.

Andriolo, K. (2002). "Murder by suicide: Episodes from Muslim history," *American Anthropologist*, ISSN 1548-1433, 104(3): 736-42.

Atran, S. (2003). "Genesis of suicide terrorism," *Science*, ISSN 1095-9203, 299(5612): 1534-39.

Berman, E. and Laitin, D.D. (Jan 2008). "Religion, terrorism and public goods," NBER Working Paper No. 13725 http://www.nber.org/papers/w13725.

Coolsaet, R. (2005). *Al Qaeda: The Myth, The Root Causes of International Terrorism and How To Handle Them*, ISBN 9038206933.

Eickelman, D. F. (Winter 2000). "Islam and the languages of Modernity," Daedalus, ISSN 1548-6192, 129(1):119-35.

Elias, N. and Scotson, J.L. (1994). *The Established and the Outsiders: A Sociological Enquiry Into Community Problems*, ISBN 083979499.

Federal Research Division (September 1999). "The sociology and psychology of terrorism: Who becomes a terrorist and why?" Washington: Library of Congress.

Hamid, M. (2008). *The Reluctant Fundamentalist*, ISBN 197801510130433.

Gouldner, A. W. and Peterson, R. A. (1962). *Notes on Technology and the Moral Order*, ISBN 0672511592.

Hutchinson, W. (2007). "The systemic roots of suicide bombing," *Systems Research and Behavioral Science*, ISSN 1099-1743, 24: 191-200.

Ismail, S. (2003). "Islamic political thought," in T. Bell and R. Bellamy (eds.), *Cambridge History of Twentieth Century Political Thought*, ISBN 0521691621, pp 590-601.

Kirby, A. (2007). "The London bombers as 'self-starters': A case study in indigenous radicalization and the emergence of autonomous cliques," *Studies in Conflict and Terrorism*, ISSN 1521-0731, 30(5): 415-28.

Kuhn, S.L., Stiner, M.C., Reese, D.S. and Gulac, E. (2001). "Ornaments in the earliest upper Paleolithic: New results from the levant" *Proceedings of the National Academy of Sciences*, ISSN 1091-6490, 98(13): 7641-46.

Lawrence, B. (1995). *Defenders of God*, ISBN 9780062505392.

Malek, S. (2006). "The suicide bomber in his own words," New Statesman, 135(4799): 26-9, http://www.newstatesman.com/200607030030.

Margalit, A. and Buruma, I. (Jan 17, 2002). "Occidentalism," *New York Review of Books*, ISSN 0028-7504.

Maruyama, M. (June 1963). "The second cybernetics: Deviation by amplifying mutual causal processes," *American Scientist*, ISSN 0003-0996, 51(2): 464-79.

McCauley, C. (2009). "The psychology of terrorism," *Social Science Research Council*, http://www.ssrc.org/sept11/essays/mccauley.htm.

Nasiri, O. (2008). *Inside the Jihad: My Life With Al Qaeda, A Spy's Story*, ISBN 9780465023882.

Neumann, P.R. (2008). *Joining Al-Qaeda: Jihadist Recruitment in Europe*, ISBN 9780415547314.

Packer, G. (2006). "The moderate martyr," *The New Yorker*, ISSN 0028-792X, (September 11): 61-69, .

Pape, R.A. (2005). *Dying To Win: The Strategic Logic of Suicide Terrorism*, ISBN 1400063175.

Pedahzur, A., Perliger, A, and Weinberg, L. (2003). "Altruism and fatalism: The characteristics of Palestinian suicide terrorists," *Deviant Behavior: An Interdisciplinary Journal*, ISSN 1521-0456, 24: 405-423.

Post, J.M., Sprinzak, E. and Denny, L.M. (2003). "The terrorists in their own words: Interviews with 35 Middle Eastern terrorists," *Terrorism and Political Violence*, ISSN 1556-1836, 15(1): 171-85.

Post, J.M. (2006). "The psychological dynamics of terrorism," in L. Richardson (ed.), *The Roots of Terrorism*, ISBN 97814000684816, pp. 17-28.

Prigogine, I. (1980). *From Being To Becoming*, ISBN 9780716711087.

Prigogine, I. (1996). *La Fin des Certitudes*, ISBN 9782738109866.

Richardson, L. (2006). *What Terrorists Want: Understanding the Enemy, Containing the Threat*, ISBN 9781400064816.

Richardson, L. (ed.) (2006). *The Roots of Terrorism*, ISBN 04125954371.

Routledge, C. and Arndt, J. (2009). "Self-Sacrifice as self-defence," *European Journal of Social Psychology*, ISSN 1099-0992, 38(3): 531-41.

Roy, O. (1995). *The Failure of Political Islam*, ISBN 0674291416.

Roy, O. (2004). *Globalized Islam: The Search For a New Ummah*, ISBN 0231134983.

Sageman, M. (2004). *Understanding Terror Networks*, ISBN 9780812238083.

Sageman, M. (2008). *Leaderless Jihad: Terrorism Networks in the Twenty-First Century*, ISBN 9780812240658.

Salzman, P.C. (2008). *Culture and Conflict in the Middle East*, ISBN 9781591025870.

Stiner, M.C. Munro, M.D., Surovell, T.A., Tcherov, E. and Bar-Yosef, O. (1999). "Paleolithic population growth pulses evidenced by small animal exploitation," *Science*, ISSN 1095-9203, 283: 190-4.

Strenski, I. (2003). "Sacrifice, gift and the social logic of Muslim human bombers," *Terrorism and Political Violence*, ISSN 1556-1836, 15(3): 1-34.

Tolman, E.C. (1948). "Cognitive maps in rats and men," *Psychological Review*, ISSN 0033-295X, 55: 189-208.

Weber, M. (1904/58). *The Protestant Ethic and the Spirit of Capitalism*, ISBN 9780415254069.

Robert Artigiani earned a Ph.D. in history at American University in Washington, D.C. He taught courses in cultural history, the history of science, the history of technology, and introductory philosophy at several American colleges. Now a Professor Emeritus at the U.S. Naval Academy, he remains dedicated to exploring the possibilities raised by Prigogine's call for a "new rationality." Accordingly, he emphasizes the paradigm shift creating a science that describes nature in terms of wholes rather than parts, processes rather than things, biology rather than physics, freedom rather than determinism, and symmetry breaks rather than continuities. Aiming to strengthen the bond between natural science and the humanities, his paper uses terrorism to show that concepts borrowed from Complexity theory can provide tools for better understanding significant cultural phenomena.

Chapter H

Modelling the management of evolving operations

James S. Baldwin & Keith Ridgway

There has been a growing trend recently, in not only viewing and treating firms and their supply chains, along with individuals and even their decision-making processes, as complex systems, but in the modelling and simulation of such systems. However, in the particular subject area of Operations and Production Management this has been rather limited. This chapter attempts to address this and build on the few studies that have been published in this area whilst also aiming to develop a decision-support tool that can provide managers with guidance when contemplating, for example, a change management programme. The chapter begins with a review of the recent advances in complex systems thinking, modelling and simulation, and then turns attention to the very real practical need for decision-support tools in Operations and Production Management. After highlighting the differences between the various systems approaches, an evolutionary complex systems simulation model is presented. Following an outline of the research methods and an explanation of the model, several simulations are presented of the organizational evolution of a firms operations and production practices and policies. Organizational evolution is based on the opinions of managers from different functional areas of how these practices and policies interact with one another. A discussion of the significance of the results, in both practical and academic senses, then follows.

Introduction

Supply networks, organizations, individuals and even their decision-making processes are increasingly understood as complex systems (Choi et al., 2001; Frizelle & Woodcock, 1995; Macbeth, 2002; MacIntosh & MacLean, 2001; McCarthy, 2003). Particularly during the last decade or so there has been a growth in the modelling and simulation of such systems (Chaharbaghi, 1991; Islo, 2001; Li et al., 2003; Lim & Zhang, 2003; Nilsson & Darley, 2006; Zhou et al., 2003). However, Operations and Production Management work has been rather neglected. This research builds on the few studies that have attempted this, and tries to form a basis on which to develop a decision-support tool giving managers practical assistance, for example during a change management programme. The original research was undertaken in collaboration with Peter Allen and a coauthored paper (Baldwin, 2010) gives a mathematical technical description of the model. Here recent advances in complex systems thinking, modelling and simulation are highlighted coupled with the need for decision-support tools in the management of operations. Also attention is drawn to a particularly problem, which is the sometimes conflicting interests, motivations and concerns of different functional managers and the potential advantages and disadvantages of these diverse priorities for overall firm performance. After describing a hierarchy of systems approaches based on different modelling assumptions, an evolutionary complex systems (ECS) simulation model is described. This approach uniquely has the capability to explore the consequences of the diversity in managers decision-making processes from different functional areas and the effects on the potential evolutionary paths that a firm can take. Following an outline description of the model the outcomes of management decision-making are described and a discussion of the significance both intrinsically and for the Operations and Production Management literature. The chapter concludes with some closing remarks on the research and practical implications of this work.

With complex socio-economic systems, evolution and change are inevitable and the performance and survival of a firm is largely determined by its management of change (Jarratt, 1999; Macbeth, 2002; McCarthy, 2004); also with manufacturing firms, technological change in particular (Raymond *et al.*, 1996). However, change management is fraught with problems not least getting everyone 'singing from the same song sheet'. For example, there has long been a recognition of the differences in motivations, interests and priorities of managers from different functional areas (O'Leary-Kelly & Flores, 2002; Rhee & Mehra, 2006) which if not managed appropriately, in terms of strategic alignment (Skinner, 1969), can significantly depress firm performance (Malhotra & Sharma, 2002). This has prompted calls for decision-support tools that increase understanding of the underlying processes in the adoption and implementation of new technologies (Baldwin *et al.*, 2005; Das & Narasimhan, 2001; Klassen & Whybark, 1999), practices (Cua *et al.*, 2001; McKone *et al.*, 2001; Zhu & Cote, 2004; Zhu *et al.*, 2008) and policies (McCarthy, 2004; O'Leary-Kelly & Flores, 2002) within the context of the firm's corporate and functional strategic emphasis (Brown *et al.*, 2007; Leachman *et al.*, 2005). The role of management decision-making in organizational evolution can crucial. Raymond *et al.* (1996), for example, whilst investigating technology adoption in SMEs, identified the decision-making process as one of three main profiles of strategic advantage, along with technological expertise and organizational capabilities.

There are two approaches in the development of decision-support systems—theoretical and computational modelling—in the literature. For example, Karkkainen and Hallikas (2006) explored, via case-study, the dynamics of inter-organizational network-related decision-making under a series of scenarios relating to risk management, learning and the business environment. They pointed out that not only was there a dearth of research in management decision-making processes underlying organizational change but that a holistic and systemic approach was needed to understand both intra- and inter-organizational decision-making. To address this, Meade *et al.* (2006) applied the chaos and complexity theories to give an understanding of the decisions for formulating strategies to successfully position products in the technology adoption life-cycle. Using several case studies of firms within the ICT industry, this work demonstrated the usefulness of this approach. In terms of computational modelling, Lim and Zhang (2003), Zhou *et al.* (2003), and Nilsson and Darley (2006) applied agent-based modelling and created virtual factories to further understanding in the change management process and also to create a decision-support. This used 'what-if' scenarios where the consequences of particular decisions could be analysed and evaluated. Agents in these applications typically represented different machines, sales, operations planning, warehousing, and customers. Nilsson and Darley's (2006) work provides a rationale for the use of the complex adaptive systems (CAS) and was also the first study to produce empirically verified results of agent-based models in the subject area of Operations and Production Management. At a level of aggregation higher, Kaihara (2003), created a virtual market and explored strategies in a supply chain model based on a problem of resource allocation within a dynamic environment. The aim, largely successful, was to develop a decision-support tool to optimize supply chain performance.

Management decision-making is central to organizational change/transformation but there is also rising awareness of the impact of diversity in the decision-making process (Allen *et al.*, 2006; Poundarikapuram & Veeramani, 2004). Here diversity refers not only to the different decisions that can and are made but to the range of approaches taken when decision-making due different perspectives, beliefs, attitudes and information-processing (Allen *et al.*, 2006). Simon's (1955; 1983) notion of bounded rationality from the cognitive sciences and more recently evolutionary economics, which refers to the incomplete knowledge that people have and use (and misuse) when decision-making, is also relevant here (Nilsson & Darley, 2006). Getting a balance is key, as diversity is seen as being hugely advantageous for innovative capabilities if set in a conducive organizational culture (Jarratt, 1999). It also has the potential to affect the evolution of a company particularly in times of change, such as when introducing a new technology, practice or policy (Baldwin *et al.*, 2005; Jarratt, 1999). However, capturing this level of diversity has been a problem not just in terms of theory (but see, for example, Holland, 1995; Jantsch, 1980; Prigogine & Stengers, 1987) but also, and perhaps more evidently, in application.

ECS modelling differs from the other computational modelling techniques, such as system dynamics, agent-based and other self-organizational models, as it incorporates the role and influence of micro-diversity and experiential learning which are arguably the driving forces and impetus behind evolutionary, rather than adaptive, change (Allen *et al.*, 2006). ECS theory is a European branch of complexity thinking stemming from Prigogine's (1973) Nobel Prize winning work. The approach has now been successfully applied to ecosystems, urban systems, economic markets and, more recently, in the evolution of an entire industry (Allen *et al.*, 2005, 2006, 2007; Baldwin *et al.*, 2005).

To illustrate the differences between the different systems models consider the underlying assumptions used. Modelling assumptions create a hierarchy of models from known certainties and perfect prediction through to explorations of the unknown and the least-likely of potentialities (Allen *et al.*, 2007). All systems models have at least two assumptions: 1) that a boundary exists between the system, in this case the firm, and its environment, and 2) that the system's components, e.g., the firm's technologies, practices and policies, can be classified to produce a taxonomy.

Additional assumptions concern the system's components and their interactions. System dynamic models have components and interactions that represent the average. This gives rise to the single average 'most likely' path. Such models give the impression of complete understanding and knowledge, and resulting in complete predictability. But what do average components and interactions actually mean when applied to a firm? Take, for example, the implementation of line-balancing or empowering employees. Not every firm has the same approach. Indeed, if every firm's approach was examined, there would be a wide diversity. Furthermore, not all implementations of line-balancing and/or empowering employees have the same outcomes; what would work in one organization, or even one point in time, may not necessarily work in another. Also the 'most likely' event may not be very likely often there is a plethora of different events of almost equal likelihood.

By incorporating the non-average, the nature of the model moves from certainty to change, and from prediction to exploration and potentials. Through the inclusion of the potential types of interactions that can occur, enables models to explore and find many possible future scenarios through self-organizational processes (e.g., in the example, above the different outcomes of line-balancing or employee empowerment implementation). The few complex system computational models found in the literature are of this type (see, for example, Chaharbaghi, 1991; Islo, 2001; Kaihara, 2003; Li *et al.*, 2003; Lim & Zhang, 2003; Meade *et al.*, 2006; Nilsson & Darley, 2006; Zhou *et al.*, 2003). Although useful both practically and intellectually, these models are still limited, particularly in their application to social systems. Such models have a stochastic mechanism that generates 'noise' to represent diversity. This is perhaps more realistic than assuming only average conditions and interactions. However the components, or agents, are still of an average type that are subject to a pre-defined, 'if-then' rule-based system, even though they introduce non-average interactions. The noisy interaction of 'average' elements does not have the same outcome, or spread of outcomes as the interaction of diverse, heterogeneous individual elements. In the former we may see different configurations and regimes of operation appearing, but in the latter, new combinations of different elements leading to new, emergent capabilities and dimensions of performance may happen. This is the essential characteristic of evolutionary change—a process of qualitative, not just quantitative, change.

ECS models have heterogeneous components and it is this that distinguishes them from self-organizational models (Allen *et al.*, 2007). Representing a diversity types of components, through the introduction of internal or micro-diversity, as well as a wide diversity of interactions produces a more realistic representation of true evolutionary processes. Whilst blind adaptation is associated with self-organizational models, ECS model coevolution through experiential learning. Control devolves from the global/system level to the local/individual level and is an expression of individual behaviors and their performance and success relative to others within the system; evolution proceeds through de-centralised, rather than centralised, decision-making. In addition, evolution, being an open process, ensues through a combination of not only the determinism of the individuals' purposefulness but also by chance events. The intake of new kinds of individuals, or the changing views and thoughts that they may have is not a rational, calculable process because the implications of any particular heterogeneity is not known until after any evolutionary step has occurred in system behavior. Here, the evolution that does occur is not readily predictable but comes from the interplay of heterogeneous individuals and the different performances of the resulting organizations. This is largely characterized by an inevitable lack of pre-existing knowledge of the link between individual and system behavior and can be thought of as resulting from a degree of 'error-making' (Allen *et al.*, 2006). The role of chance is fundamental, creating a rich medium for experiential learning through continuing experiments in behavior space (Allen *et al.*, 2007). This approach incorporates diversity at all levels of description and so is appropriate when trying to better understand the role of diversity in decision-making and the impacts on a firm's evolutionary trajectories.

Research methods and preliminary results

The development of the ECS model, which simulated a firm's evolutionary trajectory reflecting management assumptions, was achieved by: a) building a profile of a firm through a case-study approach involving observation and simple semi-structured interviews; and then b) gauging, via a quantitative questionnaire, the managers' perception of how the firm's technologies, practices and policies (also referred to as 'character-states') interacted with one another in the context of their overarching operations strategy. In so doing, it was possible to compare and contrast different decision-making capacities, which enabled an exploration of decision-making consequences resulting from potentially diverse information sources and assumptions.

The case-study approach was selected to better illustrate and exemplify the utility of the ECS modelling technique (Eisenhardt, 1989; Meredith, 1998) and is consistent with similar research in Operations and Production Management research (e.g., Meade *et al.*, 2006; Nilsson & Darley, 2006). As is common with case study research, sampling was purposive (Saunders *et al.*, 2007). That is, a number of firms were pre-screened to determine whether they had a suitable profile that would help achieve the research aim. A mechanical engineering firm was approached and consented to act as the case study. The CEO, and three senior managers, responsible for Marketing, Manufacturing, and R&D, participated.

Interviews were based on a simple, semi-structured, qualitative questionnaire, which was sent to each participant prior to the interview, and was accompanied by pre-prepared paper-based check-list of common operations practices and policies, which was not made known to the participants to avoid interviewer bias. To maximise internal validity (Saunders *et al.*, 2007), the interview schedule and practice check-list was first piloted on two industrialists and an academic familiar with Operations and Production Management after which small adjustments were made. Interviewer and interviewee biases were minimized through the selection of neutral settings for interviews and through a standardised presentation of non-leading questions (Saunders *et al.*, 2007). Interviews were recorded but due to the basic nature of this part of data collection and analysis, were not transcribed. They were instead directly interpreted and coded using the pre-prepared practice list as a guide. Essentially, if the practices from the list were mentioned then it was marked for inclusion for the next quantitative questionnaire phase plus any additional practices and policies identified. This was conducted during the interview. There was then a post-interview analysis of the recordings to verify the list of practices. No further practices were added at this stage.

The interview firstly encouraged a discussion of the firm's operations strategy, by asking participants what they deemed important for the survival of the company in terms of performance, using as the basis for discussion both the Four Competitive Priorities (cost, quality, time and flexibility), from Hayes and Wheelwrights (1984) and the Five Performance Objectives (quality, speed, dependability, flexibility and cost), from Slack *et al.* (2007). These performance criteria were familiar with the interviewees and are consistent with both the literature and previous empirical research on manufacturing performance (see, for example, Brown *et al.*, 2007; Cua *et al.*, 2001; Das & Narasimhan, 2001; Fynes *et al.*, 2005). From the interviews, four main performance criteria, i.e., product quality, cost efficiency, customer relationship, and schedule adherence, were found to be relevant and there was also an indication that they had differing degrees of importance which is consistent with the literature (Hayes & Wheelwright, 1984; Skinner, 1969). Three out of the four performance criteria, i.e., product quality, cost efficiency and schedule adherence, directly mapped on to Hayes & Wheelwrights' (1984) Competitive Priorities and Slack *et al.*'s (2007) Performance Objectives. Customer relationship although alluding to aspects of flexibility and dependability, had a much more informal element relating to social relationships of trust building, which is more relevant to aligning organizational cultures and corporate strategies.

The interview process then encouraged a discussion of what 'characterized' the company, by asking questions about the firm's technologies, practices and policies covering, in a sequential manner, the workforce, scheduling, suppliers, quality, R&D and production processes. The pre-prepared check-list consisted of 47 practices (see appendix) which were elicited from the literature particularly from Womack *et al*'s (1990) and McCarthy *et al*'s (1997, 2000) work on the evolution and development of the automotive industry supported by the generic manufacturing practices alluded to in Kinni (1996), Schonberger (2008) and Slack *et al.* (2007). Although, this level of data collection is fairly basic and could have been achieved using a questionnaire survey based on generic operations technologies, practices and policies, a better understanding of the idiosyncrasies of the firm was attained through interviews. Furthermore, different descriptors, which were more relevant to that particular organization, were identified, along with additional technologies, practices and policies (i.e., not on the pre-prepared check list). The interviews also ensured both participation in the next quantitative questionnaire phase and more importantly it gave participants a better understanding of their requirements in this phase. Twenty-five character-states (listed in table 1) were identified as the most important technologies, practices and policies for continued successful firm performance providing the basis for the quantitative questionnaire.

The quantitative questionnaire was designed to gather the managers' views of how the 25 character-states interacted with one-another in relation to the overall performance of the firm. To achieve this, the questionnaire had three parts. The first asked participants to rank the overall importance of the four performance criteria to the company, for example, 1st: customer relationship; 2nd: schedule adherence, and so on. Three out of the four (the CEO, and the marketing and R&D managers) ranked customer relationship first followed by product quality, schedule adherence and cost efficiency. The Manufacturing Manager indicated a ranking of product quality, schedule adherence, customer relationship followed by cost efficiency. The second part asked participants to rank the impact or strength of association of each of the character-states on each of the performance criteria; for example, CS1, R&D investment, may be associated 1st with product quality, 2nd with customer relationship, 3rd with cost efficiency and 4th with schedule adherence, etc. These first two sections determined whether different practices contributed to some performance criteria more than others, i.e., whether some practices were more important than others, and provided a basis for weighting mechanisms in the ECS model enabling a reflection of the character-states' impact on the overall operations strategy. However, due to the informal nature of the customer relationship performance criterion, the majority of technologies, practices and policies, did not directly contribute, i.e., scored poorly, and as such was excluded from the weighting mechanism.

The final part of the questionnaire asked participants to gauge the interactions between each of the character-states in terms of the overall performance of the firm. The answer options were based on a 7-point Likert scale (-3 to +3) determining the degree of positive/neutral/negative interactivity. That is, for example, participants could indicate say a moderately synergistic interaction between CS1 and CS2 as '+2' and a strongly antagonistic interaction between CS1 and CS3 as '-3' and so on. These scores were weighted in accordance with their impact on the ranked performance criteria. The ECS model drew directly from these weighted scores. Piloting for the questionnaire, involving two academics and two industrialists, was in two stages after which minor/incremental adjustments were made to the questionnaire to ensure construct validity (Saunders *et al.*, 2007).

The simulation model

With ECS modelling, structures and the organization of different practices may be explored. The work presented here traces its origins back to the insights expressed in the works of Prigogine (1973), colleagues (Glansdorff & Prigogine, 1971; Kondepudi & Prigogine, 1998; Nicolis & Prigogine, 1977; 1989; Prigogine & Stengers, 1987), and others (Allen, 1982; 1984; Haken & Mikhailov, 1993; Jantsch, 1980) who have all demonstrated how complex systems evolve through the emergence of fluctuations and instabilities within a system. Prigogine (1973) developed a simple model, known as the 'Brusselator' (after the Brussels' School of Thermodynamics), which described how non-equilibrium systems become unstable and begin oscillating. The conditions to be met are that the system is open, that the gradient (i.e., flow of matter and energy) creates a far-from-equilibrium state, and that there are autocatalytic steps in the reaction chain.

Building on these principles, Allen (1976) developed a mathematical expression describing the introduction and growth of new 'behaviors' into a system, such as new species in a natural ecosystem. The ECS model developed for this study was adapted from this and was designed to simulate the interaction between the firm's character-states drawing directly from the four decision-makers' questionnaire data. In terms of manufacturing, the behaviors/species and their interactions represent the manufacturing firm's technologies, practices and policies and how they work together. Biological evolution, through selection, surrounds the diffusion and proliferation of innovative behaviors determined by their success, or relative performance, in birth and death rates. Birth and death rates represent, for example, the performance in the competition for resources, mating success, avoiding/catching prey, and rearing offspring. In terms of manufacturing, the success of character-states reflects the importance of the character state to the organization in terms of, for example, product quality, schedule adherence and cost efficiency. Successful bundles of practices and behaviors will experience positive feedback and growth when their particular characteristic performances correspond to that which the selection environment requires.

The model was based on the equations given in Allen *et al.* (2007) and describes the growth in the total health of a manufacturing firm, which is seen as the sum of the activities of its constituent practices. It is the synergy, neutrality or conflict be-

No.	Label	Description
1	R&D investment	Significant R&D investment on product quality and process efficiency improvements.
2	Continuous production	Three production shifts ensure non-stop production.
3	Cells with automated equipment	Factory layout is based on the cell principle with a significant substitution of human labour with mechanized labour.
4	Setup time reduction	Setup processes for machinery/equipment are analysed in order to reduce time between setups for different production runs.
5	Setup automation	Set-ups between production runs are largely automated rather than manual.
6	Preventive maintenance	Operators and in-house engineers perform routine maintenance on a regular basis on workstations and process machinery including, cleaning, oiling/greasing, adjustment, and parts replacement to avoid breakdowns.
7	Outsourced corrective maintenance	All but the simplest machine/equipment breakdown repairs are largely outsourced.
8	MRP system (material replenishment)	Materials Requirement Planning software system to aid production planning and inventory control.
9	ERP system (organise/monitor resources)	A number of software based Enterprise Resource Planning modules are currently being implemented to organise and monitor resources with the aim of replacing the existing MRP system.
10	Full resource visibility	ERP manufacturing module enhancing the visibility of resources through stock to production.
11	Resource priority control	A Pareto based control system to prioritize purchasing, stocking and allocation of resources.
12	ERP supply chain integration	ERP supply chain management module to facilitate communication and information sharing to reduce costs and enhance both responsiveness and quality.
13	Supplier cooperation	The organization has an open book, cooperative relationship with suppliers and customers.
14	TQM sourcing	Vendors are vetted according to stringent quality standards to ensure consistently high quality sourced components and raw materials.
15	Quality systems/standards	A series of standards for quality management systems maintained by the International Organization for Standardization and administered by accreditation and certification bodies.

No.	Label	Description
16	5S's programme	The 5S's programme involves: 1. Sort/Segregate (keeping only essential equipment/tools/materials at workstations); 2. Simplify/Straighten (workstations anthropometrically designed to improve efficiency of movements); 3. Shine/Sweep (workstation cleanliness); 4. Standardise (removing variations in movement/flow); 5. Sustain/Self-discipline (periodic motivational reviews of employee performance).
17	Decentralised error detection and correction	Operators are largely responsible for detecting and rectifying quality problems as and when they occur at their workstations.
18	100% inspection	To ensure the highest quality, each and every product is inspected.
19	Production process traceability	A quality system to enhance traceability along the production process.
20	Line-balancing	Tasks are assigned to workstations to level overall time requirements and fluctuations.
21	Job rotation	Operators regularly work on other qualitatively different tasks.
22	Flexible workers	Flexibility is achieved mainly through multi-skilling but also includes both time flexibility (PT/FT, specific working times, and to cover variable demand) and location flexibility for 'indirect' manufacturing jobs (i.e., occasional home-working to fully mobile).
23	Empowering employees	Employees are empowered through both suggestion involvement (i.e., suggest process improvements) and 'job involvement' (i.e., redesign processes to improve efficiency/quality).
24	Employee multi-skilling	Operators develop a set of skills to enable work on qualitatively different tasks.
25	Proactive annual training	Operators are intensively trained annually in current and future practice to support multi-skilling and flexibility.

*Note: 'character-states' will be abbreviated in the text to 'CS' when referred together with a particular technology, practice or policy; for example 'CS1, R&D investment'

Table 1 *Firm character-states (CSs): Number, label and description*

tween its practices that affects the size of each one, and therefore the total output or sum of them all. The model uses a pair matrix defined from the questionnaire data from the four senior managers concerning their view of the synergy, neutrality or conflictual nature of the 25 practices, which defines how each of the 25 practices impinges on each other (a 25 by 25 matrix). For a firm with a given set of practices the synergy - conflict matrix gives the net synergy for each of the particular practices in the presence of the others. This in turn gives an indication of the overall survival or health of the system/firm. The limits to 'health' however will be set by the size of the practices already present. The model then calculates the growth or decline of each practice in the presence of the others.

During a model run after each time step the size of the different practices are updated and the total health of the system/firm is calculated. Several other variables in the ECS model can be manipulated and calibrated the running of the model. Of which three are of particular interest. The first is the running time of the simulation. This may be adjusted to permit finding stable solutions which are typically found within 10-50,000 arbitrary time units. The second variable is the number of character-state initiations in the model that allows exploration of particular organizational forms. The third is the performance value of the character-state. Performance may lie between 0-30 arbitrary units. The higher the value, the better the performance and importance within the organization. The simulations presented here launched the character-states with a starting value of 5 performance-units.

Before the results are presented there are two qualifications. The first is that the particular simulations presented here are among many possible trajectories dependent on initial conditions. All simulations of the different managers, however, begin with identical initial conditions. However, selecting particular solutions is somewhat problematical when dealing with evolutionary systems as an infinite number of possible evolutionary trajectories are possible. Nonetheless, the simulations presented were deemed a fair representation of a series of repeat simulations. The second qualification is that, unlike reality, the simulations show an evolution of the firm with all character-states starting equally with 5 performance-units each. Nonetheless, the results from this procedure do highlight differences (and similarities) between the different decision-makers.

Simulation results and discussion

To appreciate the effects of diverse decision-making capacities, five simulations are discussed. Figure 1 depicts, in simplified form, five separate evolutionary trajectories of the firm. The first line, to the front of the graph, represents the results of a simulation in which all the managers' scores were averaged out—the group simulation. The second, third, fourth and fifth lines represent the results of simulations based individually on the CEO's, Marketing, R&D, and Manufacturing Managers' scores, respectively. Each point of the line represents a character-state's 'performance'; the number indicated below each point on the line corresponds to the character-state numbers in the list of practices in table 1. The height of the line in the line graph is an indication of the performance or the value to the firm of that particular character-state relative to the other character-states in terms of the performance criteria (i.e., customer relationship, product quality, schedule adherence and cost efficiency).

There are several general points here. The first is that the simulations reveal several potential management concerns, both collectively (in terms of group decision-making) and individually, those that can then be flagged up for further discussion and exploration, similar to the 'what-if' scenario building of Nilsson and Darley's (2006) work. Taking the collective concerns as an illustration, when the managers' opinions are aggregated (the first line, to the front of the graph in Figure 1), CS7, outsourced corrective maintenance, fails, indicating a consensus that this practice is problematical. The simulation also flags up concerns over CS6, preventive maintenance, CS20, line-balancing and CS10, full resource visibility, which were low performers relative to other character-states. Overall, the integrity of the organization was very good, however, signifying that as a decision-making group the managers are more complementary than not.

The second point concerns the nature of the model, the micro-diversity that has been captured and the potential insights that this gives. As can be seen from the research methods, the model draws directly from the opinions and views of managers rather than from logical 'if-then' rules and averaged mathematical representations of agents that characterise agent-based and other self-organizational models, particularly those proposed by Lim and Zhang (2003), Zhou et al. (2003), Li et al. (2003), and Poundarikapuram and Veeramani (2004). Whereas here the micro-diversity is represented here by the individual managers' opinions, which are not fully represented in the other simulations.

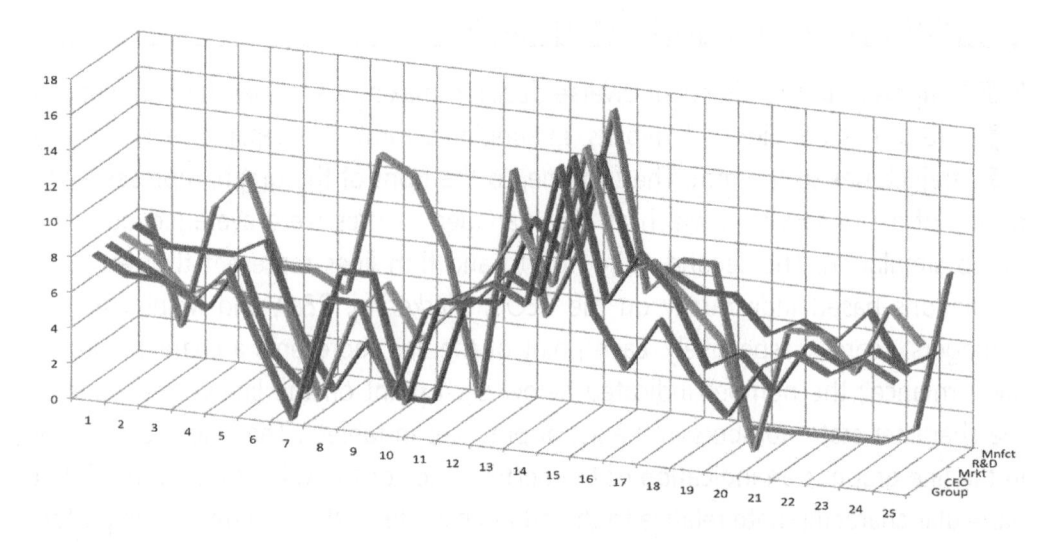

Figure 1 *Simulations of firm evolution: 1) Managers' average; 2) CEO; 3) Marketing Manager; 4) R&D Manager; and, 5) Manufacturing Manager.*

This limitation exists in previous research (see, for example, Baldwin *et al.*, 2005) where opinions of manufacturing managers, operations managers, CEOs and company managers were averaged out. As such, significant information is lost. In Baldwin *et al*'s (2005) study, there was an indication that the informants had very diverse views of how technologies and practices interacted with one another. Unfortunately, the methodology prevented a thorough analysis of views of the individual respondents. This was due to the large numbers of characteristics; that is, the survey instrument had to be divided into four parts with one informant only giving their opinions on a quarter of the total number of characteristics. To further illustrate this limitation, the means and standard deviations of the character-state performances resulting from the grouped managers' opinions are presented in table 2.

Figure 2 graphically displays the results from 100 simulations of the performance of individual character-states giving an indication of variability among the managers' scores whilst also the degree of character state failure throughout the simulations. As can be seen, character-states with the most variability, in descending order are: CS9, ERP system; CS8, MRP system; CS10, full resource visibility; CS20, line-balancing; CS24, employee multi-skilling; CS23, empowering employees; CS19, production process traceability; CS22, flexible workers; CS21, job rotation; CS11, resource priority control; and, CS12, ERP supply chain integration. Character states that had high failure rates, in descending order, are: CS7, outsourced corrective maintenance failing in 23% of simulations; CS20, line-balancing (14% failure rate); CS10, full resource visibility (13%

CS	Mean	SD	CS	Mean	SD	CS	Mean	SD
1	8.00	0.60	10	3.25	3.59	18	8.50	0.58
2	5.75	2.50	11	4.75	3.20	19	4.75	3.40
3	6.75	1.71	12	8.50	3.11	20	3.00	3.46
4	6.75	1.50	13	9.00	2.83	21	4.50	3.32
5	7.25	2.63	14	8.50	0.58	22	5.00	3.37
6	4.25	1.50	15	16.00	0.82	23	4.75	3.40
7	1.50	2.38	16	8.25	0.96	24	5.50	3.42
8	7.25	4.35	17	9.75	0.96	25	7.25	1.89
9	6.75	4.99						

Table 2 *Means and standard deviations of the character-state performances.*

failure rate); CS21, job rotation (12% failure rate); and CS22, flexible workers, CS9, ERP system, and CS11, resource priority control, all failing in 10% of simulations.

An important point to make here is that the degree of failure rate in the simulations is not a sufficient indicator of a problem in actual practice, if, for example, it is accompanied with high variability which is an indicator of management disagreement. However, if a high failure rate in combination with low variability is found then the likelihood of a genuine problem in practice is high. This is most evident with CS7, outsourced corrective maintenance, which has the highest failure rate together with only low-to-moderate variability.

This limitation is addressed here in that the opinions of different managers in a firm are considered fully and can be explored individually. Thus, to explore this variability in more detail, the next step was to analyse the opinions of the different decision-makers separately and compare and contrast the results. Figure 1 also shows a simulation of the CEO's opinions at the stable solution (the second line in the graph). Interestingly, this simulation largely mirrors the first simulation where all the managers' opinions are aggregated. This raises several questions. Does the CEO have a more overarching model of the 'mechanics' of the firm reflecting the consensus of the group of managers? Or does the CEO have the ability to project his understanding/influence onto the other managers in their particular fields of functional expertise? This may reflect in some respects, for example, what Jarratt (1999) sees as achieving the right balance between centralised and de-centralised systems in the management of diversity. An important observation is that the CEO's results did not have any character-state failures. This gives a strong impression that the CEO does indeed have at least a healthy view of the organization and, in contrast to the other managers, sees how all the character-states work and fit together. However, the CEO did appear to have concerns over CS6, preventive maintenance, CS7,

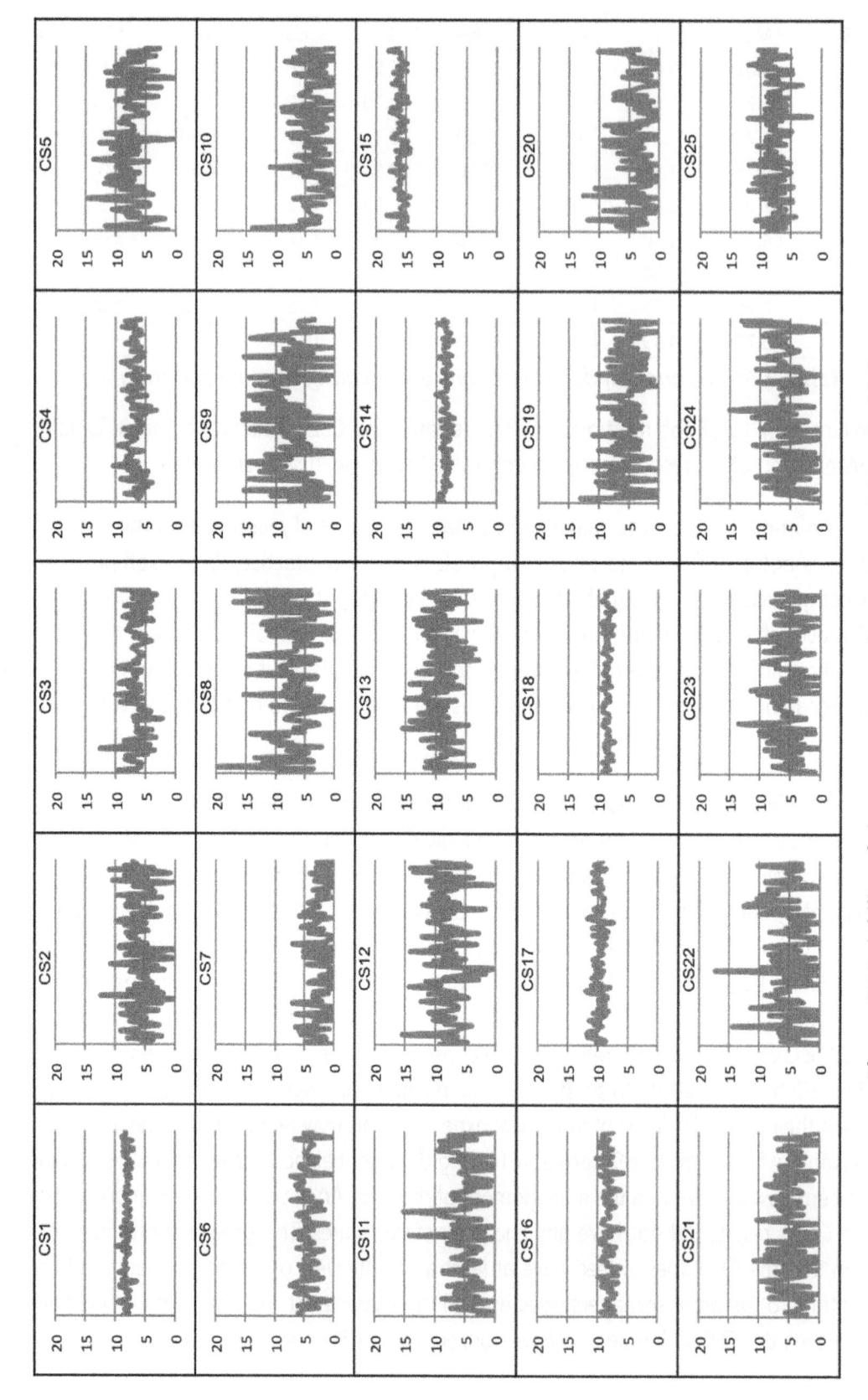

Figure 2 *Variability of character-state performance and degree of failure.*

outsourced corrective maintenance, CS21, job rotation and CS10, full resource visibility, which, apart from CS21, job rotation, reflects the simulation of the aggregated opinions.

Figure 1 portrays the Marketing Manager's simulation at the final solution (the third line in the graph). This begins to demonstrate the significant role that individual differences or diversity has in the management decision-making process and adds support to the arguments of Allen et al. (2006), Poundarikapuram and Veeramani (2004), and O'Leary-Kelly and Flores (2002). In the Marketing Manager's simulation there are several differences that need to be highlighted. The first is that only seventeen of the twenty-five character-states improved on their starting value. This is in contrast to the first two simulations where there were improvements for twenty of the aggregated scores and nineteen of the CEO's scores. However, seven of the nine character-states that surpassed 8 performance-units agreed with the CEO. The main difference, particularly in terms of the CEO's results, was that two character-states, CS7, outsourced corrective maintenance, and CS20, line-balancing, failed altogether. Although the performance of the former character-state reflects the aggregated results and to a degree the CEO's simulation, the latter is opposed to the opinion of the CEO. The Marketing Manager's negative impression of line-balancing may be a symptom, for example, of functional barriers (Malhotra & Sharma, 2002; Rhee & Mehra, 2006) and is another area for further management analysis.

The R&D Manager's simulation at closing is shown in Figure 1 (fourth line in the graph). The results largely agreed with the latter two managers but with obvious exceptions. Agreement surrounded CS15, quality systems/standards, CS17, decentralised error detection and correction, CS16, 5S's programme, CS13, supplier cooperation, and CS1, R&D investment, which all had good performances with end values of over 8 performance-units (that is, eight out of ten character-states in agreement with the CEO). Furthermore, both CS7, outsourced corrective maintenance and CS10, full resource visibility, failed for the R&D Manager—the CEO and Marketing Manager's simulations also resulted in low scores for the latter with 1 and 4 performance-units, respectively. The point of divergence concerns both CS9, ERP system, which had no place in the final solution, and, CS8, MRP system, which had a very weak performance relative to most other character-states. In this instance, a plausible explanation is the fact that the firm has both ERP and MRP systems running simultaneously, where the former should be in replacement of the latter. However, no other managers' simulation flagged this.

The final simulation at conclusion (see Figure 1; the fifth line in the graph), based on the Manufacturing Manager's opinion scores, took on the most extreme final configuration and was in stark contrast to the rest of the decision-makers' simulations and is the best example of the significant role and impact of diversity among decision-makers alluded to by Nilsson and Darley (2006), Allen et al. (2006) and Jarratt (1999). The most obvious difference was that six character-states failed with an additional two barely surviving, finishing with less than 2 performance-units. Of the eight character-states that failed or underperformed, six surrounded policies concerning employees (i.e., employee multi-skilling, line-balancing, job rotation, flexible workers, empowering employees, and continuous production). This pattern indicates that the Manufacturing Manager has issues with the way the workforce is utilised. Suggested reasons could be that the employee policies are not working as intended or that the manager has different preferences. With only fifteen character-states gaining on the original values, it was, however, interesting that twelve of these reached or exceeded 8 performance-units, which was the most out of all simulations. With both the failures and high scoring character-states, this simulation represents the most extreme potential evolutionary trajectory of the firm out of the five presented here. This simulation when compared to the other simulations also lends significant support for a long standing call voiced by O'Leary-Kelly and Flores's (2002) and Malhotra and Sharma (2002) for more integration and understanding between functional areas and particularly between operations and other functions.

On a general reflection, a consensus is evident among all managers surrounding the importance of a good proportion of character-states including (indicated by 8 performance-units or above in the majority of simulations): CS15, quality systems/standards, CS14, TQM sourcing, CS18, 100% inspection, CS17, decentralised error detection and correction, CS7, 5Ss programme, CS5, setup automation, CS12, ERP supply chain integration, CS13, supplier cooperation, and CS1, R&D investment. The practical usefulness perhaps lies more in the exploration of the more problematical areas (Nilsson and Darley, 2006). Character-states in need of review and discussion (signified by multiple failures) include CS7, outsourced corrective maintenance (3 failures) and CS20, line-balancing (2 failures). The results suggest that the policies concerning employees may also need revisiting, as a good proportion faired relatively poorly rarely breaching 8 performance-units. In terms of methodology, the findings also strengthen the consistency/reliability of the data collection procedure adapted from previous work (Allen et al., 2005, 2006, 2007; Baldwin et al., 2005).

Closing remarks

In terms of practical value, ECS models and simulations, along with other similar tools such as those advocated by Nilsson and Darley (2006) and Meade et al. (2006), offer a more realistic decision-support tool for management with which to explore the strengths, weaknesses and consequences of different decision-making capacities within the firm. In this research, for example, out of the four decision-makers, the CEO appeared to have the most balanced view of the organization as all character-states successfully survived. Both the Marketing and R&D Managers had similar simulation outcomes to the CEO but with two to three character-state failures. The Manufacturing Manager's simulation took the most extreme evolutionary trajectory and highlighted the potentially disastrous effects of diversity in decision-making.

On a more academic note, further case studies are still required; firstly, to strengthen the reliability and validity of the methods employed; and, secondly, to encompass more management decision-making scenarios. In future research, clarification could be sought into the potential underlying reasons and consequences for the successes and failures of particular technologies, practices and policies. Unfortunately, in this instance, the empirical setting could not be re-visited. Not long since the main investigation was conducted the case study firm ran into difficulties and ceased operations, approximately a year after the questionnaire survey (late 2006). There are several other avenues for future research that builds on and can extend this work. Firstly, the approach may be used to explore underlying opinions, beliefs and attitudes along with their potential consequences on the evolutionary trajectory of a firm when introducing an entirely new manufacturing technology, practice or policy. At the time of this study, a new ERP system was being implemented and the simulations revealed particular synergies as well as conflicts with other practices. Ideally, this modelling approach should have been applied prior to implementation, perhaps with the input of external experts, and would have perhaps highlighted the most prevalent issues and potential pitfalls. Alternatively, firms may explore a significant change in operations strategy, say from low cost strategy to a high quality or differentiation strategy. The ECS model could then explore the performance of current practice and how new practices could further help (or hinder) the firm. There is also a possibility to model at a level of aggregation above, i.e., the supply chain or perhaps an industrial sector. With the former, supply chain practices in the context of supply chain strategies may be simulated highlighting both what practices (and individual firms) would help or hinder the overall performance of the supply chain.

To conclude, this research aimed to provide insights into the potential evolutionary effects of the diversity in management decision-making and attempted to add to both the theoretical development of complex systems thinking and to the application of computational models and simulations which is still arguably lacking in the particular area of Operations and Production Management (Macbeth, 2002; McCarthy, 2004; Nilsson & Darley, 2006).

References

Allen, P.M. (1976). "Evolution, population dynamics, and stability," *Proceedings of the National Academy of Science*, ISSN 0027-8424, 73(3): 665-668.

Allen, P.M. (1982). "The genesis of structure in social systems: The paradigm of self-organization," in C. Renfrew (eds.), *Theory and Explanation in Archaeology*, ISBN 0125869606, pp. 347-374.

Allen, P.M. (1984). "Self-organization and evolution in urban systems," in R. Crosby (eds.), *Cities and Regions as Non-Linear Decision Systems*, ISBN 0865315302, pp. 29-62.

Allen, P.M., Boulton, J., Strathern, M. and Baldwin, J.S. (2005). "The implications of complexity for business process and strategy," in K.A. Richardson (eds), *Managing Organizational Complexity: Philosophy, Theory and Application*, ISBN 1593113188, pp. 397-418.

Allen, P.M., Strathern, M. and Baldwin, J.S. (2006). "Evolution, diversity and organizations," in E. Garnsey and J. McGlade (eds.), *Complexity and Coevolution*, ISBN 184542140X, pp. 22-60.

Allen, P.M., Strathern, M. and Baldwin, J.S. (2007). "Complexity and the limits to learning," *Journal of Evolutionary Economics*, ISSN 0936-9937, 17(4): 401-431.

Atkins, P.W. (1984). *The Second Law*, ISBN 071675004X.

Baldwin, J.S., Allen, P.M. and Ridgway, K. (2010). "An evolutionary complex systems decision-support tool for the management of operations," *International Journal of Operations & Production Management*, ISSN 0144-3577, 30: 7.

Baldwin, J.S., Allen, P.M., Winder, B. and Ridgway, K. (2005). "Modelling manufacturing evolution: thoughts on sustainable industrial development," *Journal of Cleaner Production*, ISSN 0959-6526, 13(9): 887-902.

Brown, S., Squire, B. and Blackmon, K. (2007). "The contribution of manufacturing strategy involvement and alignment to world-class manufacturing performance," *International Journal of Operations and Production Management*, ISSN 0144-3577, 27(3-4): 282-302.

Chaharbaghi, K. (1991). "DSSL II: A powerful tool for modelling and analyzing complex systems," *International Journal of Operations and Production Management*, ISSN 0144-3577, 11(4): 44-88.

Choi, T.Y., Dooley, K.J. and Rungtusanatham, M. (2001). "Supply networks and complex adaptive systems: control versus emergence," *Journal of Operations Management*, ISSN 0272-6963, 19(3): 351-366.

Cua, K.O., McKone, K.E. and Schroeder, R.G. (2001). "Relationships between implementation of TQM, JIT, and TPM and manufacturing performance," *Journal of Operations Management*, ISSN 0272-6963, 19(6): 675-694.

Das, A. and Narasimhan, R. (2001). "Process-technology fit and its implications for manufacturing performance," *Journal of Operations Management*, ISSN 0272-6963, 19(5): 521-540.

Eisenhardt, K.M. (1989). "Building theories from case study research," *Academy of Management Review*, ISSN 0363-7425, 14(4): 532-550.

Frizelle, G. and Woodcock, E. (1995), "Measuring complexity as an aid to developing operational strategy," *International Journal of Operations and Production Management*, ISSN 0272-6963, 15(5): 26-39.

Fynes, B., Voss, C. and de Burca, S. (2005). "The impact of supply chain relationship dynamics on manufacturing performance," *International Journal of Operations and Production Management*, ISSN 0272-6963, 25(1): 6-19.

Glansdorff, P. and Prigogine, I. (1971). *Thermodynamics of Structure, Stability and Fluctuations*, ISBN 0471302805.

Haken, H. and Mikhailov, A. (eds.) (1993). *Interdisciplinary Approaches to Nonlinear Complex Systems*, ISBN 3642510329.

Hayes, R.H. and Wheelwright, S.C. (1984). *Restoring Our Competitive Edge: Competing Through Manufacturing*, ISBN 0471051594.

Holland, J. (1995). *Hidden Order: How Adaptation Builds Complexity*, ISBN 0201442302.

Islo, H.E. (2001). "Simulation models of organizational systems," *International Journal of Technology Management*, ISSN 0267-5730, 21(3-4): 393-419.

Jantsch, E. (1980). *The Self-Organizing Universe: Scientific and Human Implications of the Emerging Paradigm of Evolution*, ISBN 0080243118.

Jarratt, A. (1999). "Managing diversity and innovation in a complex organization," *International Journal of Technology Management*, ISSN 0267-5730, 17(1-2): 5-15.

Kaihara, T. (2003). "Multi-agent based supply chain modelling with dynamic environment," *International Journal of Production Economics*, ISSN 0925-5273, 85(2): 263-269.

Karkkainen, H. and Hallikas, J. (2006). "Decision making in inter-organizational relationships: implications from systems thinking," *International Journal of Technology Management*, ISSN 0267-5730, 33(2-3): 144-159.

Kinni, T.B. (1996). *America's Best: IndustryWeek's Guide to World-Class Manufacturing Plants*, ISBN 0471160024.

Klassen, R.D. and Whybark, D.C. (1999). "The impact of environmental technologies on manufacturing performance," *Academy of Management Journal*, ISSN 0001-4273, 42(6): 599-615.

Kondepudi, D. and Prigogine, I. (1998). *Modern Thermodynamics: From Heat Engines to Dissipative Structures*, ISBN 0471973947.

Leachman, C., Pegels, C.C. and Shin, S.K. (2005). "Manufacturing performance: evaluation and determinants," *International Journal of Operations and Production Management*, ISSN 0144-357, 25(9-10): 851-874.

Li, S.L., Loulou, R. and Rahman, A. (2003). "Technological progress and technology acquisition: Strategic decision under uncertainty," *Production and Operations Management*, ISSN 1937-5956, 12(1): 102-119.

Lim, M.K. and Zhang, Z. (2003). "A multi-agent based manufacturing control strategy for responsive manufacturing," *Journal of Materials Processing Technology*, ISSN 0924-0136, 139(1-3): 379-384.

Macbeth, D.K. (2002). "Emergent strategy in managing cooperative supply chain change," *International Journal of Operations and Production Management*, ISSN 0144-357, 22(7-8): 728-740.

MacIntosh, R. and MacLean, D. (2001). "Conditioned emergence: researching change and changing research," *International Journal of Operations and Production Management*, ISSN 0144-357, 21(9-10): 1343-1357.

Malhotra, M.K. and Sharma, S. (2002). "Spanning the continuum between marketing and operations," *Journal of Operations Management*, ISSN 0272-6963, 20(3): 209-219.

McCarthy, I.P. (2003). "Technology management: A complex adaptive systems approach," *International Journal of Technology Management*, ISSN 0267-5730, 25(8): 728-745.

McCarthy, I.P. (2004). "Manufacturing strategy: understanding the fitness landscape," *International Journal of Operations and Production Management*, ISSN 0144-357, 24(1-2): 124-150.

McCarthy, I.P., Leseure, M., Ridgway, K. and Fieller, N. (1997). "Building a manufacturing cladogram," *International Journal of Technology Management*, ISSN 0267-5730, 13(3): 269-286.

McCarthy, I.P., Leseure, M., Ridgway, K. and Fieller, N. (2000). "Organizational diversity, evolution and cladistic classifications," *The International Journal of Management Science*, ISSN 0305-0483, 42(28): 77-95.

McKone, K.E., Schroeder, R.G. and Cua, K.O. (2001). "The impact of total productive maintenance practices on manufacturing performance," *Journal of Operations Management*, ISSN 0272-6963, 19(1): 39-58.

Meade, P., Rabelo, L. and Jones, A. (2006). "Applications of chaos and complexity theories to the technology adoption life cycle: case studies in the hard-drive, microprocessor, and server high-tech industries," *International Journal of Technology Management*, ISSN 0267-5730, 36(4): 318-335.

Meredith, J. (1998). "Building operations management theory through case and field research," *Journal of Operations Management*, ISSN 0272-6963, 16(4): 441-454.

Nicolis, G. and Prigogine, I. (1977). *Self-Organization in Nonequilibrium Systems: From Dissipative Structures to Order through Fluctuations*, ISBN 9780471024019.

Nicolis, G. and Prigogine, I. (1989). *Exploring Complexity*, ISBN 0716718596.

Nilsson, F. and Darley, V. (2006). "On complex adaptive systems and agent-based modelling for improving decision-making in manufacturing and logistics settings - Experiences from a packaging company," *International Journal of Operations and Production Management*, ISSN 0267-5730, 26(11-12): 1351-1373.

O'Leary-Kelly, S.W. and Flores, B.E. (2002). "The integration of manufacturing and marketing/sales decisions: impact on organizational performance," *Journal of Operations Management*, ISSN 0272-6963, 20(3): 221-240.

Poundarikapuram, S. and Veeramani, D. (2004). "Distributed decision-making in supply chains and private E-marketplaces," *Production and Operations Management*, ISSN 1937-5956, 13(1): 111-121.

Prigogine, I. (1973). "Irreversibility as a symmetry-breaking process," *Nature*, ISSN 0028-0836, 246(5428): 67-71.

Prigogine, I. and Stengers, I. (1987). *Order out of Chaos*, ISBN 0553340824.

Raymond, L., Julien, P.A., Carriere, J.B. and Lachance, R. (1996). "Managing technological change in manufacturing SMEs: A multiple case analysis," *International Journal of Technology Management*, ISSN 0267-5730, 11(3-4): 270-285.

Rhee, M. and Mehra, S. (2006). "Aligning operations, marketing, and competitive strategies to enhance performance: An empirical test in the retail banking industry," *Omega: International Journal of Management Science*, ISSN 0305-0483, 34(5): 505-515.

Saunders, M., Lewis, P. and Thornhill, A. (2007). *Research Methods for Business Students*, ISBN 1405886137.

Schonberger, R.J. (2008). *World Class Manufacturing: The Lessons of Simplicity Applied*, ISBN 0029292700.

Simon, H.A. (1955). "A behavioral model of rational choice," *Quarterley Journal of Economics*, 6(1): 99-118.

Simon, H.A. (1983). *Reason in Human Affairs*, ISBN 0804711798.

Skinner, W. (1969). "Manufacturing: missing link in corporate strategy," *Harvard Business Review*, ISSN 0017-8012, 47(3): 136-45.

Slack, N., Chambers, S. and Johnston, R. (2007). *Operations Management*, ISBN 0273708473.

Womack, J.P., Jones, D.T. and Roos, D. (1990). *The Machine that Changed the World*, ISBN 0892563508.

Zhou, Z.D., Wang, H.H., Chen, Y.P., Ai, W., Ong, S.K., Fuh, J.Y.H. and Nee, A.Y.C. (2003). "A multi-agent-based agile scheduling model for a virtual manufacturing environment," *International Journal of Advanced Manufacturing Technology*, ISSN 0268-3768, 21(12): 980-984.

Zhu, Q. and Cote, R.P. (2004). "Integrating green supply chain management into an embryonic eco-industrial development: a case study of the Guitang Group," *Journal of Cleaner Production*, ISSN 0959-6526, 12(8-10): 1025-1035.

Zhu, Q.H., Sarkis, J. and Lai, K.H. (2008). "Confirmation of a measurement model for green supply chain management practices implementation," *International Journal of Production Economics,* ISSN 0925-5273, 111(2): 261-273.

Appendix

Pre-prepared check list of operations management technologies, practices and policies (adapted from McCarthy *et al.,* 1997; Womack *et al.,* 1990).

✓	TECHNOLOGIES, PRACTICES AND POLICIES	✓	TECHNOLOGIES, PRACTICES AND POLICIES
	1. Standardization of parts		31. Individual error correction; products are not re-routed to a special fixing station
	2. Assembly time standards		
	3. Assembly line layout		
	4. Reduction of craft skills		32. Sequential dependency of workers
	5. Automation (machine paced shop)		33. Line balancing
	6. Pull production system		34. Team policy (motivation, pay and autonomy for team
	7. Reduction of lot size		
	8. Pull procurement		35. Groups Vs teams
	9. Operator based machine maintenance		36. Job enrichment
			37. Manufacturing cells
	10. Quality circles		38. Concurrent engineering
	11. Employee innovation prizes		39. ABC costing
	12. Job rotation		40. Excess capacity
	13. Large volume production		41. Flexible automation for product versions
	14. Suppliers selected primarily on price		
			42. Agile automation for different products
	15. Exchange of workers with suppliers		
	16. Socialization training (master/apprentice)		43. Insourcing
			44. Immigrant workforce
	17. Proactive training programmes		45. Dedicated automation
	18. Product range reduction		46. Division of labour
	19. Autonomation		47. Employees are system tools and simply operate machines
	20. Multiple sub-contracting		
	21. Quality systems (tools, procedures, ISO9000)		48. Employees as system developers; value adding
	22. Quality philosophy (TQM, culture)		49. Product focus
	23. Open book policy with suppliers; cost sharing		50. Parallel processing
			51. Dependence on written rules; unwillingness to change rules as the economic order quantity
	24. Flexible multi-functional workforce		
	25. Set-up time reduction		
	26. Kaizen change management		52. Further intensification of labour; employees are considered part of the machine to be replaced by machines`
	27. TQM sourcing; suppliers selected on quality		
	28. 100% inspection/sampling		
	29. U-shape layout		
	30. Preventive maintenance		

Keith Ridgway CBE, the AMRC Group Research director, worked in industry before moving to academia in 1980. He joined the University of Sheffield as a lecturer in 1988, and became Professor of Design and Manufacture in 1997. He worked with local businessman Adrian Allen to launch the AMRC with Boeing in 2001. In 2009, he helped launch the Nuclear Advanced Manufacturing Research Centre. Keith was awarded an OBE in 2005 and CBE in 2011. He became a Fellow of the Royal Academy of Engineering in 2006.

James Baldwin received his PhD and pursued a Postdoctoral Fellowship under the newly founded Advanced Manufacturing Research Centre (AMRC) with Boeing. He now works in the AMRC's Rolls-Royce Factory of the Future where he is the technical co-ordinator of a large, €7.6million, Framework 7 project, partly funded by the European Commission, which is a collaboration between several European universities and industrial partners, including Rolls-Royce and Electrolux. Between these times at the AMRC, James was a lecturer of project, operations and supply chain management for five years at the University of Sheffield's of School of Management. He has published over 80 papers in journals, books and conference proceedings.

Chapter Θ

The evolution of industries in diverse markets: A complexity approach

Elizabeth Garnsey, Simon Ford & Paul Heffernan

The question of how industries emerge and develop over time is one that has drawn attention from academics across a range of disciplines, leading to a dominant narrative around the concept of the product life cycle. In this chapter we take a grounded approach to the subject, exploring cases of industrial development in a number of product-based industries: instrumentation, semiconductors, microcomputers, and the Network Computer. We analyze these cases through the coevolutionary lens of complexity. Our analysis illustrates some of the limitations of linear product life cycle perspective, while the complexity lens highlights how the complex unpredictable nature of evolving industrial systems, which differentiate as they coevolve and are integrated through processes of self-organization and entrainment.

Introduction

In this chapter we address the question of how industries emerge and develop over time, and show how a complexity perspective can throw light on this issue. We start with the problem of multi-faceted evidence that is difficult to sum up and make sense of: how can we characterize the unfolding of industrial sectors that are highly distinctive in their producer histories, technologies and users? We begin with life cycle approaches to explaining the development of four industry sectors, viewed as exemplars, and then consider variations in their experience. We go on to describe how a complexity perspective provides a meta-theoretical orientation that can identify commonalities in the face of diversity, an approach that Peter Allen pioneered in his studies of innovation.

The interconnectedness of industrial activity suggests that new industries do not arise de novo but emerge from existing activity and knowledge. Just as new technologies depend on recombinations of knowledge and practice, new industries depend to a large extent on recombinations of knowledge and practices from prior industries. New experimental activity in economic life draws on the endowments of past activity, in the science base and in existing industry.

The simplest form of evolving industrial activity can be seen in product class industries that originate in the invention of a new product and proceed through to commoditization of the innovative outcome of the invention. In the field of industrial economics, the work of Abernathy and Utterback (1978), and of Klepper and Graddy (1990), have been particularly influential in addressing the development of product class industries. The Abernathy and Utterback model addressed product innovation, process innovation and competitive structure both at the level of the organization and in relation to the life cycle of the industry. The work of Klepper and Graddy (1990) supports their model of industrial development. Initially the number of firms in the industry is said to expand, then at a later point there is a shakeout in the number of firms followed by a period in which the number is stable (Klepper & Graddy, 1990). The early entrants into the industry are typically small and have experience in related technologies. Sometimes they are users of the new product, while in other instances they are spin-offs of incumbent firms. They often introduce major product innovations based on information about users' needs and/or the technological means available to satisfy them (Klepper & Graddy, 1990).

The life cycle model of industry development presents an analytical distinction between product and process innovation at the industry level. In the case of assembled products, basic product concepts take form in the early phase of the industry and once the dominant design emerges, opportunities for radical product innovation recede. Firms in the industry tend to produce similar products and the competitive basis rests on process innovation to lower the cost of production. Industries develop a division of labour, arising from firm diversification and specialization, while also developing value chains and industry governance institutions, processes which may also cause rigidities to arise.

Focusing on product class industries, researchers in this field have drawn a necessary boundary in defining the extent of the subject of study. Yet there are also a number of alternative ways of defining an industry that go beyond the core product around which innovation is described as occurring. These include descriptions of industries as:

- "[A] group of firms serving the same group of buyers and selling products that are

close substitutes for each other" (Dean, 1951: 151);
- "[D]efined by identifying the knowledge structures shared among key stakeholders ... that provide the inter-organizational dimension of identity against which individual firms can define themselves" (Clegg et al., 2007: 498);
- "[T]he arena in which firms compete" (Tikkanen, 2008: 183), and;
- "[A] population of organizations of the same form" (Low & Abrahamson, 1997: 440).

These alternative definitions demonstrate the difficulty that arises in attempting to define the boundaries of an industry. While many industry definitions, like those above, are supply-oriented, we seek to encompass both supply and demand relations. Demand and user inputs are key features of industry evolution. In place of modern economists' definition of industries as made up of competing firms that produce a common class of products, the basis for the product life cycle approach, we returned to Adam Smith. He was concerned in *The Wealth of Nations* (1776) with how the provision of goods and services meets the "wants of the people." We define an industry as a system of activities in which producers and suppliers meet specific user needs and preferences, while customers and users seek suppliers to provide for these needs. These interactions are traditionally coordinated by market exchange. However the industry boundary, like all system boundaries, is set for purposes of inquiry and in response to the evidence. Industries are open systems subject to the dynamic of new firm entry and exit, alongside entry by firms from other sectors and other parts of the value chain, just as consumer requirements and demography alter as the industry develops.

How can we make sense of such developments? We take a grounded approach to the subject by first exploring cases of industrial development before going on to analyze these through the coevolutionary lens of complexity. We begin by reviewing accounts of product-based industry development, turning then to investigate more diversified industrial development in instrumentation, semiconductors, personal computers and the Network Computer. We focus on those diversified industries that comprise clusters of related activity around complementary products and services. These have a more varied evolutionary history than those based around a single product and are more difficult to analyze.

Industrial development in product-based industries

Industrial evolution

According to product life cycle models, an industry emerges as unmet needs and new productive possibilities draw producers into providing for demand in new ways. Producers experiment with a new product in the hopes of meeting future demand. The potential for meeting needs in ways that consumers had not previously considered is often inherent in technological advances. Supply may stimulate demand by meeting unmet needs with innovative products or services. Over time, supply and demand continually move out of alignment and readjust, as productive potential improves and perceived needs are reflected in purchases. Adjustments set off a chain of further reactions, propelling the industry along its cycle. There are almost infinite variations on this pattern. Moreover, because industries are interrelated, as one industry expands it provides or curtails opportunity space for other industries.

Some industries show a definite pattern of evolution. This is found in particular among industries where there is a dominant design and economies of scale are possible, reducing costs as markets expand. Various depictions of the industry life cycle exist. Abernathy and Utterback (1978) define three phases of industry evolution: the "fluid", "transitional" and "specific" phases. Klepper and Graddy (1990) adopt a similar model with their stages 1, 2 and 3. Grant (2002) offers an industry development model using introduction and growth as equivalents to Utterback's fluid and transitional phases. Moore (1996) proposes similar stages in the development of a "business ecosystem": "pioneering", "expansion", "authority", adding a fourth stage of "renewal" or "death".

In the initial, or "fluid" phase (Abernathy & Utterback, 1978), as a result of access to a new technology, or new organizational or marketing capacity, a few firms become capable of producing something new. Often these are new firms with no vested interests in prevailing conditions. At first the quality and reliability of the pioneer products are limited, and only a few potential users show interest in them. However if the product or service affords benefit to users, demand grows, and new entrants are attracted to the area of activity, particularly if barriers to entry are low. Entrepreneurial teams (or R&D teams in established firms) refine the design and technology, and a variety of alternative designs and production methods emerge. This period is characterized by uncertainty, concerning, for example, user preferences, technology and standards, and industry structure. However, over time, different versions of the product coalesce into a limited set of technological alternatives as the industry progresses into the "transitional" phase.

Consumer demand begins to catch up with the new technical capability. Those producers who can come up with a design that has incorporated consumer preferences and past learning, may find their market growing rapidly if they have the necessary distribution channels and can obtain and payment for the necessary supplies. They may catch a wave of consumers moving to share in the perceived benefits of the new product. But competitors soon move in to defend a substitute product by upgrading, or to gain a share of the profits by imitating the new product. There may be a struggle to establish the dominant design, with alternative technologies backed by different firms.

The start of the "specific phase" is marked by the emergence of a dominant design, particularly in industries where complementarity of products and components can reduce costs and improve performance. This is not necessarily the best of all possible solutions, indeed it is often inferior to certain leading edge designs. However the emergence of an accepted design solution makes it possible to concentrate the major efforts of participants in the industry onto refining that solution and bringing down its cost. A further period of refinement of the design and improvements in quality and performance then takes place in order to reach a wider consumer base. This may involve intense competition and an industry "shake-out," in which certain firms come to dominate. The division of labour becomes progressively more complex . The industry grows until all who want the new product at the going price have purchased it, and the industry has to survive on demand for replacements. New ways to persuade other consumers of the benefits of the product must be found. This usually requires a fall in price, and thus a fall in production costs. Pressure is brought to bear on costs in the production process and new methods of production introduced. New markets are sought among new consumers, domestic or foreign. The industry may renew itself through reductions in price and upgrading of products. Finally, however, unless some form of renewal occurs, the product will be displaced and the industry will decline.

Variations in the product life cycle

The product life cycle (PLC) provides a means of exploring the processes involved in the maturation of industries made up of similar products, and a basis for comparison between industries. There is empirical evidence (e.g., Utterback & Suarez, 1993) to support the existence of PLC features in a number of key industries, including electric lamps, typewriters, automobiles and TV sets, but they are less obvious elsewhere.

The reasons why the product life cycle is more evident in certain industries relate to the nature of the production process and the needs the product meets (Garnsey & Minshall, 2002). A simple classification of production distinguishes between unit production, small batch, large batch, mass production and continuous flow production (Woodward, 1970). A product life cycle with distinctive phases is found to a greater extent in large batch and mass production of assembled products than in cases where production is not carried out in volume, or where there is continuous flow production, e.g., paper making or plate glass. Unit production does not allow of economies of scale, and in continuous flow production, variations in design are more limited. On the other hand, if there is great scope for variations in design and requirements, as in electronic components, design is less likely to standardize.

According to Klepper and Graddy (1990), two factors appear to have an important effect on the pace of the evolutionary process: product technology and buyers' preferences. Products characterized by limited opportunities for technological change tend to be subject to less uncertainty and to reach maturity faster where products are characterized by considerable diversity in buyers' preferences, it is more difficult for dominant designs to emerge, which tends to lengthen the time it takes to reach maturity.

The concept of the product life cycle is most relevant where the industry coincides with a given product which can be scaled up to volume production and which has a finite scope for variation once a dominant design has emerged. The standardization of design depends on the extent to which variants on the product are possible, and also on the extent to which users' needs vary, calling for product differentiation. In fragmented industries, where levels of competition are high, there is often product diversity and a proliferation of overlapping product life cycles rather than one dominant cycle.

Where there is a reduced number of components and materials, the possibilities for variation of design are fewer. The transition from product to process innovation applies to products in which production can be scaled up gradually. In the case of products where minimum scale requirements are quite high (e.g., plate glass), production process costs are relevant from early on. In process industries there is an early shift of effort and experiment onto the production process. Some products are never produced in volume, including large turbine generators and aircraft, where radical product innovations remain more important than incremental process innovations.

The PLC model helps us to understand the kind of processes that shape industrial evolution, but it has limitations as a practical tool. Some have viewed it as of little value; Porter claimed that the PLC models lack a rationale and predictive capacity: "... nothing in the concept allows us to predict when it will hold and when it will not" (1980: 162). However Porter did not examine the effects of differences in production processes by industry in this respect. Utterback has shown how differences between industries in the extent to which an identifiable life cycle has occurred can be related to differences in production processes (Utterback, 1994).

But though the model can be applied retrospectively, during the unfolding of the processes it is never clear how long a phase will continue. The product life cycle is a learning model, a heuristic device for exploring the processes involved in change, not a predictive model. It has limitations when applied to some of the sectors discussed below.

Industrial development in instrumentation

The instrumentation industry is a multi-sector, multi-product industry which faces many different market conditions. The medical, research and industrial markets are highly diverse. Instruments include: scientific analytical instruments, affected by funding for science and R&D; medical instrumentation shaped by developments in health care and affected by the nature of demand, health insurance and public health schemes; electronic Instruments, include standard instruments which have been commoditized and are increasingly the province of large international companies, while specialist and precision instruments continue to provide niches for small entrepreneurial firms; production control instruments—generally more rugged and automated than laboratory products—where costs of entry are often a deterrent to small enterprises, but diversity in requirements creates niches which provide openings for entrepreneurs.

The path of new instrumentation started with specific needs in a particular discipline, but the successful development of the instrumentation in turn led to its applications in other areas of science. Computers were first developed in the 1940s and 1950s by scientists whose research required very elaborate calculations. Interest in computational capability was initially limited to certain specialist domains (Ritchie, 1986). The availability of new instrumentation also resulted in interdisciplinary collaboration, as when University of Cambridge physicists migrated into microbiology to pursue investigations made possible by the electron microscope. Similarly, nuclear magnetic resonance was developed by physicists at Harvard and Stanford to measure the magnetic moments of atomic nuclei, first developed as a tool of analytical chemistry before being transferred into biology and medicine. Indeed it is worth noting that most innovation in medical instrumentation has come from universities and medical schools and not from the medical device industry (Rosenberg, 1994)

Emerging applications required further work to improve performance and new lines of investigation led to further basic research. In the 1940s and 1950s, investigations into the potential of semiconductors and the transistor (which we discuss further in the next section) stimulated a massive research effort in solid-state physics and the science of surface states. Similarly, while initially the laser was no more than a scientific curiosity at Bell Laboratories, with the firm failing to see its potential in the communications industry, its subsequent development led to research in neglected topics in optics and encouraged the emergence of optoelectronics. This in turn saw the laser find applications in semiconductor fabrication in the 1960s (von Hippel, 1988).

It was not long before new instrumentation moved into industry. Many other techniques used in the manufacturing of semiconductors, for example, ion implantation techniques to control the deposit of impurities, along with the use of interferometers to produce precise patterns on a chip in lithography, originated as techniques in high energy particle physics. In a similar manner, cryogenics, the science of freezing, also moved rapidly from basic to applied research and into industry (Langrish, 1972).

The tacit knowledge acquired by scientists in the laboratory was often exploited by entrepreneurially-inclined individuals. In chromatography, for example, knowledge was available in published form which could be accessed by established companies. In the post second world war period until the 1980s. "The majority of instrument innovation involved the movement of technical leaders to form their own companies or join recently established firms. The scientific base and essential enabling technologies existed in university, government or large industrial laboratories, while new instrument products were largely the work of small new firms" (Shimshoni, 1970: 85).

Entrepreneurial start-ups often have innovative customers in the period of experimentation and diversity at the outset of a new industry. As knowledge about the new instrumentation advances, wider opportunities to enter commercial markets arise. Production industries are continually developing instrumentation to streamline and improve control over production processes, with these developments opening up new opportunity spaces for innovative instruments.

Small focussed teams are frequently the most efficient innovators, whatever their base. However, some instrumentation is beyond the scope of small or new firms to commercialize and requires the capital available to established firms. In the electron microscope industry for example, interaction took place mainly between R&D teams in established companies and university researchers. There has been a three way interaction between mobile scientists formed in leading science laboratories, small entrepreneurial firms (often founded by technical experts) and established firms with technical capabilities.

Instruments which are more accessible both in price and usability can reach a much larger research community, greatly extending the research capability of laboratories of all kinds. Oxford Instruments, for example, was an entrepreneurial small firm founded to make available leading edge equipment to centres of semiconductor research and development. Instruments which are more accessible both in price and usability can reach a much larger research community, greatly extending the research capability of laboratories of all kinds.

For some instruments, larger firms were at an advantage because of the capital requirements; e.g., electron microscopy required the resources of a large technology-based company because of the array of sophisticated and costly technical procedures that were required to develop this instrument (Shimshoni, 1970; Agar, 1996). New entrepreneurial firms were less well positioned to exploit opportunities where capital requirements were high and had difficulty competing with established companies in more mature markets. Commercial customers required equipment that did not need highly specialized technicians, with products needing to be designed for manufacture in order to reduce production costs and improve ease of use.

Instrumentation activities were subject to disruption as new technologies arose in allied fields. In the 1980s the introduction of the microcomputer and Windows-based graphical interface represented a discontinuous source of innovation, making it possible for computer power and packaged software to be applied to instrumentation in small labs and companies. This opened up new prospects but disrupted established markets. Many instrumentation companies failed to adapt and there were wholesale closures, exacerbated by a shrinking manufacturing sector which reduced revenues from older industrial instrumentation in multi-product instrumentation companies. The transition from electromechanical to electronic and software-driven instrumentation eliminated many firms in the UK that proved incapable of transforming their skills base, while creating extensive opportunities for innovative instrumentation firms. The transition from mainframe to networked microcomputers and graphical user interface created difficulties for older instruments firms with insufficient expertise in emerging technologies, as new hardware-software combinations emerged.

Reflections on industrial development in instrumentation

In this example from instrumentation there are clear similarities with the development of product-based industries; small entrepreneurial firms have given way as producers to fewer dominant firms as the market for a set of instruments has matured, with this leading to the concentration of industrial instrumentation production in mature sectors. Selection forces operated through cost pressures, eliminating firms that lack the reserves or competence to see them through fluctuations in their fortunes. The firms that were eliminated were often very new firms and those established firms that had become unwieldy in size and lacked the capital or managerial skills required to adjust to new conditions. As process innovation became cost-driven, and basic research equipment became commoditized, production then moved to emerging economies where labour costs were lower.

The genesis of an industry often involves new ways to solve problems and meet needs, often directly or at some remove the outcome of new knowledge or science. Such was the case in optoelectronics, where there were multiple knowledge flows between science and industry. The transfer from basic to applied research, along with the movement from academic lab to industry, was interactive and involved the fusion of technologies from different sources. The transfer of instrumentation did not move in a simple linear fashion from science into commercial applications. Instead, work on basic research problems yielded solutions to applied problems and findings that did not fit existing theories were identified by detailed application, raising further questions in basic science. User innovation played an extensive part in this industry's development, with much of the equipment in an electronics manufacturing plant having its origin in the research laboratory (Rosenberg, 1994).

Within the instrumentation sector new entrepreneurial firms have transformed themselves in order to address markets of this kind. Although they might not be selected and grow to become the leading firms in the industry in which they first chose to compete, they often continue to specialize in innovative instrumentation, remaining in niches that allow them to produce a stream of new products for the research market. As entrepreneurial firms have searched for these niches, new markets for instrumentation have arisen in new industries. For example, the automation of biotechnology research has created extensive openings for new instrumentation firms in new cycles of complementary innovation. This has stimulated the increasing sophistication and internationalization of the instrumentation industry, with a complex structure of distribution networks emerging. However, as a consequence of remaining in a niche, entrepreneurial firms are unable to reap the rewards of moving from early commercialization into production for a more mature and larger market sector. Hewlett Packard is the most well known example of a firm that has been able to serve both kinds of market; such firms are comparatively rare.

Despite the diversity of markets for instrumentation, there are certain common features faced by firms in the wider cluster of activities that we describe as forming the instrumentation industry. Self-renewing processes result from the role of instruments in production chains that continually need new forms of instrumentation. Underlying technology platforms affect the whole industry as instruments have become interconnected in clusters of activity, with both software and hardware compatibility operating as a requirement. Developments in information technology place an increasing premium on modes of ensuring complementarity. As a consequence, common standards of design and interchange have become increasingly important as information is shared and diffused through the internet. The self-standing instruments of previous generations are becoming more interactive forms of equipment, with implications for the coevolution of various parts of the instrumentation industry. Interchange between software and hardware can be achieved by modularity of design together with common interface protocols. Both uniformity, for information exchange, and diversity to enhance innovation may be achieved in this way.

At the same time, the increasing sophistication and internationalization of the instrumentation industry has seen the development of a complex structure of distribution networks in which distributors are often manufacturers in their own right, in competition with small instrument makers or aiming to acquire them.

Industrial development in semiconductors

The semiconductor was a form of meta-instrumentation which has transformed a vast array of other industries. It depended on advances in instrumentation and its development illustrates the complexity of analyzing a product class industry that gave rise to a plethora of other applications.

The semiconductor originated from research carried out in the laboratories of the Bell Telephone Company, where the transistor was invented in 1947. The transistor required less current, generated less heat, and was over fifty times smaller than the vacuum tubes earlier used for this purpose (Wolfe, 1983). With US government grants that continued into the Cold War, Bell Labs was funded in some respects like a public sector research institute. Within about five years, transistors were reliable enough for commercial use. They were exploited not by established companies, but by a maverick employee of Bell Labs, William Shockley, one of the coinventors of the transistor. Returning to his home town Palo Alto, he set up Shockley Semiconductors in 1955 to pursue opportunities opened up by the transistor.

This was the origin of a swarm of entrepreneurial endeavours that created the semiconductor sector in Silicon Valley, largely through multi-generational spin-outs. Shockley fell out with his employees, who left and started up their own firms. The first spin-out was Fairchild Semiconductors, originally a joint venture, and the first company to work exclusively in silicon. It was by combining a number of transistors, diodes, resistors and capacitors onto a single piece of silicon that Robert Noyce invented the integrated circuit at Fairchild Semiconductors in 1959. Ten years later, Noyce had spun out his own company, Intel, with Gordon Moore. Within two years, Noyce and Moore had developed the 1103 memory chip, the size of two letters in a line of type, each chip containing four thousand transistors.

At the end of Intel's first year in business, which had been devoted almost exclusively to research, sales totalled less than three thousand dollars and the work force numbered forty-two. In 1972, thanks largely to the 1103 chip, sales were $23.4 million and the work force numbered 1,002. In the next year sales almost tripled, to $66 million, and the work force increased two and a half times, to 2,528. (Wolfe 1983: 364)

Other semiconductor ventures were drawn by these high returns, the numbers fuelled by large numbers of engineers produced by universities in the US and by employee departures to found new start ups. By 1972 there were 330 semiconductor manufacturing firms in the United States (Freeman 1995: 234).

From the late 1960s the output of the semiconductor industry was increasingly concentrated in about a dozen companies, as economies of scale led to an increase in plant size and capital costs. The continuous lowering of prices deterred new entrants. European companies entered the semiconductor industry at a later date, by which time entry by new small firms was more difficult. Even as scale barriers set in, opportunities arose in the application of semiconductor devices in the production of new products and IT applications for new markets. In the 1970s, a significant proportion of semiconductor production was relocated from Silicon Valley to Japan and Korea, where mass production costs were generally lower and quality was higher. This relocation provoked a slump in Silicon Valley during the 1980s, until new opportunities for applications led to a renewal of the semiconductor industry in the 1990s.

The industry pioneer, Intel, played a crucial role in the growth and renewal of the semiconductor industry as it reached early maturity. While the original focus at Intel was on dynamic random access memory (DRAM), revenues from DRAM began to decline in the 1970s as commoditization set in and foreign competition increased. This renewal was driven by advances made in microprocessors, a particular configuration of integrated circuits. The first microprocessor was designed at Intel in 1968 in response to a request from a Japanese manufacturer for specialist chips for use in a desktop calculator. Following further refinements to this original design, the microprocessor was soon available for thousands of applications, including the microcomputer.

With its core business in computer memory, a sector which it dominated despite the numerous firms entering the sector (Grove, 1997), Intel did not at first recognize the importance of the microprocessor. However, local fabrication plants managers had discretion over resource allocation in their plants and elected to increase their output on microprocessors because they had higher margins. This was not a deliberate corporate-wide strategy; in the early 1980s neither the importance of the microprocessor was recognized nor the future growth of microcomputers anticipated. Nevertheless, as they realized what was occurring, senior Intel management adjusted their strategy and pursued the development of microprocessors, going on to provide the core technology for the microcomputer industry (Burgelman, 1994).

Reflections on industrial development in semiconductors

The evolution of the US semiconductor industry provides an example of the complementarities between large and small firms. Existing large firms produced much of the basic technology and the technically skilled personnel which were essential to the start up of new technology-based firms; the new firms provided the entrepreneurial drive and took the innovation into new markets. Important innovations sprang from these small firms, but Bell Telephone Labs continued to account for a high proportion of major innovations in semiconductor technology (Rothwell & Zegveld, 1982).

To date, the generic applications of the products of the semiconductor industry have prevented the onset of the kind of industry maturation cycle the product class industry model describes. Intel moved out of the memory market when prices rapidly declined and Japanese and Korean producers were in the ascendancy. However they had expertise in other technologies, notably microprocessors, and they have managed to protect their intellectual property in this area. Furthermore, as applications for this product can be found in so many markets, diminishing returns have not set in. A similar story can be found in other semiconductor devices such as RISC microprocessors. The Cambridge firm ARM has been able to successfully license its RISC chips to a variety of corporate clients who value the low energy consumption of these chips, and who in turn have developed a multiplicity of new products.

Intel has pursued strategies of alliance, but also has created defensive walls around its intellectual property in order to retain the rewards of innovation. In this way it has created significant barriers to entry. Furthermore, Intel has recognized the importance of networking products such as switches, ethernet hubs, and plug-in network cards that provided connectivity and allow PCs to be networked. Thus Intel attempted to take the lead in developing and integrating the wider convergent industry, in what has been called the extended 'business ecosystem' (Moore, 1993). Intel's experience illustrates how an industry leader can shape developments in unique ways through strategies that are path dependent, reflecting the leader's history and learning from experience (Gawer & Cusumano, 2002). Through a combination of defensive and innovative strategies, a leading firm like Intel (and ultimately the strategies and decisions of its managers) can have a major impact on industry developments.

Industrial development in microcomputers

Digital computers were developed in the 1940s, heavily funded by defence spending at IBM and elsewhere. During the 1960s and 1970s computing facilities became progressively more accessible with the introduction of timeshare systems and minicomputers. However, computers remained complex and expensive until the mid 1970s when the invention of the microprocessor made possible the development of the microcomputer.

The early days of the microcomputer provided new scope for entrepreneurs. The first ideas came largely from amateurs keen to gain access to digital computing. Entrepreneurial ventures experimented to create a variety of kits for these hobbyists, connecting up electronic components to the increasingly powerful chips available from young semiconductor companies. The first commercial microcomputer, the Altair, launched by a small business, MITS, in Albuquerque in 1975 had as its core the Intel 8080 8-bit microprocessor that had been released in 1973. Hobbyist user groups developed applications for the Altair, including games, music, databases and personal accounting software. Small third party suppliers sprung up to provide software and add-ons. A compiler for the BASIC computer language was developed by Bill Gates and Paul Allen, before they founded Microsoft.

Demand for the Altair outstripped supply and imitations were soon available based on the S-100 bus architecture and CP/M operating system of the Altair. New ventures, drawn to the emerging market, were experimenting with alternative designs and by 1977 a number of improved products were available, including the Commodore PET, Tandy TRS-80 and Apple II, each using their own operating systems and a variety of chips (e.g., Motorola 6800; Zilog Z80). The Apple II, conceived as a consumer product by Steve Jobs, attracted customers outside the hobbyist market and sales increased dramatically, from $750k in 1977, to $983m in 1983. At this early stage, the design of microcomputers included most of the now familiar elements: a microprocessor unit, a keyboard, a storage device and a monitor.

Users fuelled demand for microcomputers. In established computer companies, engineers were ordering their own microcomputers to bypass interaction with the mainframe computers which then dominated the business market. As the 'toy computers' improved their performance, they made inroads into the markets of both mainframe computers and the more recent generation of minicomputers, produced by companies like Digital Equipment Corporation and Wang[1]. Minicomputers were more accessible than mainframes but still costly and beyond the reach of ordinary consumers and small businesses.

The emerging market also attracted the attention of what was then the world's largest computer manufacturer, IBM. The first IBM Personal Computer (PC), launched in 1981, was innovative in introducing the first 16-bit microprocessor used in a microcomputer, the Intel 8088 chip. IBM's entry into the microcomputer market attracted buyers in the business sector, faithful to its brand. IBM had calculated that an open standard would encourage the production of software and IBM-compatible complementary products, enhancing their product. Users and suppliers were ready for a new standard that would allow for variation around a common format. The advent of the IBM PC contributed to the early industry shakeout and closure of many firms with 8-bit products.

IBM had outsourced the PC's operating system from Microsoft without requiring an exclusive license beyond the first 12 months. This encouraged other producers to provide peripheral complementary products that enhanced the value of the PC. However, IBM had very limited intellectual property governing the architecture of its microcomputer and this enabled competitors to produce rival products, ('PC clones'), which also worked with Microsoft's operating system using components readily avail-

---

1. Minicomputers were more accessible than mainframes but still costly and beyond the reach of ordinary consumers and small businesses.

able on the market. The most successful of the companies to introduce IBM PC compatible models, Compaq (founded in 1982), was soon in direct competition with IBM and achieved an annual revenue in excess of $1 billion in 1987.

In 1987, in an attempt to regain control of the standard, IBM introduced its PS/2 Personal System product range with proprietary logic chips and interface standards. IBM ceased production of its previous PC models, but this only encouraged sales of PC clone makers; Compaq's profits tripled in three months in 1987 as exits from the industry rose. But ultimately, it was not the hardware producers that came to dominate the PC industry. While it is the combination of hardware and software that enables the operation of the PC, the user experience is more closely associated with the software. By 1987, Microsoft's operating system, MS-DOS, was in use in hundreds of cloned products and had become the basis for a multitude of other software applications. While there were other companies that had developed operating systems that were technically superior to MS-DOS, they found to their cost that the market was locked into MS-DOS, and these variants were soon eliminated from the market.

Microsoft and Intel emerged from the industry shakedown as the dominant players in the industry. Microsoft had begun as a small supplier, but gained crucial leverage through its partnership with IBM when it was able to license the DOS operating system to PC clone producers. Microsoft's revenues grew with the rapid expansion of the PC market, taking off around 1987, the year when the dominance of the PC was established.

Reflections on industrial development in microcomputers

The emergence of a dominant design in the form of the IBM PC, along with the subsequent shakeout appears to be a classic example of a product class industry and its development is explained in these terms by Utterback (1994). Yet what we now describe as the PC industry has provided a market for a wide range of components and complementary products. The PC provides us with an example of diversity reduction through convergence around a basic product design, before giving rise to extensive variety generation around the new standard.

The industry shakeout that occurred was driven by the need for interoperability and a single standard. Apple did not make its proprietary technology available as an open standard[2]. This prevented the pioneer from gaining licensing revenues and creating an alliance of companies using its operating system. In the UK, another pioneer,

---

[2]. Apple's new CEO John Sculley had previously guarded the Pepsi recipe, though he continued a policy that had been established by Apple's founders.

Acorn Computers, also followed a proprietary strategy, not anticipating that in an industry reliant on complementary products, customers would soon shun a minority system incompatible with the industry standard.

More firms entered the industry because they were confident of a market for products compatible with the industry-leading standard established by the IBM PC. Thus, even as diversity diminished in the architecture of the PC, new generations of improvement were made in its software applications, components and peripherals. Apple co- founder Steve Wozniak explained the need for markets to set standards in terms reminiscent of complexity ideas: "You've got to let end users develop their own standards ... when a new market evolves like PCs did, there's a period of time when you've got to let the world go in random directions and eventually it will subside because it wants standardization" (Langlois, 1992: 45).

As we have already noted, user interaction played a strong role in the emergence of the industry. Hobbyists shaped the functionality of the microcomputer during its formative stages until its commercial potential was recognized, first by hobbyists themselves, such as Steve Jobs and Steve Wozniak at Apple, and then by the industry incumbent IBM (Langlois, 1992).

Descriptions of the rate of change have been enshrined in what has come to be known as 'Moore's Law'. This term is used to refer to exponential increases in computing performance and is founded on the pace of advance in semiconductors[3]. The factors supporting this empirical generalization are complex. Market forces were operative, but US government funding of information technologies and South East Asian government support for their emerging semiconductor industries were enabling factors making possible the massive improvement in the performance and yield of computer chips and increasing miniaturization.

---

3. In 1965, Gordon Moore, then a Director of Fairchild Semiconductor, observed that the number of components on a cost-effective silicon wafer had doubled each year since 1959. He predicted that this rate of increase would continue for a further 10 years. In 1975, Moore extended his prediction, but this time referred to the maximum complexity over 2 year periods. The period of the effect is frequently quoted as 18 months (e.g., EC, 2000; Cringely, 1996).

Industrial development around the Network Computer

The concept of a Network Computer (NC) was proposed by Larry Ellison, the co-founder and Chairman of Oracle in 1995. His proposal was for a simpler device that would deliver computing applications to the home over a network rather than hold all data locally, and which would cost less than $500 (Southwick, 2003)[4]. At the time, the Internet was beginning to be extended from the scientific and security communities to commercial applications and was opening up new possibilities for complementary technologies. In these terms, Ellison's suggestion that the future lay in low-cost digital devices connected to a centralized network that would provide access to the Internet and application software did not appear unreasonable (Ford & Garnsey, 2007).

Ellison commissioned the Cambridge-based Acorn Computers to develop a reference profile, based on a set-top box design that they had developed. The first NC device was unveiled at the Oracle Open World Conference in San Francisco in 1996. On its launch, Oracle announced that fellow industry heavyweights Apple, IBM, Netscape and Sun would be joining it in an alliance to develop NC devices. These devices would all be based on the architectural specification that Acorn had devised for Oracle, an open standard called the "NC Reference Profile" (NCRP). Ellison also announced the launch of a new Oracle subsidiary, Network Computer Inc (NCI), which would be responsible for handling the licensing of the NC reference profile.

In the terms of the contract between Acorn and Oracle, Acorn could not seek revenues from licensing the NC reference profile as this was NCI's role. However, Acorn could manufacture its own NC devices, so along with designing the specification, the Acorn engineers worked towards the development of their own NC, the NetStation, for release in August 1996. Although the Acorn NetStation was released on schedule, with it came the first signs that all was not going to plan. Oracle was supposed to be leading on service delivery but its servers were frequently offline for maintenance. For a computing system that was dependent on always being connected to a central server, this was unacceptable.

There were also further setbacks, with the five NC alliance members conflicting over the interpretations of the NC Reference Profile. Price wars in the memory and hard-disk drive industries, along with the rise of Dell's direct marketing strategy, were pushing PC prices below the $1000 mark, in the process making the NC's price tag of $500 appear increasingly less attractive. Further competition came in October 1996

---

4. A typical PC cost in the region of $1500 dollars in 1995.

when Microsoft and Intel announced that they were to lead a rival consortium that would develop an alternative network computing system, the NetPC. Although it was a year before this group released any hardware, this announcement introduced further uncertainty into the market, furthering benefiting the incumbent PC technology.

Meanwhile, Hermann Hauser had set up two NC ventures, both of which failed to make any headway in the market. NetChannel had been set up to provide a service to those with NC devices but were failing to attract subscribers. By early 1998 it had only attracted 10,000 subscribers in the US whereas its main competitor, WebTV (which was providing Internet access and services through a proprietary set-top box technology), had a subscription base of 300,000. Sell-offs followed in the middle of 1998, with AOL purchasing the US operation and NTL picking up the European subsidiary.

While the launch of the Network Computer had come about amid great fanfare and press attention, it failed to stimulate partners into becoming co-developers. NCI managed to sign up several East Asian manufacturers—Akai, Funai, IDEA and Proton—as licensees but these were not the big name consumer electronics manufacturers necessary to spearhead the rollout of NC devices into homes and within four years the rival to the PC disappeared from view.

Reflections on industrial development around the Network Computer

Instances of failure, here of an industry to take off, provide useful opportunities for learning. On simple inspection it may appear that the Network Computer failed to challenge the PC effectively because its technology was inferior. Yet most disruptive technologies are inferior at the time they emerge (Christensen, 1997) and to dismiss the failure of the industry to develop in such terms is to overlook the interplay of causes that contributed to the NC's weakness.

As a new paradigm in computing, Ellison's Network Computer concept represented a form of technological mutation, being conceived as a simpler distributed device than the PC and derived from the centralized mainframe terminal. The NC reference profile represented further evolution of the technology as its blueprint had been determined through Acorn's adaptation of its digital set-top box. However, the NC paradigm had not been developed in an environment in which user involvement shaped the characteristics of the final device. The 'top-down' approach to industrial development stands in stark contrast to the 'bottom-up' emergence of the microcomputer industry that had occurred through the enthusiastic response of hobbyists. The PC had grown from a resource-rich niche that had expanded over time. Touted from the start as a PC-beater, the Network Computer had no such niche from which to grow and it lacked the user involvement that has been identified in the history of successful technologies (Bijker *et al.*, 1987; Bijker, 1995)

As we saw in the example of the IBM PC, open standards can enable the replication and mutation of core technologies, along with enabling the creation of complementary technologies when the standard promotes interoperability. For an interactive innovation there is a requirement for a certain level of standardization so that other actors can participate successfully. This explains part of the success of the PC: its modularity facilitated competition across a disaggregated value chain, promoting innovation in performance improvements and cost reduction (Curry & Kenney, 1999). In theory, as an open standard, the NC reference profile would provide such an architectural blueprint. However, only a few East Asian manufacturers licensed the technology as established players in the electronics sector were part of the PC ecosystem.

Disruptive technologies, and new technologies in general, are often released with significant improvements still to be made in their design for successful commercialization. Such was the case with the Network Computer where the initial performance was inferior to the PC, with software limited to just a few applications that possessed lower functionality and aesthetic quality than comparable PC software. The lack of network infrastructure was ultimately, the most serious impediment to the uptake of the NC (Ford & Garnsey, 2007).

Discussion: The organization of industrial activity

Complex systems and coevolution

We turn now from the multi-faceted processes of change revealed in the evidence above to ways of making sense of complex dynamic processes of this kind. In the development of industrial ecosystems, certain evolutionary processes can be detected beyond the diversity of detail. These are processes of variety generation, selection and retention operating to favour certain elements in the system—here products, technologies and firms—while eliminating others. At the composite level, systemic evolution takes place as resources are attracted into a new industrial ecosystem, where variety is engendered and exchange relations become more complex, before entrainment sets in, together with other forms of system integration through collaboration and regulation.

While traditional models of structures focused on change and evolution through 'top-down' centralized control, a complexity approach emphasizes the ability of uncoordinated, 'bottom-up' dynamics to generate coherent structure. The notion of 'structure' is used here to describe the internal mechanism developed by the system to receive, encode, transform and store information, and to react to this information through some form of output. Complexity informs us that internal structure can evolve without the intervention of an external designer or the presence of some centralized form of internal control. If the capacities of the system can satisfy a number of constraints, it can develop a distributed form of internal structure through a process of self-organization. The structure that becomes apparent through this process is neither a passive reflection of the external environment, nor a deterministic result of active, pre-programmed internal factors. So while a firm differs from an industry in being a purposive system, and an industry self-organizes around market exchange, there is some common logic that underpins their evolution. Complexity analysis applied to human activity recognizes that the way a system is perceived and represented shapes motivation and action. In the case of an emerging industry, a sense of identity among participants arises as the system self-organizes but self-organization may in turn promote deliberate forms of organization at the industry level, including industry bodies, trade and employers' institutions.

An emphasis on coevolution points to the connections between related evolving phenomena. It questions the idea that the environment is external, that a system of human activity is determined by its environment and the changes imposed thereby. As Capra (1996) comments "... evolution cannot be limited to the adaptation of organisms to their environment, because the environment itself is shaped by a network of living systems capable of adaptation and creativity ... through a subtle interplay of competition and cooperation, creation and mutual adaptation" (p. 222). Accordingly, we move towards a view of the firm operating within a complex adaptive system, with the appropriate level of analysis being the network of interactions between the firm and its partners, suppliers and customers. Whether an industry is viewed as made up of competing firms producing similar products (product class definition) or as a web of interacting firms providing complementary goods and services to meet common consumer wants (business ecosystem definition), the industry can be seen to operate as a complex adaptive system. It is up to the observer to draw boundaries around the system for purposes of analysis.

Coevolution takes place through signals and responses between interacting elements in the system of exchange. Whether price signals or other factors dominate in stimulating activity depends on a variety of conditions. Where innovation continues to renew the industry, agents are engaged in a continual transformation of resources into outputs that meet continual readjustments in demand. This kind of dissipative structure, continually in flux or 'disorderly order' (e.g., Hayles, 1990, Stacey, 2003), requires high levels of energy to maintain self-organization. As learning increases throughout the production chain, there are further rises in productivity and reductions in costs. Rates of return improve in the new industry while competition is still limited and the rents of innovation can be retained.

Coevolution can involve both interdependence—cooperation—and competition, both of which can lead to positive feedback effects which improve the fitness of individual participants (Kaufmann, 1995). In the newly forming industrial system, cooperation between firms can be adaptive if, for example, it reduces uncertainty with respect to standards or technology. Similarly, competition between firms may lead them to adopt designs or features which have been shown to be well-received by consumers. In either case, there is a reduction in design variety, leading the system towards a dominant design.

Kaufmann also points, however, to two outcomes of coevolution that he considers to be undesirable. The first, "chaos" or the "Red Queen effect", results when coevolving agents engage in an "arms race" as they each attempt to outdo the adaptation of the other. In the second case, "stability", agents settle into a state of "stable equilibrium" where they do best by not changing their strategy so long as others maintain theirs. Kaufmann considers the first case to be so chaotic and changeable that it is not possible to achieve and maintain high levels of fitness, while in the second, rigidity leads to resistance to change, i.e., learning cannot be achieved where change is too rapid for participants to see cause and effect at work. However systems in a third state, the "edge of chaos", enjoy a combination of order and flexibility which make it possible to achieve higher levels of fitness for given environments (Kauffman, 1995).

The emergence of a dominant design, and the transition to what Abernathy and Utterback refer to as the "specific" stage, could correspond to a state of "stable equilibrium", the outcome of which is the decline of the system, unless it is renewed. The reduction in variety depicted in the early stages of life cycle model can be adaptive if it moves the system away from chaos, but could lead to a state of excessive stability later in the system's evolution, in other words, there is a danger of "lock-in".

The industrial systems considered here do not conform to simple life cycle models, but have seen the effect of reducing variation over time, followed by renewed product innovation or disruption. The self-organizing processes result in an extended system of exchange that goes beyond a single product-based industry and which makes up a cluster of connected activity.

We need not confine analysis to product class or business ecosystem approaches—there are a variety of types of clustered activity, integrated by economic exchange, which operate as complex adaptive systems. These clusters of industrial activities have various dimensions, depending on the aspects of exchange, so that we can identify a number of different types of clusters. These include:

1. Market-based clusters. Firms producing similar products are serving common markets, meeting needs of users with related requirements. These may be local, regional, national or global.
2. Production chain clusters. The firms in the production or value-adding chain that contribute to a final product are interrelated. These too may have extensive spatial spread.
3. Local clusters. Firms in proximity to each other which draw on a common local resource pool are interrelated, e.g., through their uses of labour and funding they make up local clusters, as in industrial districts. They may have local markets for their output, but may well serve a wider customer base.
4. Clusters of complementary activity. These link related production chains when products are complementary. A few products are self-contained or autonomous and have an inherent use value, such as knives or ladders. Many more products require complementary products, processes and infrastructure to function, as do electric lamps, telephones or aircraft. To attract custom, they must relate to customers' previous product purchases and connect with complementary products in current use. Complementarities increase over time, and are a key feature of networked activity in communication and IT industries.

As we have seen, the evolution of clusters of industrial activity is less regular than that of product class industries that evolve from invention through to commodity. These clusters resemble the business ecosystems described by Moore (1996), which, emphasizing coevolution, draw on a complex-systems narrative. Any given firm will be part of more than one of these clusters, and will be affected by (and will affect) the evolution of each of them (evident in all of the cases considered here). Hence, even the product class industries described by Utterback (1994) will overlap with other systems or clusters, each of which both influences and is influenced by changes in the others[5].

---

5. Utterback acknowledges that organizations may operate in several "industries" as a result of his product-based definition, but the suggestion here is that single product organizations too will operate

While industries show recurrent evolutionary features, they are never precisely repeated. During industrial evolution, once certain key products and processes are standardized, complementarity draws related activity into synchrony. Elements of the system become increasingly connected and specialization increases interdependencies. Complex supply networks form, with an increasingly specialized division of labour, creating new opportunities for system shapers and integrators. In advanced economies, system integrators are often the winners among the participating firms in an industrial cluster or ecosystem—this is clear from the role of Microsoft and Intel in the evidence discussed above. Leading supermarkets and leading aircraft companies provide other examples of firms that are able to capture a disproportionate share of the value generated in the wider business ecosystem.

Selection processes extend beyond rewarding fitness to survive and diffuse in a given environment. Fitness can be understood not only as capacity to withstand competition but also in terms of a firm's capacity to fit into complementary activity. Some firms produce unique products and have no direct competitors. They must however be complementary to other activities in the exchange processes in which they participate. Along with competitive selection, where resources are scarce and exchange is a zero sum game, there are processes of cooperative exchange and symbiosis that favour certain elements of the system, promoting their survival and propagation. Together these processes shape the experience of individual units, that in turn affect the system's evolution. As linkages become more elaborate, closer synchronization of activity leads to entrainment, or common periodicity, including industry-wide production and business cycles. For example the marked business cycles in the semiconductor industry result from amplified demand and supply effects working through the production chain.

Self-organization through interaction

In economic life, transferable units of practical knowledge can be shared through partnerships, imitated and renewed, providing new firms with access to a problem-solving repertoire of technologies and administrative procedures developed elsewhere. These can be viewed as providing the information endowment in an industry accessed by collective learning and used to create further knowledge and to form new firms in a similar mould. This process of knowledge accumulation can be identified within and between the industries considered here.

As process and product innovation involves solving problems, self-organization is about a continual search for knowledge and procedures that will contribute to prob-

---

in several different systems

lem solving. Firms and clusters that are accomplished at learning can rapidly identify necessary problem solving knowledge and procedures wherever they may be located. This search capability has been characterized as "learning about learning" or "knowledge of knowledge" (Faulkner & Senker, 1995). Learning, therefore, is in part aimed at identifying new core capabilities and complementary assets, and targeting new network members that possess these capabilities and assets. The early history of the PC industry is a particularly good example of this type of behavior, with early developments relying on the collaboration of enthusiasts, and the IBM PC itself drawing on capabilities from outside IBM itself (e.g., Microsoft and Intel), facilitated by its modular design. Difficulties that prevented the development of a similar collaborative network of complementary activities in the 'premature' Network Computer case were responsible for its demise.

For any firm or cluster, the learning that has taken place in the past is a good indicator of where learning is likely to take place in the future; its history comprises a path dependency that both restricts and amplifies the learning possibilities and the potential for accessing new complementary assets (Rycroft & Kash, 2002). Therefore, most tend to learn 'locally', by engaging in search and discovery activities close to their previous learning 'neighborhood'. While localized learning can build upon itself and become a major source of positive feedback, it can also lead to localized optimization. Non-local alternatives, which may be optimal at a global level, are often ignored, fail to be recognized or dismissed as too difficult to attain (Kauffman, 1993). A cluster's path dependencies and selection environment thus both restrict and amplify the self-organization process (De Vany, 1996). In semiconductors, Intel's change of emphasis from memory to microprocessors represents a case of self-organization within the firm, and the ability of its senior managers to recognize the benefits of pursuing a broader strategy (Grove, 1991).

In many industries, local networks are the key to shared learning. A form of direct inheritance occurs through a firm's employees founding spin-outs, whereby knowledge is directly transmitted from one firm to a successor. The process through which existing firms give rise to new ones via offshoots or spin-outs is a major source of the multiplication of new firms in an emerging industry and local economy.

Access to industry-based knowledge occurs through membership of the industrial community, often with shared training and joint activity between members of an industry (Van de Ven & Garud, 1993). People employed in an industry often share a common base of experience, knowledge and training. They may share a common commitment to a new technological approach or unite in defence of a threatened technology. Industry-wide institutions and common training are among the factors that create a sense of identity and a boundary. As the industry establishes its activities and its members develop a sense of identity, interactions within the industry become denser than with the rest of the economy. This network can extend to customers, as was the case in instrumentation, where not only "wants", but technical solutions were generated by users.

Adaptation and attraction

As pointed out earlier, the dominant design that emerges in an industry need not be the best possible solution, from a technical perspective (e.g., VHS Video Tape Recorders). Even if the design is appropriate at a given point in time, positive feedback effects may lead to its being retained to a point at which it is not (e.g., QWERTY keyboard). The dominant design might be considered to be a local peak in the fitness landscape, which is sub-optimal, but very difficult to escape (Lissack, 1997). Since the fitness landscape is continually deforming (Kaufman & Macready, 1995)—changed by the activities of those operating within the system, and in interacting systems—optimal fitness requires a change in strategy, which may be difficult to implement in the context of a densely connected network, and relationships of coevolution. The case of instrumentation firms in the UK appears to illustrate the difficulty or responding to changes in the landscape, with the failure to respond to the transition to digital solutions and the use of microcomputers. Changes in the landscape were also evident in the case of Network Computers, where the reducing cost of PCs, and direct action by Microsoft and other incumbents reduced the attractiveness of their "low-cost" offering.

Bak and Boettcher (1997), drawing on Schumpeter's (1934) notion of "Creative Destruction", and Eldredge and Gould's (1972) concept of "Punctuated Equilibrium", argue, that periods of stasis are important since they allow the system to settle and develop, while disruptive events permit change in the system. But only where industry participants respond appropriately to the disruption do they avoid "lock-in" or the increasing hold of inertia. The dangers of lock-in were evident in the UK instrumentation sector. In contrast, the experience of the PC industry, following the launch of the IBM PC, demonstrates the benefits of a period of predictability, emerging, in this case, as a result of the establishment of an industry standard. Thus it is not stability as such that ossifies an industry, indeed some industries have remained stable over long periods, but rather inability to adapt to changing conditions.

Reflections on theory and evidence

In addressing how industries emerge and develop over time, we have explored some limitations of the dominant approach to understanding industrial change, the product life cycle. Our four examples transcend this type of simple linear model, highlighting how industrial systems are complex and unpredictable, differentiate through coevolution, and are integrated through processs of self-organization and entrainment.

Our case studies of industry evolution highlight the coevolutionary nature of industrial change. In the instrumentation industry we saw how a highly diversified range of products was created, frequently being generated through user development, and finding multiple markets and applications. Advances in instrumentation gave rise to semiconductors, which were a form of meta-instrument. The emergence of semiconductors allowed new applications to be developed that renewed a number of existing industries and led to the emergence of new activities. The PC was one of the new industries to which semiconductors, in the guise of the microprocessor, gave rise. Commenting on the nature of this type of process, Metcalfe (2000) notes that "...knowledge accumulation is an unfolding process in which the realization of possibilities makes possible the specification of new possibilities. Like any autocatalytic process the output of knowledge becomes the input into new knowledge and one idea results in another" (p. 5).

On first inspection, the PC industry appears to be a classic example of a product life-cycle showing the familiar trajectory of coalescence around a dominant design followed by commoditization. However, the pioneers who launched the dominant design did not capture a commensurate share of the value they generated. Instead, the system integrators engaged in chip design and software emerged as winners, favoured by the open architecture of the PC. Later, the Network Computer emerged as a competitive threat to the PC. This early challenge was unsuccessful because the nascent industry failed to develop the complementarities that were required, in part because of path dependent resource investments that firms had already made as part of the open PC system. The story increases in complexity following the convergence between mobile phones, PCs, personal digital assistance and tablet computers; these developments require telling elsewhere.

In 1943 Thomas Watson, then Chairman of IBM, said that he thought that there was a market for only five computers. He made this statement before the emergence of semiconductors, when computers were powered by vacuum tubes and were big enough to fill a room. Watson had little idea of the dramatic effects that the miniaturization of semiconductor devices would have on the computing industry; his view failed to take into account discontinuous innovations, the future pace of technological advance and changes in user requirements. The continuing miniaturization of semiconductor devices made it possible for new computing devices such as the Network Computer to challenge the supremacy of the PC. However, across the examples we have seen that users frequently play a part in this 'unfolding', stimulating innovation through early work to meet their own needs (von Hippel, 1988). In the case of the Network Computer, the failure to involve users at an early stage prevented the type of early experimentation that had occurred in the hobbyist days of the microcomputer. The unpredictability of industrial change stems in large part from the positive feedbacks that occur that can amplify and accelerate learning and knowledge accumulation.

A complex systems perspective can explain the processes depicted in the life cycle models, but can also illuminate development of industries which do not follow that predicted path, showing how and why industrial evolution is subject to common processes with diverse outcomes.

Acknowledgements

This chapter builds on seminars and working papers by Elizabeth Garnsey on industry evolution, prepared for the ESRC Nexsus project on Complex Dynamic Systems and the Social Sciences 2000-2004. Professor Peter Allen was inspiring as leader of this inter-university project. The authors of the Cambridge team would also like to thank Stan Metcalfe, James McGlade and Mark Strathern and other members of Nexsus for stimulating and enjoyable interaction on the Nexsus project.

References

Abernathy, W. and Utterback, J. (1978). "Patterns of innovation in industry," *Technology Review*, ISSN 1099-274X, 80(7): 40-47

Agar A. (1996). "The story of European commercial electron microscopes," *Advances in Imaging and Electron Physics*, ISSN 2041-172, 96: 415-584.

Bak, P. and Boettcher, S. (1997). "Self-organized criticality and punctuated equilibria," *Physica D: Nonlinear Phenomena*, ISSN 0167-2789, 107(2-4): 143-150.

Bijker, W.E. (1995). *Of Bicycles, Bakelite and Bulbs: Towards a Theory of Sociotechnical Change*, ISBN 0262023768.

Bijker, W.E., Hughes, T.P. and Pinch, T.J. (1987). *The Social Construction of Technological*

Systems: New Directions in the Sociology and History of Technology, ISBN 0262022621.

Burgelman, R.A. (1994). "Fading Memories: A Process Theory of Strategic Business Exit in Dynamic Environments," *Administrative Science Quarterly*, ISSN 0001-8392, 39(1): 24-56.

Capra, F. (1996). *The Web of Life*, ISBN 038547675.

Christensen, C.M. (1997). *The Innovator's Dilemma: When New Technologies Cause Great Firms to Fail*, ISBN 142219602X (2013).

Clegg, S.R. Rhodes, C. Kornberger, M. (2007). "Desperately seeking legitimacy: Organizational identity and emerging industries," *Organization Studies*, ISSN 0170-8406, 28(4): 495-513.

Cringely, R. (1996). Accidental Empires, ISBN 0140258264.

Curry, J. and Kenney, M. (1999). "Beating the clock: Corporate responses to rapid change in the PC industry," *California Management Review*, ISSN 0008-1256, 42(1): 8-36.

De Vany, A. (1996). "Information, chance, and evolution: Alchian and the economics of self-organization," *Economic Inquiry*, ISSN 0095-2583, (34): 427-443.

Dean, J. (1951). *Managerial Economics*, Prentice Hall, Inc, Englewood Cliffs N.J.

Eldredge, N. and Gould, S.J. (1972). "Punctuated equilibria: An alternative to phyletic gradualism," in: T.J.M. Schopf (ed.), *Models in Paleobiology*, Freeman, Cooper and Company, San Francisco, pp. 82-115.

Faulkner, W. and Senker, J. (1995). *Knowledge Frontiers: Public Sector Research and Industrial Innovation in Biotechnology, Engineering Ceramics, and Parallel Computing*, ISBN 0198288336.

Freeman, C. (1995). "The national system of innovation: Historical perspective," *Cambridge Journal of Economics*, ISSN 0309-166X, 19(1): 5-24.

Garnsey, E. and Minshall, T., "High tech enterprise in evolving industries", Centre for Technology Management, University of Cambridge, Working Paper no CTM 2000/02

Gawer, A. and Cusumano, M. (2002). *Platform Leadership: How Intel, Microsoft, and Cisco drive Industry Innovation*, ISBN 1578515149.

Grant, R. (2002). *Contemporary Strategy Analysis: Concepts, Techniques, Applications*, ISBN 0631231366.

Grove, A.S. (1997). *Only the Paranoid Survive: How to Exploit the Crisis Points That Challenge Every Company and Career*, ISBN 0002558106.

Hayles, N.K. (1990). *Chaos Bound: Orderly Disorder in Contemporary Literature and Science*, ISBN 080149701.

Kauffman, S.A. (1993). *The Origins of Order: Self-organization and Selection in Evolution*, ISBN 0195079515.

Kauffman, S.A. (1995). *At Home in the Universe: The Search for Laws of Complexity*, ISBN 0140174141.

Kauffman, S.A. and Macready, W.G. (1995). "Technological evolution and adaptive organizations," *Complexity*, ISSN 1099-0526, 1: 26-43.

Klepper, S. and Graddy, E. (1990). "The evolution of new industries and the determinants of market structure," *RAND Journal of Economics*, ISSN 1756-2171, 21(1): 27-44.

Langlois, R.N. (1992). "External economies and economic progress: The case of the microcomputer industry," *Business History Review*, ISSN 0007-6805, 66(1): 1-50.

Langrish, J. (1972). *Wealth from Knowledge: Studies of Innovation in Industry*, ISBN 0333120078.

Lissack, M. (1997). "Mind your metaphors: lessons from complexity studies," *Long Range Planning*, ISSN 0024-6301, 30(2): 294-298.

Low, M.B. and Abrahamson, E. (1997). "Movements, bandwagons and clones: Industry evolution and the entrepreneurial process," *Journal of Business Venturing*, ISSN 0883-9026, 12: 435-457.

Metcalfe, J.S. (2000). "Restless capitalism, experimental economies," in W. During, R. Oakey and M. Kipling (eds.), *New Technology-Based Firms at the Turn of the Century*, ISBN 0080437915, chapter 2.

Moore, J.F. (1993). "Predators and prey: A new ecology of competition," *Harvard Business Review*, ISSN 0017-8012, (May-June): 75-86.

Moore, J.F. (1996). *The Death of Competition: Leadership and Strategy in the Age of Business Ecosystems*, ISBN 0887308090.

Porter, M.E. (1980). *Competitive Strategy: Techniques for Analyzing Industries and Competitors*, ISBN 0029253608.

Ritchie, D. (1986). *The Computer Pioneers: The Making of the Modern Computer*, ISBN 067152397X.

Rosenberg, N. (1994). *Exploring the Black Box: Technology, Economics and History*, ISBN 0521452708.

Rothwell, R. and Zegveld, W. (1982). *Innovation and the Small and Medium Sized Firm*, ISBN 0898380995.

Rycroft, R.W. and Kash, D.E. (2004). "Self-organizing innovation networks: implications for globalization," *Technovation*, ISSN 0166-4972, 24: 187-197.

Schumpeter, J. A. (1934). *The Theory of Economic Development*, Cambridge, MA: Harvard University Press.

Shimshoni, D. (1970). "The mobile scientist in the American instrument industry," *Minerva*, ISSN 0026-4695, 8(1-4): 59-89.

Southwick, K. (2003). *Everyone Else Must Fail: The Unvarnished Truth About Oracle and Larry Ellison*, ISBN 0712621482.

Stacey, R.D. (2003). *Strategic Management and Organizational Dynamics: The Challenge of Complexity*, ISBN 0273725599.

Tikkanen, I. (2008). "Innovations, exports, and Finnish electrical industry life cycle 1960-2005," *Journal of Euromarketing*, ISSN 1936-6426, 17(3/4): 183-197.

Utterback, J.M. (1994). *Mastering the Dynamics of Innovation*, ISBN 0875847404.

Utterback, J.M. and Suárez, F.F. (1993). "Innovation, competition and industry structure," *Research Policy*, ISSN 0048-7333, 22(1): 1-21.

Van de Ven, A.H. and Garud, R. (1993). "Innovation and Industry Development: The case of Cochlear Implants," *Research on Technology Innovation, Management and Policy*, ISSN 0737-107, 5: 1-46.

von Hippel, E. (1988). *The Sources of Innovation*, ISBN 0195040856.

Wolfe, T. (1983). "The tinkerings of Robert Noyce: How the sun rose on the Silicon Valley," *Esquire Magazine*, ISSN 0014-0791, (December): 346-374.

Woodward, J. (1970). *Industrial Organization: Behavior and Control*, ISBN 019859805X.

Elizabeth Garnsey is Emeritus Reader in Innovation Studies, University of Cambridge Engineering Department. She has experience on the university-industry interface, on support for new business in incubation centres and science parks and on acquisition by and joint ventures with larger corporations. She has been monitoring the progress of high tech firms in Cambridge since 1988 and was a founder and academic organiser of the first Cambridge Enterprise Conferences. She was seconded to St John's Innovation Centre as an advisor and researcher 2000-2002. She has been advisor to the Bank of England, the Treasury and the Confederation of British Industry on high tech enterprise and an Expert Witness to Parliamentary committees. She has worked on the Shell Springboard Programme for young environmental companies. Her research on Daylight Saving has been cited in Parliament. She is a member of the Board of Trustees of Camfed. In 2011 she was a Visiting Professor at the National University of Singapore (NUS) and is engaged in on-going work with NUS colleagues.

Simon Ford joined the Centre for Technology Management as a Research Associate in January 2007. Prior to this he completed his PhD at the centre on the subject of technological obsolescence in the PC industry.

Paul Heffernan is an independent consultant with Bobnick. Previously he was Director, Centre for Smart Infrastructure and Construction and Lecturer in Human Resource Management, Institute for Manufacturing, Department of Engineering, University of Cambridge. His interest in human factors developed from nearly 20 years industrial experience in aerospace, railways and construction. Paul trained at British Aerospace, including an apprenticeship as an aircraft fitter, before working in production planning, management and projects. He later broadened his horizons with a move to railways and construction, joining the Jubilee Line Extension project. He continues to consult on project management in the construction industry.

Chapter I

The evolution of complex exchange dynamics in a pre-market economy: A case study from north-east Iberia

James McGlade & Mark Strathern

This paper focuses on a 'prestige goods' exchange system in a pre-market economy and its transformation after the founding of the Greek colony at Empúries in north-east Catalonia, and the region's exposure to wider Mediterranean markets. We shall argue that the dynamics of such an exchange network and the inherent nonlinearity of its interactions, is best viewed within the compass of a complex evolutionary system. A model is proposed to study the nonlinearities resident in such a complex exchange system and in particular to address questions such as: 1) The capacity of highly connected systems to generate emergent properties that cannot be deduced from an analysis of the system's constituent parts; 2) The system's long-run behavior and its relation to stability/resilience criteria, and 3) The degree of predictability resident in complex exchange networks.

Introduction

Over the last few decades, the computer simulation of coupled social, economic and environmental processes has made substantive contributions to our understanding of evolutionary dynamics of complex systems (Epstein & Axtell, 1996; Conte et al., 1997; Gilbert & Troitzsch, 2005). A key aspect of this work is that it foregrounds the role of emergent, self-organizing processes in the creation of evolutionary structuring (e.g., Nicolis & Prigogine 1977; Allen & Sanglier, 1981; Kauffman, 1993). The insights provided by theses pioneering studies into the world of complexity has lead to the construction of increasingly complex models of social systems (e.g., Helbing, 2012) and importantly, these have focused on the nature and role played by models, not so much as explanandum, but as exploratory, investigative tools. This idea is contrary to the frequent perception of models as maps of reality a common epistemological gaffe, hindering constructive debate in disciplines such as archaeology, anthropology and geography. Rather than being seen as one to one mappings of real world phenomena, model building is essentially a creative process that allows us to experiment: to track nonlinear causalities, to explore counterintuitive dynamics and potential futures—in essence, to increase the range of investigative methods that can be brought to bear on the interpretation of complex problem sets (McGlade, 2003, 2005). It is in this spirit of enquiry that we shall approach the question of societal dynamics and specifically the dynamics displayed by competitive networks of power as it is played out in a pre-market economy.

In what follows, we shall construct an evolutionary model to examine the behaviour of a proto-market, prehistoric system driven by motives of power and territorial domination. Specifically, we shall focus on the socio-spatial dynamics of a "prestige goods economy" in Iron Age Iberia, Our primary objective is to construct a dynamical model of the transaction dynamics so as to explore the sets of interactions and nonlinear feedbacks underpinning the evolution of a complex trade/exchange system. In addition, we shall also seek to shed light on questions related to the stability and resilience of such economic networks. Before proceeding, however, we shall first outline the nature of prehistoric exchange and prestige goods economies generally, followed by a brief overview of the socio-political system which flourished in the proto-historic Iberian landscapes of north-east Catalonia.

Exchange processes in pre-market economies

Exchange and commodities

The exchange of commodities, whether as gifts, for the acquisition of power and status or simply for commercial gain, has assumed a fundamental role in the evolution of human societies. This can be traced from the earliest prehistoric times; indeed, from the Palaeolithic, the accumulation and circulation of artefacts and subsistence resources over long distances, fulfilled a variety of socio-political functions, including the creation of alliances—vitally important aspects of survival in difficult environments governed by climatic uncertainty and fluctuating animal populations. Anthropologists generally make a fundamental distinction between gift and commodity exchange (Mauss, 1925; Hyde, 1983). That is, they distinguish between those social, symbolic transactions, based on the reciprocal exchange of gifts as a display of personal wealth and status, often as a prelude to the creation of alliances. In such highly ritualized gift economies, there are usually prohibitions against the conversion of gifts to exchange capital. By contrast, the exchange of goods as commodities, initiates an entirely different set of relationships that heralds a world of commerce and the acquisition of goods that is based—not on any ritualized idea of balanced reciprocity—but on a simple profit motive. In effect we see a gradual decoupling, or to use Polanyi's term, a disembedding of objects from their cultural and symbolic context. In this paper, we shall examine a system that in many ways is on the cusp of this transition—from artefacts as prestige items, to artefacts as commodities.

Prestige goods economies

In pre-market economies, social position is maintained through access to valuables, which confer prestige and are essentially the currency of power (Johnson & Earle, 2000). Commodity transactions in pre-market economies essentially can be thought of as a two tier system, comprising local domestic trade/exchange of everyday items such as agricultural produce, cloth, leather, tools and locally made pottery. The second tier involved the movement of rarer types of commodities such as metals (gold, silver, bronze), ivory, high status ceramics, jewellery, perfumes and wine. Not surprisingly, this latter category of luxurious, status enhancing goods was highly coveted, forming one of the key sources of competition and even warfare in prehistoric societies. For millennia, the acquisition of rare, exotic artefacts formed the basis by which ambitious members of society sought to gain political power and status. Prestige items thus became a signature of social dominance by aristocratic elites. In this sense, the emergence of prestige goods economies is synonymous with the rise of inequality and from an anthropological perspective, is the hallmark of ranked and stratified societies. Crucially, the control of prestige goods networks is the basis of power and authority, particularly in the types of prehistoric society we are dealing with here. Indeed the power of elites is maintained through increasing levels of consumption that is based on competitive emulation (Friedman & Rowlands, 1977; Godelier, 1999).

From an anthropological perspective, there has been much speculation on just how prestige and its acquisition work in ranked societies. For example, both Helms (1993, 1994) and Johnson and Earle (2000) point out that the possession of luxury commodities plays a central role in the creation of status and the ability to acquire political office; thus, valuables are marks of social status that function as a way of define an individual's political and economic position. In this way, the control and distribution of prestige items becomes a form of "political currency.

Essentially, prestige goods are an embedded socio-symbolic category that is exclusively designed to enhance the status and power of the ruling elite. Such systems work well in situations where there is a limited supply of exchange items and thus the transaction landscape can be controlled by a few dominant individuals or entrepreneurs. However, if this type of closed system is exposed to wider international markets, the influx of new goods has the effect of flooding the market and renders exclusivity of control more difficult, as new entrepreneurs emerge to challenge the authority and control of elites to orchestrate the movement of high status goods. This process amounts to a 'democratization' of the exchange system, as an increasing number of individuals compete for access. More fundamentally, it represents a move from a world of exchange as a socio-symbolic system, to a world of commerce and market economics.

In what follows, we shall argue that this transformation is precisely what occurred in north-east Iberia as a consequence of the colonial incursions of first, Phoenician, and later Greek traders who were responsible for the region's exposure to wider Mediterranean markets. In effect, we are dealing with an economy, that is on the cusp of a transition from a system of prestige goods transactions, to that of a fully fledged commercial system. Moreover, we shall argue that the dynamics of such an exchange network and the inherent nonlinearity of its interactions, is best viewed within the compass of a complex evolutionary system; thus, we are particularly interested in:

1. The capacity of highly connected systems to generate emergent properties that cannot be deduced from an analysis of the system's constituent parts;
2. The system's long-run behavior and level of stability/resilience, and;
3. The degree of predictability resident in complex exchange networks.

It is these properties and their contribution to evolutionary dynamics, that will form the conceptual basis of our model building exercise, but before embarking of our model construction, we shall outline the socio-political context of the exchange system we wish to study.

A case study from proto-historic Iberia

The Iberian Iron Age landscape

The exchange dynamics we shall model is focused on the Iron Age landscapes of northeast Catalonia, specifically the Empordà plain and its Mediterranean littoral (Figure 1). Chronologically, we are dealing with the centuries preceding the Roman invasion of the Iberian Peninsula; that is, a period beginning c. 600 BC with Greek colonization and the founding of the city of Emporion, until the arrival of the Roman legions in 218 BC at the start of the 2nd Punic War.

The indigenous groups occupying these landscapes, in common with many Iron Age societies in the Euro-Mediterranean zone, were hierarchical in structure, being ruled by aristocratic warrior elites (Sanmarti, 2004). Power, wealth and status were maintained through territorial domination (warfare) and especially through the control of agricultural resources and access to prestige goods networks. The landscape can thus be viewed as a socio-spatial network whose power nodes are represented by a series of politico-administrative centres, or oppida, each controlling a territory occupied by a group of smaller dependent settlements (Figure 1). A relatively high level of social, economic and political integration is apparent, and this was the landscape, which the first Greek colonists were to encounter (Picazo et al., 2000).

Colonial encounters: The coming of the Greeks

Around 575BC, the colony of Emporion was founded as part of the expansion of Greek commercial interests (Figure 2). From their capital, Massilia (Marseilles), merchants from Phocaea established a chain of small ports and factories around the Gulf of Lyon and the northeast coast of the Iberian Peninsula. Emporion was situated at a strategic point on the Gulf of Roses, between two rivers, the Fluvià and the Ter, which provided ideal communication with the landscapes of the interior.

The arrival of the Greeks in the Empordà in the first half of the 6th century can best be seen as the harbinger of profound changes in the economic life and social organization of the indigenous population (e.g., Martín, 1987; Picazo et al., 2000). Archaeologically, what we see is the beginning of a large-scale influx of Mediterranean products into first, elite circles, and eventually into domestic contexts. This was to eventuate a gradual transformation of indigenous economies as they became part of the wider world of Mediterranean commerce; thus the societies of the north-east became embedded for the first time in market exchange systems in contrast to transactions that previously were dominated by domestic consumption within local or regional tribal contexts (Cabrera, 1998).

One of the immediate consequences of increased access to Mediterranean networks was the rapid adoption of new systems of production that, especially following the introduction of iron agricultural tools, lead to the intensive exploitation of previously marginal environments, new methods of crop rotation and the construction of silos for grain storage. All these developments enabled surplus to be produced for commercial exchange in the new colonial markets. In fact, when the Greeks arrived in the Empordà region, the local populations already had commercial relations with the Phoenicians and, on a smaller scale, with the Etruscans. In addition, the Greek colonial

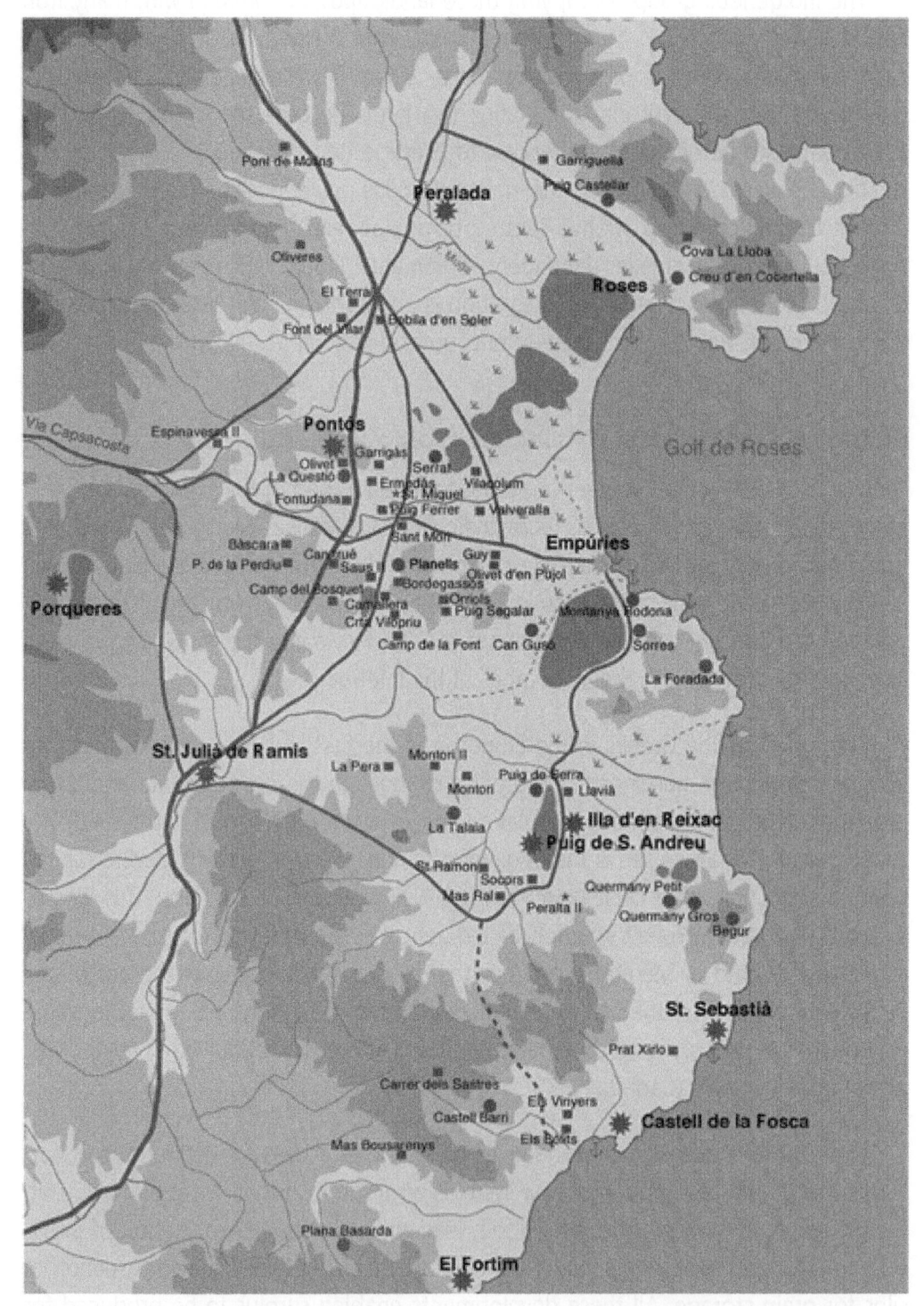

Figure 1 *Map showing the network of sites on the Empordà plain during the middle Iberian period (IV-IIIc BC). The principal territorial capitals (oppida) are marked by an asterisk.*

Figure 2 *the Greco-Roman colony of Emporion*

city at Emporion was close to the ancient basin of the river Fluviá which provided access to La Garrotxa and the Pyrenees zone, both rich in mineral resources. Thus, from c. 600 BC the Empordà plain can be considered as an expanding commercial market, but, notably, one displaying evidence of the role of indigenous agency in commercial transactions (Cabrera, 1998). Further evidence of indigenous market manipulation comes from the large quantity of Phoenician amphorae, which circulated in Catalonia between c.650 and 575 BC. It has been argued that such a marked increase in wine production was not simply the result of speculative marketing by entrepreneurs, but may have been the consequence of a real demand on behalf of the indigenous population; thus the local demand in Catalonia may, in a sense, have been driving the Phoenician economy (Cabrera, 1996, 1998; Sanmarti, 2004: 31). Further, it is interesting to observe that the archaeological record shows a decline in the popularity of table-ware, perhaps also reflecting a deliberate selection, (i.e., disinterest), on the part of the native populations (ibid.). Here we have a good example of the active role played by indigenous groups in the selection of Mediterranean trade goods and the way that the market was driven by local demand.

During the 5th-4th centuries BC, Emporion developed an important commercial network that reached the Midi de la France, Rousillon, the entire Catalan littoral, the

Levant, as well as the south-east and the Balearic islands. Mostly this is to be seen from the presence of Greek imports, especially of Attic pottery. It seems that in return, Emporion exported cereals to Athens and through this connection it was consolidated as a redistribution centre of Attic pottery for the northwest Mediterranean (Ruiz de Arbulo, 1992; Sanmartí-Grego, 1992). Thus, the stimulus for much of the expansion we see at Emporion was probably driven by the needs of a growing Athenian population and its demand for wheat.

Ultimately, what was to emerge from this interaction with the Greek world was the appearance of a more homogeneous material culture—aided particularly by the adoption of wheel-thrown pottery types to feed a growing market that also demanded high quality production (Figure 3). These new commercial opportunities were to have profound effects on social structure as access and control of prestige goods from the wider Mediterranean, increased the wealth and power of existing elites, and thus heightening existing socio-political tensions in an increasingly competitive landscape (Buxo et al., 1998; Picazo et al., 2000). However, it needs to be reiterated that the Greek presence by itself was not responsible for the substantive changes that occurred in social, political and economic structures. Rather, these dynamics were inherent in the social cultural entanglements that accompany all native/colonial encounters (Dietler, 1997a,b) and tend act to reinforce inequalities that were already embedded in the pre-existing exchange landscape (Cabrera, 1998).

Above all, the most radical change wrought by the Greek presence was in the sphere of social relations. What we see is a radical disruption in the normative world of kinship relations that acted to cement social group. These kin-based dynamics were gradually replaced by the acquisitive nature of a new class-oriented society; a society whose membership and ultimate status is no longer based on ancestral precedent, but on the capacity to create wealth (and hence status) from the opportunities presented by participation in Mediterranean commerce. Thus, of paramount importance

Figure 3 *An example of Greek prestige ceramic imports*

was the ability of individual native entrepreneurs to negotiate successful commercial transactions that stimulated more intensive levels of production. In this way, they acted literally as agents of change.

Emerging polities: The new oppida landscape

But perhaps the most remarkable thing about these developments was that they were occurring within the context of a radically changing socio-political landscape—a landscape of rapid urbanization focused on powerful political centres, or oppida. The origins of this process lie in the socio-political upheavals during the Late Bronze Age, with the appearance of characteristically proto-urban settlements layouts. This, in itself, suggests the hand of an emerging individual or group exercising a deliberate form of spatial planning or alternatively, perhaps even political control by an aristocratic elite. This urbanization shows a marked increase after the establishment of the Greeks at Emporion and suggests the emergence of an increasingly competitive landscape. For example, archaeological evidence shows a restructuring of the principal settlements with a new emphasis on defensive structures, such as the construction of elaborate fortifications at Ullastret, a site where we also have evidence of destruction through burning, perhaps signifying warfare.

A not unsurprising outcome of the radical changes that accompanied the colonial Greek presence was to accelerate and amplify existing social inequalities. As we have noted above, these had emerged during the Late Bronze Age as a consequence of the power and prestige acquired through control of previously established trading networks. Thus we see an increasing territorialization of the landscape: what previously was a landscape divided along ancestral lines and organized for the mutual benefit of the population, gives way to a new conception of land ownership and authority. From a geographical perspective, we begin to see the landscapes of the northeast as being divided with a view to controlling the primary river systems—the primary axes of trade—and it is in this context that we see the expansion of fortified oppida as powerful central places. The rapid urbanization that occurred during this period was also accompanied by the increasing presence of large-scale agricultural storage facilities (silos), such as at Pontós and other sites in the Empordà, as one of the main characteristics of expanding demographic and economic growth across the region. In a sense this is the direct consequence of a remaking of the landscape based on the type of coercion that usually accompanies the establishment of political hegemony. Effectively, this involved the appropriation of labour so as to increase production for exchange purposes. In this sense, the process of "colonial entanglement" was instrumental in further accentuating social division across the landscape

Territoriality and its expression as politico-economic hegemony, was to reach its apogee between c. 400-200 BC, with the exchange network and the flow of prestige commodities at its most intense. However by c. 200 BC, the entire socio-political system was brought to an abrupt halt—not by catastrophic warfare, or the collapse of the prestige goods exchange network—but by a set of contingent events triggered by the onset of the 2nd Punic War between Rome and Carthage. Thus, during the summer of 218 B.C., a Roman army commanded by the consul Scipio, landed in the port of Emporion, with the intention of driving out the Carthaginians from the peninsula. It was this action, and the subsequent conquest of the Iberian populations throughout the peninsula, which marked the beginning of the dismantling of the oppida system and the end of a distinctive Iberian culture, as it submitted—like so many other circum-Mediterranean societies—to the yoke of Rome.

The dynamics of socio-political relations

Peer polity interaction

From the perspective of socio-political organization, the landscape we have described above has similar properties to so-called 'peer polity interaction' (PPI) models (Renfrew and Cherry 1986). The term was devised to account for networks of interaction between roughly equal polities and is intended as an explicit alternative to core-periphery models with their strong centralist dynamics and subordinate periphery. The term 'polity' is used here to describe the highest order politically autonomous unit. As defined by the authors, peer polity interaction encompasses:

> "...the full range of exchanges taking place—including imitation, emulation, competition, warfare, and the exchange of material goods and information—between autonomous (self-governing) socio-political units, generally within the same geographic region".
>
> (ibid. 1986: 1)

Importantly, PPI models highlight the relationships between socio-spatial organization and the role of power in articulating societal systems at the landscape or regional scale. The essential difference separating PPI from core-periphery models is that unlike the latter, peer polity models are focused on the settlement landscape as a network. What this means is that in the case of the Empordà landscapes, attention must be shifted from a conventional single site focus e.g., privileging sites such as Ullastret, Peralada or Pontós, as the basis of understanding socio-spatial interaction, to one that encompasses the socio-political landscape as a single networked entity. As a result, the landscape can be conceived as a distributed system of semi-autonomous nodes forming a 'flat' hierarchy, or heterarchy, rather than a `top-down` vertical type of organization. Importantly, two distinct types of organization operate simultaneously: the landscape (regional) scale functions as a heterarchy, while at the local, settlement scale what we observe is a hierarchy of relations (see Figure 4).

A key characteristic of PPI is the presence of 'structural homologies` between polities. In our present case, this is manifest in the structural and material culture patterns visible in the oppida; that is, in terms of their similar geo-topographic location: a) on small elevated hills, or in easily defensible locations, b) the presence of similar types of fortifications, c) stone built architecture, d) high density of grain storage pits, e) presence of foreign trade goods, f) shared symboling systems, g) similar writing systems.

In relating the PPI model to socio-political change, Renfrew distinguishes three primary types of interaction:

1. Competition (including warfare) and competitive emulation;
2. Symbolic entrainment and the transmission of innovation, and;
3. Increased flow of exchange goods.

In the case of the Empordà plain, with respect to competition, we assume that this was a ubiquitous element of social interaction that in some cases, precipitated warfare. Symbolic entrainment, the process whereby a symbolic system is adopted when it comes into contact with a "less developed" one, is manifestly evident in the region under study and can be seen in the presence of ritual funerary accoutrements such as Greek Corinthian perfume jars which may have accompanied funerary rituals. The third factor in the PPI triad, relating to an increased flow of exchange goods is most visible in the spread of Greek ceramics (e.g., Attic pottery) across the landscape and visible in the main oppida sites described above.

The peer-polity perspective is particularly useful, since it provides us with a regional perspective that stresses the landscape, not as the locus of a series of autonomous sites, but rather as a regional network. We shall now turn our attention to how we might understand the evolutionary dynamics of such a network.

Complex networks and emergent behavior

The foregoing description of the Iron Age Iberian societies of north-east Catalonia outlining their socio-spatial organization, provides us with a picture of a politico-

Landscape level: heterarchical organization

***Oppidum* level**: hierarchical organization
control is top down

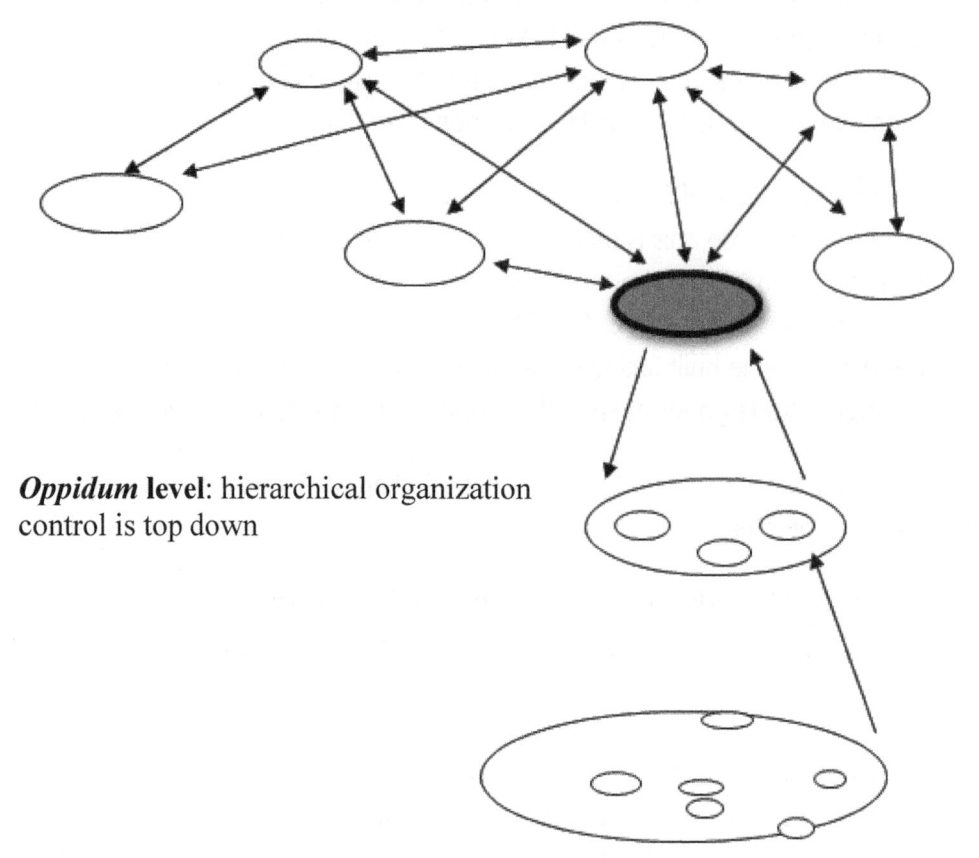

Figure 4 *Different forms of oppida network organization depending on observational scale*

economic system that is driven by the control of prestige goods. There is, however, another level of description we can usefully add to generate a more complete representation of the structural dynamics and emergent properties resident in the prestige goods exchange system. What this entails is that we view the exchange landscape within the perspective of complex dynamic systems; that is, as a complex distributed network. Over the last two decades, a number of researchers have argued for the utility of a complexity perspective for archaeological problems, and in particular, the way that nonlinear processes at the heart of societal systems can generate emergent, unanticipated behaviors (e.g., McGlade, 2006; McGlade & van der Leeuw, 1997; Bentley & Maschener 2003).

A key aspect of this theoretical focus is that in underlining the essentially nonlinear nature of complex human-environment relations, it foregrounds the importance of emergent, self-organization as a fundamental evolutionary mechanism (Allen &

Sanglier, 1981; Allen, 1997). Properties such as instability and discontinuity are thus viewed, not as aberrant processes, but rather as fundamental attributes of change in social systems. These properties are particularly visible in the dynamical behavior of networks (e.g., Barbassi & Albert, 1999, Adamic & Huberman, 2000) where research on network dynamics across a number of disciplines (e.g., computer science, economic geography, ecology) has demonstrated that it is precisely these processes which are resident in distributed networks; that is systems whose organization is not hierarchical, but heterarchical. Thus organization and decision making is decentralized across a number of nodes. Viewed from this perspective, we can conceive of the landscape in our study as a complex network of connectivities that resembles the distributed architecture of computer systems with nodes acting as semi-autonomous agents alternating control across the network of connectivities.

The relevance of this work to archaeology and broadly to the study of complex human situations lies in its provision of a framework within which the behavior and variegated decision-making capacities of individual agents can be studied simultaneously at both the micro and the macro levels; it presents unique insights into the evolutionary structure of the long-term, showing how time-delays and incomplete knowledge act to promote nonlinearity in the information structures which govern the global evolution of the system. Moreover, in investigations of these nonlinearities, a number of researchers (e.g., Axelrod, 1994, Huberman & Adamic, 1998, Barbasi & Albert, 1999; Barbasi, 2002) have shown that extremely complex behaviors and even chaotic solutions can appear in models of cooperation and competition for finite resources—a point that may have important implications and indeed applications for the kinds of commodity exchange interactions we are currently dealing with. These studies provide valuable analytical insights into network evolution in general and highlight the role of instability as a crucial source of order.

In order to explore this perspective on network complexity and the dynamics of connectivity, we shall set up a model experiment designed to explore the core dynamics of an exchange system in which both prestige items and everyday commodities circulate across a landscape in which 'competitive emulation' is a dominant feature of socio-political interaction.

Elements of a model

Model assumptions and boundaries

There are two primary attributes shared by successful models in any field of enquiry—transparency and parsimony. As was pointed out many centuries ago by William of Ockham, elegance and simplicity of structure are preferable to the density of excessive detail; simplicity rather than complexification, is the key to insight and understanding. Consistent with this principal of parsimony, our point of departure is to seek a reduced description of a prestige goods economy, i.e., one that captures the key processes and dynamics, without falling into the trap of trivial description. Above all, model building should be viewed as a branch of experimental science (McGlade 2003, 2005), a notion that is echoed in the words of the eminent evolutionary biologist, the late C.H. Waddington, who declared that models are "tools for thought", nothing more, nothing less. Our present study is precisely aligned with this spirit of exploration.

Perhaps the most important point about the modelling exercise we shall embark upon is that the settlement landscape will be viewed as a complex co-evolutionary system. This perspective will allow us to conceive of population dynamics, food production and the nature of commercial exchange, not as separate systemic entities, but rather as inexorably entwined aspects of a single dynamic. To do otherwise would be to reify entities such as "economy", "production" or "population" as if they were independent variables. Rather, if we are to understand the true evolutionary nature of this Iberian Iron Age system, then we must confront the indivisibility of these elements.

In the interests of clarity, we shall outline the primary model assumptions, thus:

1. The politico-economic landscape is made up of 10 large-scale settlements acting as political central places (oppida), and each of which is served by smaller dependent sites.
2. The domain of trade/exchange is divided into two spheres: domestic goods and commodities and prestige items.
3. Domestic exchange items include, wool, textiles, leather, cattle, goats, pigs, salt, fish.
4. Mediterranean Prestige goods emanate from the Greek colony at Emporion on the Bay of Roses and include:

Etruscan amphorae, Greek Attic pottery, Greek grey ware, Massalian pottery, Phoenician/Punic amphorae, Olive oil, Wine, Perfume, Textiles, Jewellery, Iron tools, Bronze weapons, Silver, Copper, and Slaves

5. Each oppidum comprises two populations: a) an aristocratic elite and commoners; these are coupled in a symbiotic dynamic, expressing their interdependence.

6. The evolutionary behavior of each oppidum is represented by three semi-autonomous, coupled sectors: population dynamics, food production and commercial exchange transactions. Collectively, the structural dynamics of these three coupled sectors may be said to form a complex system that is characterized by nonlinear interactions.

7. Documentary evidence from inscriptions on a lead plaque (Ampurias 3), suggest that overall control of commercial activities was in the hands of a chief merchant, the naukleros, a ship owner and orchestrator of all exchange transactions. These were carried out through a series of agents—Greeks, Phoenicians and Iberians—who acted as private entrepreneurs. These agents, based in Emporion, acted independently, and not on behalf of the city. It is known that these merchants traded directly with indigenous groups who had privileged access to the trade in prestige goods, on account of their being the first to initiate trade relations. We shall assume that this group is synonymous with the Indiketes who inhabited the oppidum at Ullastret, the closest geographically to the Greek colony at Emporion. Significantly, from an archaeological perspective, this site boasts the greatest number of Mediterranean imports. Ullastret, thus, is the gateway to the final stage in the diffusion of prestige goods to the other oppida in the landscape.

8. Effectively, the landscape is crisscrossed by two spheres of exchange, representing a) the movement of local domestic consumables and b) the flow of prestige imports. The former accounts for the flow of locally produced items such as pottery, hides, metalwork, salted fish, cheeses, meat, wool, cloth, metals and some cereals—though these latter were generally reserved for the exchange of prestige goods.

9. The transaction value of prestige items is calculated as a function of:
    - The scarcity/rarity of the artefact or commodity
    - The production costs, accounting for labour input and time involve manufacture.

- The transportation cost, accounting for the distance travelled by objects and commodities.
- A demand factor that takes account of fluctuations in the market as demand for a particular prestige item waxes and wanes in accordance.

Model description

Figure 5 shows the basic model architecture, expressed as a series of causal loops that present the three sectors of the model (population dynamics, food production and transaction dynamics). The figure shows the basic causal connections and influences that operate at the level of an individual oppidum. At its core is situated the dynamics of two interacting (coevolving populations: a ruling aristocratic Elite group and a dependent labouring, or Commoner population.

The model assigns separate demographic profiles to each group, so as to account for different birth and mortality rates that inevitably separate the life expectancy of privileged and non-privileged members of society. These demographic profiles are derived from archaeological and palaeo-demographic data obtained from sites around the Mediterranean and are representations of typical small scale agrarian societies.

The diagram (Figure 5) illustrates the key necessity of surplus agricultural production as a prerequisite for entry into the wider Mediterranean commercial networks. The acquisition of prestige or luxury items is seen as a vital factor in the maintenance of elite power and territorial control. The aggrandizement of individual oppida provokes or increases existing competition and may lead to warfare. Indeed, we are describing warrior aristocracies whose raison deter is the waging of war as a means of acquiring increased prestige. Thus achieved, it is commonplace among prehistoric elites to use feasting as a means of conspicuous display of wealth and as a demonstration of social superiority (Dietler & Hayden, 2001). This is represented in the model diagram as a process that impinges on food stocks, requiring their continuous replenishment—a circumstance that necessitates an even greater increase in agricultural production. That is in addition to normal surplus storage as an insurance against drought, crop disease and/or bad harvests. Herein lies one of the critical instabilities in the system, for surplus deficit compromises the ability of an individual opidum to participate in the prestige goods economy.

Essentially the model is constructed around the dynamics of two symbiotically linked populations (elite and commoner) and explores their interaction over a period of c. 400 years. This represents the time span between the arrival of the Greeks at

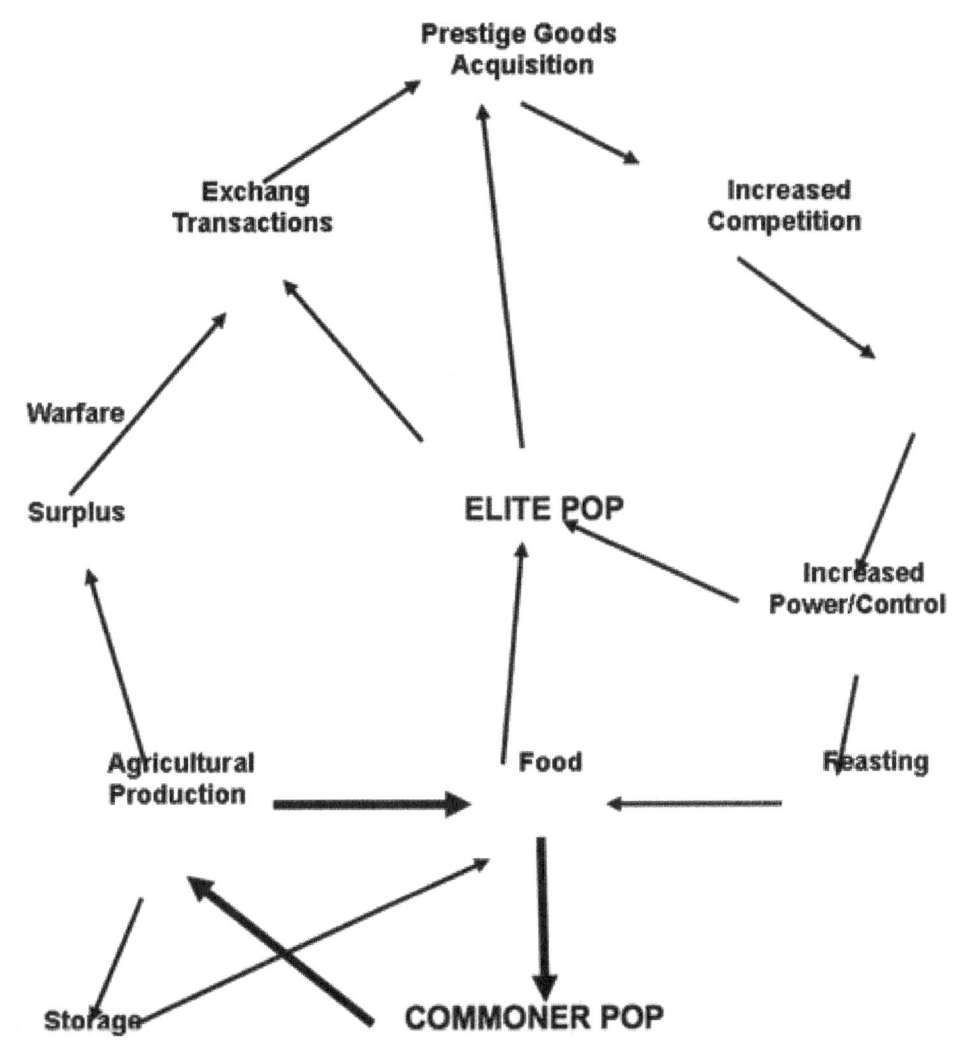

Figure 5 *Relationships describing the internal dynamics of each oppidum and its connection to external exchange networks.*

Emporion at c. 600 BC and the dissolution of the Iberian world with Roman conquest in 218 BC. Figure 4 shows that, the growth of elite power rests on the ability of the dependant commoner population to generate sufficient labor (food production) to support the non-productive elite sector. Production, itself, is modelled as an expression that accounts for available labour and technology, with the latter accounting for periodic improvements in agricultural methods (e.g., the introduction of iron), crop and animal breeding and innovations in manufacturing technology, such as the potters wheel.

The causal feedbacks linking the two populations' (see Figure 4) are, however, further complicated by the presence of environmental and cultural/biological factors which act to inhibit growth and may even precipitate collapse of the socio-political system. These are, 1. Environmental fluctuations, 2. Disease vectors and 3. Natural mortality.

1. **Environmental fluctuations**: The first of these accounts for natural fluctuations in temperature and precipitation and, especially, the frequency of extreme events that are visited upon the Empordà region. These latter include episodes of flooding as well as prolonged summer drought; will clearly have a significant impact on crop yields and the survival of livestock. The need for crop storage—not simply for exchange, but as a form of insurance—is thus included as in the model as a vital component of the Iberian economy.

2. **Disease vectors:** Infectious diseases and epidemics are ubiquitous in prehistoric populations since they have limited immunity and/or medical treatments to control them. This susceptibility is abundantly evidenced in historical and paleoanthropological studies of prehistoric and early historic societies (e.g., Schedel, 2003; Hansen, 2006; Bocquet-Appel, 2006). The reality is that small-scale agrarian populations can be decimated by the speed with which disease can spread through a settlement. In general, the most obvious consequences of infectious diseases is their impact on human nutrition, reducing fertility levels in women and hence in the population birth rate.

3. **Natural mortality**: The third mitigating factor constraining growth concerns the role of natural mortality. The calculation of this factor draws on palaeo-demographic studies from around the Mediterranean (Schedel, 2003) and provide important data relevant to the Iron Age Iberian populations we are modeling. Additionally, we need to account for an increase in mortality rates as a consequence of the endemic nature of warfare in warrior-based societies.

In summary, at the most basic level, Figure 4 shows that the socio-political system is dependent upon population growth in the commoner population, which provides the labour force needed for production. The level of production itself, determines the number of non-productive members of society (the elite) that can be supported. Increased production creates surplus, firstly as an insurance against crop failure and secondly permits entry into exchange networks. In turn, the growth of exchange increases the flow of prestige goods and the power of the aristocratic elite and it is this increase in the power of individual oppida which generates more competition across the transaction network, increasing the probability of warfare as well as the desire for territorial expansion. It is in this sense that the internal dynamics of each oppidum generates a high degree of instability that in turn, contributes a significant degree of unpredictability to the global behavior of the transaction network.

Model results

Population dynamics

Population dynamics are, of course, governed by the behavior of birth and mortality rates. These play a particularly crucial role in prehistoric populations and their inherent variability—for a variety of environmental and cultural reasons—contributes to the long-term instability of prehistoric social groups and hence the persistence and survival of populations generally. Realistic data on vital rates was obtained from previous research on prehistoric and ancient demographic profiles (e.g., Schendel, 2003). Based on these data, model runs were done with a birth rate of 0.055. The mortality was variable with an average of 0.05, and a standard deviation of 0.015. These values can be considered relatively conservative as opposed to extreme. A normal distribution was used for the variability of the mortality rate. Additionally, it is important to remember that population dynamics—and particularly mortality rates—are affected by the warfare that characterized these Iron Age Iberian societies, and this has been factored into the mortality rate.

The evolutionary trajectory of each oppidum traced over a 400-year period, can be seen in Figure 6, which represents a randomly derived output from 100 simulation runs. The population trajectories of each site demonstrate constant fluctuations in population size, reflecting the effects of high infant mortality rates that accompany small-scale warrior societies. However, this appears to be compensated by the birth rate profile, hence there is a trend to population growth over the long term.

The bar charts show how often the particular oppidum took a particular population rank over the hundred runs. On the scale represented in the bar graphs of Figure 8, 10 is the lowest placed site over the time trajectory. For instance the figure shows that the oppidum of Porqueres occupied the lowest position, i.e., bottom in terms of population, four times and the highest position six times with an average position of 4.27. The figure shows this spread across all of the oppida graphically.

Disease vectors

The tendency for prehistoric and early historic populations to be vulnerable to the onset of disease has been discussed above. In order to simulate such effects and to observe their long-term effects, a 'plague' scenario was explored. Initially, there is a general outbreak of the disease across the landscape in year 200 which spreads from Emporium via Ullastret over a 3-year period. The outbreak last for 3 years at each

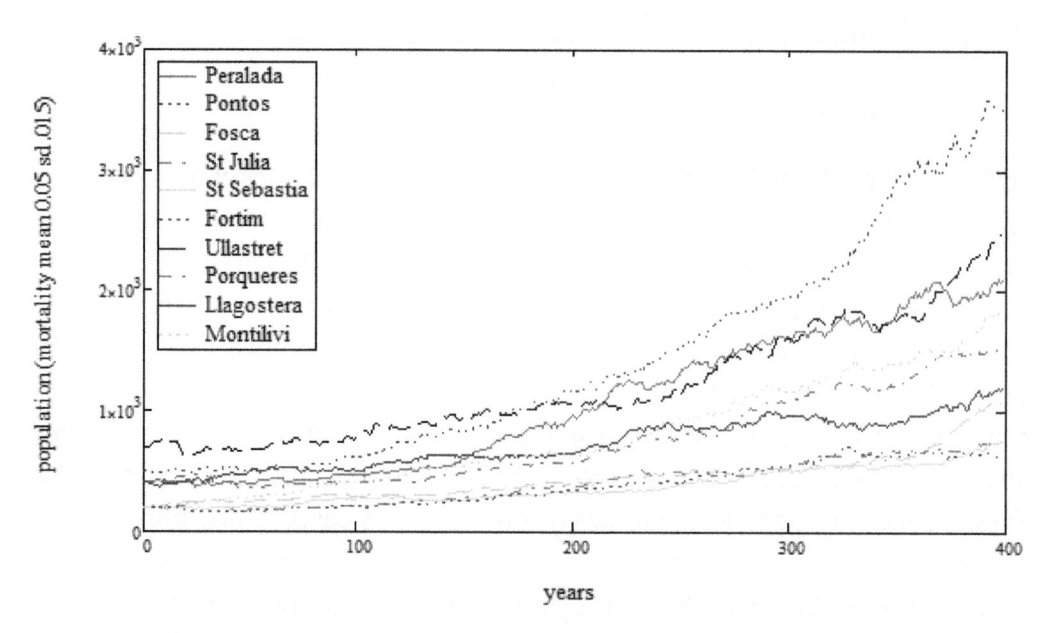

Figure 6 *Population dynamics over the long-term (400 years).*

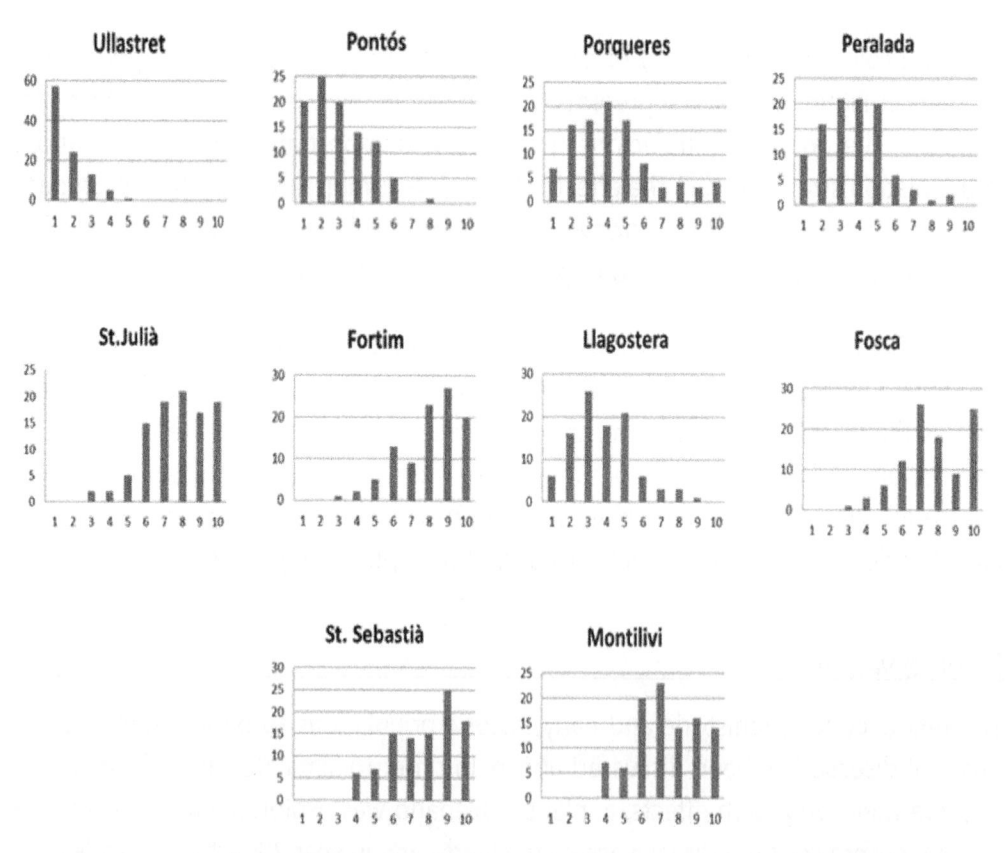

Figure 7 *The final rank places for each oppidum over of the hundred runs.*

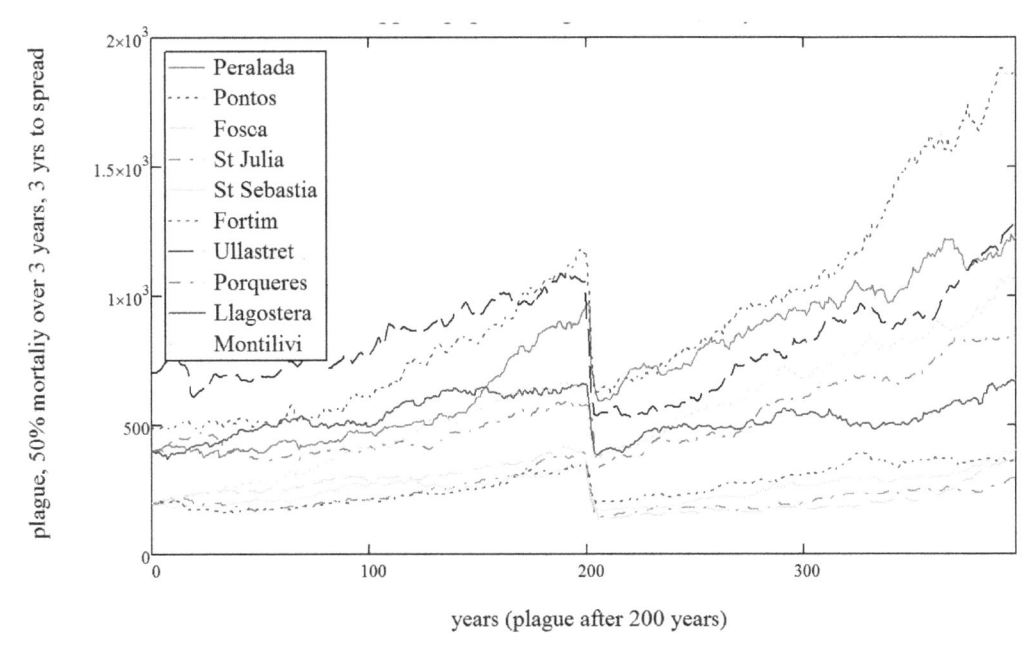

Figure 8 *Simulation results of a plague visitation on the oppida system*

oppidum. In the first year the average mortality is set to 21%, in year one, 20% in the second year and 19% in the third. This gives an average overall mortality of 50% (not unreasonable for a virulent disease in a prehistoric population). Of interest here is the recovery time in a decimated population and we see that it takes some 4 generations—100 years—for the population as a whole to return to its pre-plague level (see Figure 8).

Prestige goods exchange dynamics

The flow of prestige goods across the landscape is essentially demand-driven and reflects the changing nature of agricultural production and particularly the quantity of surplus available for exchange transactions.

The graphs in Figures 9a and 9b show the changing levels of wealth accumulation for each oppidum. What we see is that differentials in accumulation reflect, not just the capacity of individual sites to increase their wealth is not only a consequence of changing levels of surplus production, but is also a consequence of geographic distance from the source of prestige goods at Emporion. For example, the relative proximity of Ullastret to the Greek colony is reflected in its higher levels of wealth, as is the case with Peralada, which is positioned, 3rd in the overall long-term wealth index. By contrast, sites such as Montilivi and Llagostera, being far from the main river systems are characterized by lower levels of accumulation. Notably it is possible to see that locations closer to the Greek colony at Emporium tend to be wealthier regardless of the size.

Long-term versus short-term dynamics

One of the most striking aspects of the model is the intrinsic resilience it displays over the long-term. Here, we are referring to the capacity of the system to deal with an array of environmental and demographic perturbations that constantly buffet the sys-

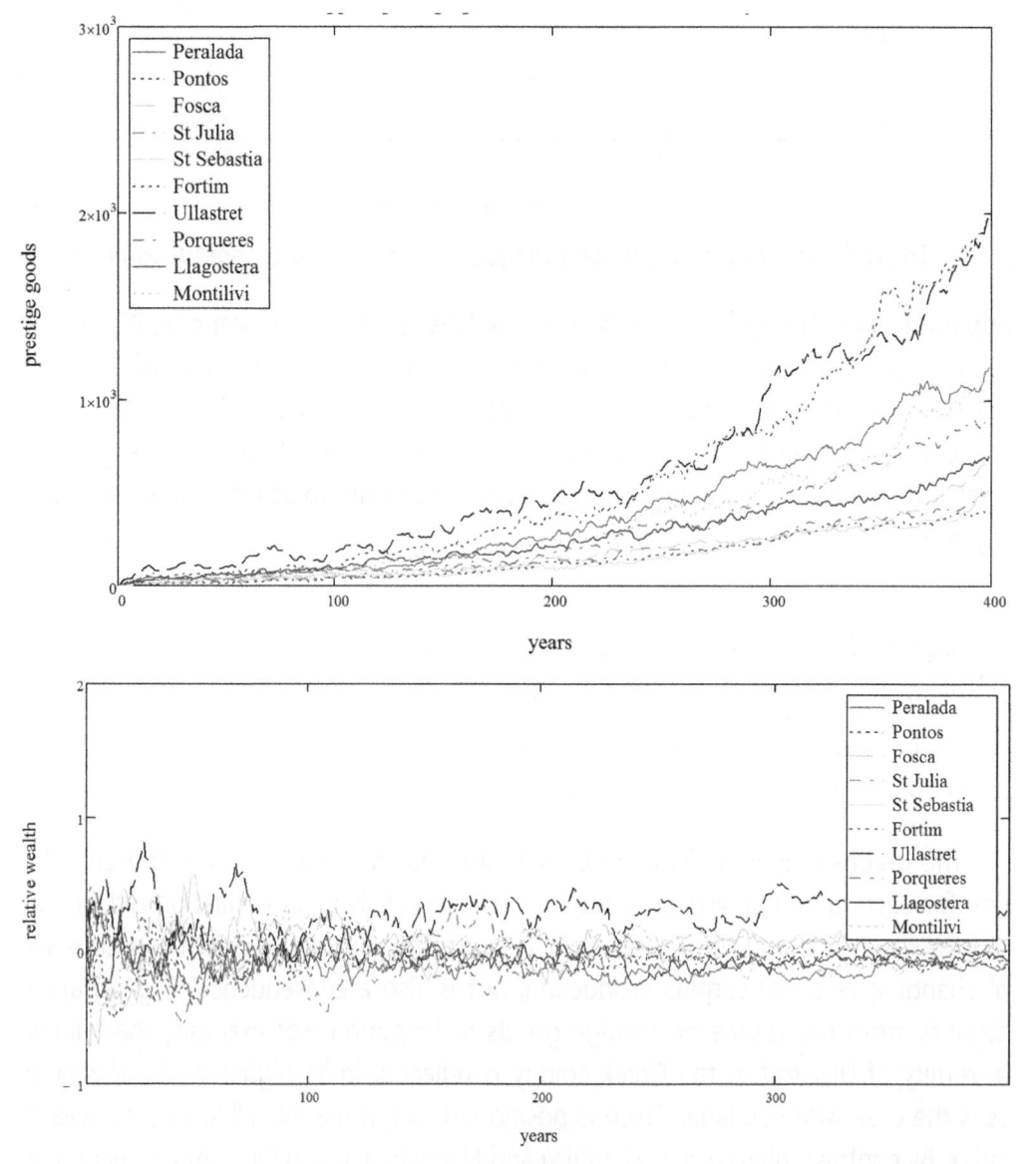

Figure 9a, 9b *Prestige goods accumulation & cumulative wealth index over 400 years*

tem. For example, if we examine the trajectory of population over the short-term, by taking a random slice of the population dynamics of one oppidum (Ullastret) over two generations (50 years) we can observe something of this resilience. What we see from Figure 10 is a highly variable trajectory as the population absorbs the effect of natural variability in birth and death rates, and the way that mortality rates are amplified by periodic disease vectors reducing the population and pushing it to the edge of collapse, followed by a gradual recovery. Such demographic fluctuations have a strong ripple effect through the socio-economic system, affecting labour input for production as well as interrupting and /or reducing the capacity of the group to participate in exchange transactions. Over the long-term, we see substantial fluctuations in the degree of prestige and power each site was able to maintain. Above all our simulation demonstrates a dynamic and constantly changing economic landscape. Notably, no single site appears to be able to accumulate vast quantities of wealth and hence acquire pre-eminence on the political stage. Thus, our model provides a good example of the operation of a distributed network, where the landscape is represented as a heterarchy of overlapping circuits of power.

In summary, what the results show is the impossibility of disaggregating the three model sectors (population dynamics, agricultural production, exchange transactions) and tantalizing them as discrete entities. This is simply because we are dealing with a complex, nonlinear structure driven by feedbacks and whose evolution in any one

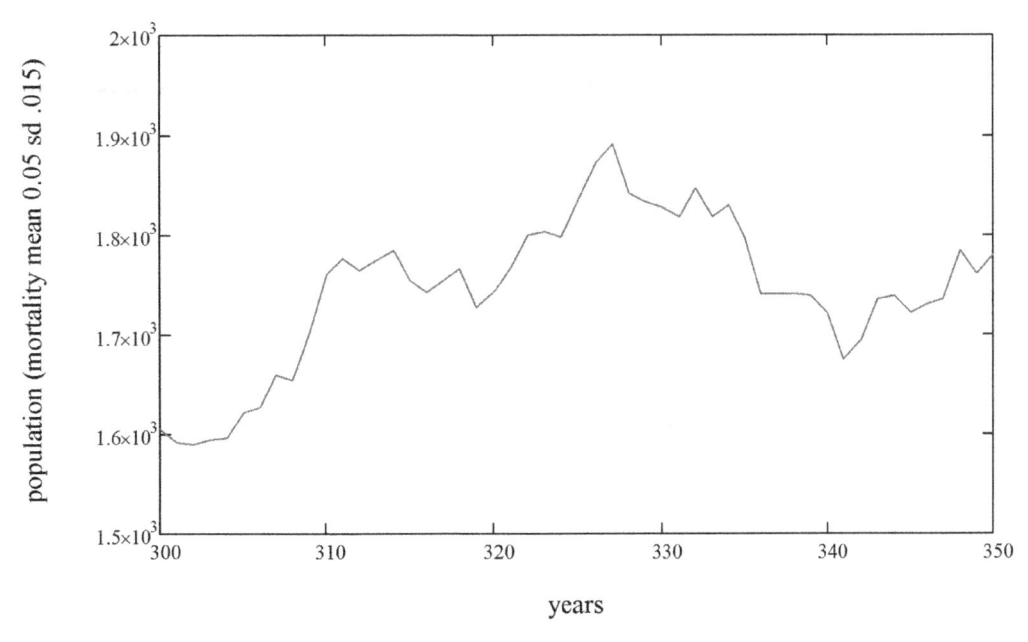

Figure 10 *Population dynamics from Ullastret traced over two generations (50 years)*

moment cannot be understood by a simple examination of its constituent parts. Additionally we see the important role played by contingent processes and particularly micro-contingent events, i.e., extreme fluctuations in birth and death rates, inclement weather, bad harvests, the emergence of unanticipated events such as disease pathologies in human, plant and animal populations. These contingent events play a significant part in the persistence and evolution of society and not least in influencing the production and accumulation of wealth and hence in the maintenance of power by the ruling elite.

Discussion

From the perspective of socio-political organization, the landscape can be understood as comprising two distinct scales of operation:

1. At the local level of each individual oppidum, we see a top-down hierarchical system at work, headed by an aristocratic elite, controlling all aspects of society. All sectors of society work to enhance the prestige and power of the ruling class—whether by acquiescence or coercion.
2. At the landscape (regional) level, each oppidum forms a node in a complex network of competitive emulation—similar to that envisioned in the peer polity interaction model (PPI) model by Renfrew and Cherry (1986). This network operates as a heterarchy, with no single site assuming dominance or control in the operation of exchange transactions.

In addition, the socio-political landscape is made up of a network of interdependencies that is a common feature of the competitive/cooperative relations we see in small scale warrior aristocracies. These, however, do not conform to any smooth, homogeneous dynamic; rather they are discontinuous, characterized by abrupt change and transformation on account of the inherent nonlinear interactions that articulate the sequence of cross-regional commercial transactions. From a stability perspective, it should be noted here that these spatio-temporal dynamics are further complicated by the internal political struggles that are frequently present among competing aristocratic elites in pursuit of power and prestige.

What our model underlines is that we need to take account of the fact that we cannot arrive at any meaningful understanding of the nature and complexity of exchange systems by isolating something termed the 'economic sector' and isolating it for study. Exchange systems are, first and foremost, intrinsic aspects of prehistoric—and in the present case, protohistoric—societies; they are embedded in social/

symbolic practice and cannot easily be understood outwith this context. In this sense, unlike our contemporary world, there are no purely 'economic' decisions that can be made, for we are not discussing commodities or the simple movement of consumables. Rather, as we have sought to emphasise above, they form elements in a discourse of power that is focused on the maintenance of prestige, which forms a central part in the construction of aristocratic/elite identity. This process of identity construction and the competitive dynamics accompanying it is the basis of the peer polity interaction (PPI) model described above. It is this model with competitive emulation as its driver, that is responsible for the heterarchical structuring that best describes the emergent socio-political landscape. What is of crucial interest here, is that the competitive emulation which forms the core of our model of a prestige goods economy, acts to prevent the emergence of any single site as the sole controlling power over the competitive landscape. It is precisely these structural dynamics and the part played by human agency as an enabling or constraining force, that require attention in future research.

Conclusion

What this paper has sought to underline is the dynamic behavior of tightly-coupled networks of power, played out through the acquisition of and movement of prestige items in the type of warrior aristocracies that inhabited north-east Iberia, as well as many parts of the Mediterranean region. As we have underlined above, the landscape functions as a distributed network held together by the mechanism of competitive emulation. It is interesting to note here that the competitive emulation, which is driving the system, is responsible for the resilience of our prestige goods model; that is, it functions as a self-organizing property, structuring and restructuring the socio-political landscape.

What this foregrounds is the central problem of structural stability in complex societal systems. By employing a complexity perspective, we can reasonably argue that it has taken our understanding of the evolutionary dynamics of peer polities a step further; i.e., the descriptive nature of the PPI model has been given a more analytical treatment, allowing us to experiment with a variety of simulated outcomes that can form the basis of a new set of hypothetical questions. Our model architecture has made it possible to investigate a number of structural and functional properties of socio-political networks that are driven by economic transactions. Perhaps most important of all, we have been able to show how the nonlinear couplings at the heart of complex peer polity networks can generate emergent, self-organizing behaviors, and thus severely compromises our ability to predict long-term evolutionary outcomes.

Finally, if it is clear that the acquisition of high status goods plays a central role in the maintenance and reproduction of social-political hierarchies and hence, of power itself, it is equally true to say that the existence of this power is ephemeral, precisely because it is based on social competition and the ebb and flow of political alliances, all of which are fundamentally unstable. What this points to is the inherent fragility of prestige-goods economies and hence the subsequent effects on the maintenance of social control in elite dominated societies will always be prone to collapse (McGlade, 1997). This is precisely the situation that characterized the Iron Age Iberian societies on the Empordà plain during the centuries preceding the advent of the Roman conquest.

References

Adamic, L. and B.A. Huberman (2000). "Power law distribution of the World Wide Web," *Science*, ISSN 0036-8075, 287(5461): 2115.

Allen, P.M. (1997). *Cities and Regions as Self-Organizing Systems: Models of Complexity*, ISBN 9056990713.

Allen, P.M. and Sanglier, M. (1981). "Urban evolution, self-organization and decision making," *Environment and Planning A*, ISSN 0308-518X, 13(2): 167-183.

Barbassi, A-L. and Albert, R. (1999). "Emergence of scaling in random networks," *Science*, ISSN 0036-8075, 286(5439): 508-512.

Bats, M. (1988). "Les inscriptions et graffitessur vases ceramiques de Lattara protohistorique (Lattes, Herault)," *Lattara* 1, ISSN 0996-6900.

Bocquet-Appel, J.-P. (ed) (2008). *Recent Advances in Palaeodemography: Data, Techniques, Patterns*, ISBN 1402064241.

Brumfiel, Elizabeth M. (1994). "Factional competition and political development in the New World: An introduction," in E.M. Brumfiel and J.W. Fox (eds.), *Factional Competition and Political Development in the New World*, ISBN 0521384001, pp. 3-13.

Buxó, R., McGlade, J., Palet, J.M. and Picazo, M. (1998). "La evolución del paisaje cultural: la estructuración a largo plazo del espacio social en el Empordà," *Arqueología Espacial*, ISSN 1136-8195, pp. 1-20: 399-413.

Cabrera, P.(1996). "Emporion y el comercio griego arcaico en el nordeste de la Peninsula Ibérica" in R. Olmos and P. Rouillard (eds), *Formes Archaiques et Arts Ibériques*, ISBN 8486839734.

Cabrera, P. (1998). "Greek trade in Iberia: the extent of interaction," *Oxford Journal of Archaeology*, ISSN 1468-0092, 17(2): 191-206.

Conte, R., Hegselmann, R., and Terna, P. (eds.) (1997). *Simulating Social Phenomena*, ISBN 3540633294 (2013).

Crumley, C. L.(1995). "Heterarchy and the Analysis of Complex Societies," in R.M. Ehrenreich, C.L. Crumley, and J.E. Levy (eds.), *Heterarchy and the Analysis of Complex Societies*, ISBN 0-913167-73-8, pp. 1-6.

Epstein, J.M., and Axtell, R.L. (1996). *Growing Artificial Societies: Social Science from the Bottom Up*, ISBN 0262550253.

Friedman, J. and Rowlands, M.J. (1977). "Notes toward an Epigenetic Model of the Evolution of 'Civilization'," in J. Friedman and M.J. Rowlands (eds.), *The Evolution of Social Systems*, ISBN 0822911337, pp. 201-78.

Godelier, M. (1999). *The Enigma of the Gift*, ISBN 0226300455

Gilbert, G. and Troitzsch, K. (2005). *Simulation for the Social Scientist*, ISBN 335216005.

Hansen, M. (2006). *The Shotgun Method: The Demography of the Ancient Greek City-State Culture*, ISBN 9780826216670.

Helms, M. (1993). *Craft and the Kingly Ideal: Art, Trade and Power*, ISBN 0292730780.

Helms, M. (1994). "Chiefdom rivalries, control, and external contacts in lower Central America," in E.M. Brumfiel and J.W. Fox (eds.), *Factional competition and political development in the New World*, ISBN 0521545846, pp. 55-60.

Helbing, D. (ed) (2012). *Social Self-Organization: Agent-based Simulations and Experiments to Study Emergent Social Behavior*, ISBN 97836422400034.

Henrich, J. and F.J. Gil-White (2001). "The evolution of prestige: freely conferred deference as a mechanism for enhancing the benefits of cultural transmission," *Evolution and Human Behavior*, ISSN 1090-5138, 22(3): 165-196

Hyde, L. (1983). *The Gift: Imagination and the Erotic Life of Property*, ISBN 0394715195.

Johnson, G. and Earle, T. (2000). *The Evolution of Human Societies: From Foraging Groups to Agrarian State*, ISBN 0804740321.

Kauffman, S. (1993). *Origins of Order: Self-Organization and Selection in Evolution*, ISBN 0195079515.

Kranton, R. (1996). "Reciprocal exchange: a self-sustaining system," *American Economic Review*, ISSN 0002-8282, 86 (4): 830-835.

McGlade, J. (1997). "The limits of social control: coherence and chaos in a prestige goods economy," in S.E. van der Leeuw and J. McGlade (eds.), *Archaeology: Time and Structured Transformation*, ISBN 0415589096, 8: 1-23.

McGlade, J. (2003). "The map is not the territory: Complexity, complication and representation," in A. Bentley and H.D.G. Maschner (eds.), *Complex Systems and Archaeological Research: Empirical and Theoretical Applications*, ISBN 9780874807554, pp. 111-119.

McGlade, J. (2005). "Systems and simulacra: Modelling, simulation and archaeological representation," in H.D.G. Maschner and C. Chippindale (eds.), *Handbook of Theories and Methods in Archaeology*, ISBN 9780759100336, pp. 554-601.

McGlade, J. (2006). "Ecohistorical regimes and la longue durée: An approach to mapping long-term social change," in E. Garnsey and J. McGlade (eds.), *Complexity and Coevolution: Continuity and Change in Socio-Economic Systems*, ISBN 184542140X, pp. 77-114.

McGlade, J. and van der Leeuw, S.E. (1997). "Archaeology and nonlinear dynamics: new approaches to long-term change," in S.E. van der Leeuw and J. McGlade (eds.), *Time, Process and Structured Transformation in Archaeology*, ISBN 1134525028, pp. 1-22.

McGlade, J. & Garnsey, E. (2006). "The Nature of Complexity," in E. Garnsey & J.McGlade (eds.), *Complexity and Coevolution: Continuity and Change in Socio-Economic Systems*, ISBN 184542140X, pp. 1-21.

Picazo, M., McGlade, J. and Buxó, R. (1998). "Caminos del Empordà: el impactode las redes de comunicación en el paisaje y en la organización del territorio," in S. Torrent i Masip (ed.), *Comerç i Vias de Comunicació, XI Col.loqui Internacional d´Arqueologia de Puigcerdà*, ISBN 8460092887, pp. 295-303.

Picazo, M., McGlade, J., Buxó, R.,and Curià, E. (2000). "Continuidad y transformación del paisaje: mil años de ocupación humana del Empordà," *Revista d'Arqueologia de Ponent*, ISSN 1131-883X, 9: 41-56.

Plana, R. (1994). *La Chora d'Emporion: Paysage et Structures Agraries dans le Nord-Est Catalan a la Period Pre-Romaine*, ISBN 2251605444 (2004).

Renfrew, C. and Cherry, J. (1986). *Peer Polity Interaction and Socio-political Change*, ISBN 9780521112222.

Ruiz de Arbulo, J. (1992). "Emporion. Ciudad y territorio (s. VI-I a.C.). Algunas reflexiones preliminaries," *La Revista d'Arqueologia de Ponent*, ISSN 1131-883X, 2: 59-74.

Sanmartí, J. (2004). "From local groups to early states: the development of complexity in protohistoric Catalonia," *Pyrenae*, ISSN 0079-8215, 35 (1): 7-41.

Sanmartí-Grego, E. (1992). "Massalia et Emporion: une origine commune, deux destins différents. Marseille grecque et la Gaule," *Etudes massaliètes*, ISSN 0986-3974, 3: 27-41.

Sanmartí-Grego, E. (1993). "Els ibers a Emporion," *Laietania*, ISSN 0212-8985, 8: 85-101.

Schendel, W. (2003). "The Greek demographic expansion: methods and comparisons," *Journal of Hellenic Studies*, ISSN 0075-4269, 123: 120-40.

Appendix

Prestige goods exchange model: Model equations

$$p_{op,t} = p_{op,t-1} \cdot \left(1 + b - op\_m_{op} \cdot rnd\_factor_{op,t-1}\right)$$

$$el_{op,t} = el_{op,t-1} \cdot \left[1 + e_g \cdot \left(e2w - \frac{el_{op,t-1}}{p_{op,t-1}}\right)\right]$$

$$c_{op,t} = p_{op,t} \cdot cons + el_{op,t} \cdot e\_xs$$

$$st_{op,t} = c_{op,t} \cdot stg\_fact$$

$$new\_st_{op,t} = st_{op,t} - st_{op,t-1} \cdot stg\_decay$$

$$surp_{op,t} = prod_{op,t} - c_{op,t} - new\_st_{op,t}$$

$$prod_{op,t} = \left(p_{op,t} - el_{op,t}\right) \cdot \left(base_{prod} \cdot tech_t + rnd2\_factor_{op,t}\right)$$

if $surp_{op,t} < -1$

$$op\_m_{op} = op\_m_{op} + stv\_fact \cdot \left(\frac{surp_{op,t}}{prod_{op,t}}\right)$$

$$dur_{op,t} = dur_{op,t-1} \cdot (1 - dur\_decay)$$

if $surp_{op,t} \geq -1$

$$op\_m_{op} = m$$

$$dur_{op,t} = dur_{op,t-1} \cdot (1 - dur\_decay)$$

$$dur_{op,t} = dur_{op,t} + \frac{surp_{op,t} \cdot dur\_decay}{dur\_cost_{op}}$$

Key to parameters and variables

t = time in years; p = the population (including elites); b = the birth rate; op_m = the oppida mortality rate; el = the elite population; e_g = the elite growth rate factor; $e2w$ = the ideal ratio of elite to workers; c = the consumption; $cons$ = the consumption rate per head of population; e_xs = the excess consumption by the elite per member of the elite; st = the desired storage; st_fact = the storage rate factor; $prod$ = the production of base goods; $base_{prod}$ = the base production rate per worker in year 0; $tech$ = the technology increase factor; $surp$ = the surplus of base goods after consumption and storage; stv_fact = the starvation response to shortfalls of base goods; dur_decay = the decay rate of durable goods; $d2c\_ratio$ = the ratio of traded durables to traded consumables; dur_cost = the cost of durables.

James McGlade was born in Scotland and studied Fine Arts in Edinburgh where he obtained an undergraduate and masters degree, subsequently lecturing for a number of years at Edinburgh College of Art, Heriot Watt University. After moving to Canada, he taught at the University of Guelph and was, for a number of years, Assistant Professor at the Nova Scotia College of Art and Design, Halifax. While in Canada, he gained a BA (honors) in Archaeology and Anthropology from Wilfrid Laurier University. James then returned to the UK to complete a Ph.D. in archaeology at the University of Cambridge. This was followed by successive lectureships at Cranfield University (International Ecotechnology Research Centre) and University College London (Institute of Archaeology). In 2005, James moved to Barcelona and currently lectures at Univesitat Oberta de Catalunya and works as a consultant with an archaeological heritage management company. James's research interests focus on two primary areas: a) Mediterranean archaeology, particularly the Iberian Peninsula and b) the evolution and management of cultural landscapes.

Mark Strathern was for many years a colleague of Professor Peter Allen's in the Complex Systems Centre at Cranfield University where he specialised study and modelling of complex systems in open human systems. He is now retired but still maintains an active interest in these areas.

Chapter K

Complexity and economic evolution from a Schumpeterian Perspective

Stan Metcalfe

This essay is written in honour of Professor Peter Allen to reflect the significance of economic complexity to the study of economic development, in general, and the link between innovation, technology and economic transformation, in particular. The work of Joseph Schumpeter is the lens through which to focus the skein of relations that we call complex economic evolution, for, among the leading economists of the last century, it is Schumpeter who has had most to say about the role of emergent phenomena in the functioning of Western capitalism. The study of emergence, novelty and innovation has been one of the principal themes of Peter Allen's work over many years. His work has provided us with conceptual and computational tools to understand the role of emergent novelty in economic progress and to this extent it is work in a Schumpeterian spirit. Along with Marx and Marshall, Schumpeter's great achievement was to formulate an evolutionary, open system perspective on modern capitalism, to explain why it could never be at rest and to link its emergent properties to the capacity to change from within. In terms of appraisal, I shall focus on three aspects of Schumpeter's scheme: the link between knowledge and enterprise, the nature of the competitive process in the presence of innovation, and the transient, out of equilibrium nature of modern capitalism. The central point is that capitalism is inherently unstable in its inner workings even though it is stable in its framing rules and ethos, as Peter Allen has always insisted, its dynamics are fundamentally grounded in the fact that its structures are open to invasion from emergent novelties, the innovations that Schumpeter places at the heart of his economics

Introduction

This essay is written in honour of Professor Peter Allen to reflect the significance of economic complexity to the study of economic development, in general, and the link between innovation, technology and economic transformation, in particular. I have chosen the work of Joseph Schumpeter, as the lens through which to focus the skein of relations that we call complex economic evolution, for, among the leading economists of the last century, it is Schumpeter who has had most to say about the role of emergent phenomena in the functioning of Western capitalism. The study of emergence, novelty and innovation has been one of the principal themes of Peter Allen's work over many years. His work has provided us with conceptual and computational tools to understand the role of emergent novelty in economic progress and to this extent it is work in a Schumpeterian spirit (Allen, 2001; Allen & McGlade, 1987). The central point is that capitalism is inherently unstable in its inner workings even though it is stable in its framing rules and ethos, as Peter Allen has always insisted, its dynamics are fundamentally grounded in the fact that its structures are open to invasion from emergent novelties, the innovations that Schumpeter places at the heart of his economics (Allen, 1976).

Emergent phenomena are novelties that cannot be explained entirely by reference to particular causes so emergence is necessarily connected to indeterminacy and uncertainty, to surprise and wonder, to the language of history. Expressed so, emergence leads to the creativity of systems, that there are emergents that did not exist in the past, which from the perspective of the past could not have been predicted[1]. The systems dimension serves to open up the possibility that development is a consequence of emergent novelty, novelty that is not imposed from outside the operation of the system but arises from its internal operation. That is to say complexity is relevant because economic systems are self-transforming as well as self-ordering, and the manner of their transformation is a consequence of the nature of their ordering processes.

It may be helpful to trace the outlines of what this means for an evolutionary research programme and to use Schumpeter's great work *The Theory of Economic Development* as the means to do so. Along with Marx and Marshall, Schumpeter's great achievement was to formulate an evolutionary, open system perspective on modern capitalism, to explain why it could never be at rest and to link its emergent properties to the capacity to change from within. In terms of appraisal, I shall focus on three aspects of Schumpeter's scheme: the link between knowledge and enterprise, the nature

---

1. Cf. Lovejoy (1927) for a thorough discussion of the concept of emergence. Shackle(1966) also makes the connection between the possibility of emergence, interpreted as non-deterministic history, and the non-illusory nature of human decision making, " a history that is to say, in which we need not regard every situation or event as the inevitable, sole and necessary consequence of antecedent situations or events"(p.107). As he goes on to say, emergence is not chaos but rather a consequence of the particular form of order that is an economy. This is very much the theme of this essay, that structure and emergence are squabbling twins. See also Shackle (1961) Part 1.

of the competitive process in the presence of innovation, and the transient, out of equilibrium nature of modern capitalism.

Schumpeter's dynamics and complexity

The *Theory of Economic Development* is Schumpeter's signature work, it encapsulates the contours of an economic vision that he never thought necessary to revise in any fundamental way[2]. Its topography is repeated with only minor amendment in *Business Cycles*, albeit with greater resort to historical illustration, and refined and restated in *Capitalism Socialism and Democracy* to take account of the changing economic sociology of enterprise and the emergence of a corporate economy. My view is that *TED* is a deeply evolutionary piece of work, a dramatic illustration of the power of language unencumbered with formulae or data, and I can only marvel at its capacity to stimulate new thoughts at every fresh reading. It is also an early example of what has come to be known as the analysis of complexity. His understanding of capitalism is that it is an adaptive system that responds too and induces the emergence of novelty from within, so that every pattern of order contains within it the unknowable seeds of its own destruction, it is a scheme in which unpredictable human creativity is the *sine qua non* of change to prevailing structures. In capitalism, every position is open to potential challenge, it is an open system in which the ordering of economic society is ever changing and it never is and never can be at rest. At the most basic level, this raises fundamental questions about the tempo and directions of change in modern capitalism, change expressed in terms of the occurrence of novel events and the subsequent adaptation of economic structures to realize the possibilities immanent in economic novelty. It is not to population growth and capital accumulation that we are directed in order to comprehend the economic record, for they are grey, derivative phenomena, but rather to enterprise, innovation and economic leadership, the vibrant colors that introduce qualitative transformation in its most fundamental terms—the doing of things that have never been done before. As in Marx and Marshall, the scheme is evolutionary in a very precise sense, in that it reflects the uneven, selective response of an economic system to the localized generation of variation from within. Capitalism develops because it stimulates and allows individuals to dare to be different but it does not and perhaps cannot require everyone to behave in this way; the few are sufficient to establish the outlines of an ensuing history in which the many add the fine details. From a complexity perspective this raises a chal-

---

2. Compare two of his last essays, Schumpeter, 1947a and 1947b with the 1928 essay, the latter being one of the first papers to bring Schumpeter's ideas to the English speaking world.

lenging question. How is it that the rules of the game that give coherence and pattern to economic life are the same rules that induce and accommodate to the transformation of economic life? If we can answer this question we will comprehend the depth of Schumpeter's theory of open-ended economic change and why he was an early and perhaps unwitting student of complexity.

We must confront at the outset Schumpeter's apparent coolness towards evolutionary methods as expressed in *TED*. Darwinian schemes are rejected not least because of the hasty generalizations associated with the word "evolution" (p. 58) but this hesitancy is more apparent than real. Leaving aside the fact that evolution is a mode of thought that is domain free; the whole structure of his argument is evolutionary in form. His innovations are the novelties, the variations that invade an existing population of production methods and the system responds by having the new displace the old, competition in tooth and claw. Schumpeter's economics is variation cum selection economics if it is anything. Moreover, it is most certainly not a description of a system in equilibrium, equilibrium has no place for any distinction between the ancient and the modern, it simply is what it always has been. You can see how museums make sense in Schumpeter's world; they are there to remind us that, whatever conventions we adhere to, the future will be different from the present. By the time he is writing *Capitalism, Socialism and Democracy* this is self evident, capitalism is by nature an evolving system that never can be stationary. One can meaningfully speak of the "old" and the "new" because capitalism operates as an out of equilibrium system in which the challenges to the function and form of the prevailing order are induced by that order. As an aside, when Schumpeter wrote his semi-centennial appraisal of Marshall's *Principles* he made it quite clear that he considered Marshall's economics to be entirely evolutionary in form and method[3]. I suspect only Schumpeter could take that view precisely because he, of all economists of that generation, understood what economic evolution amounted to[4]. This is so despite the fact that, in the *Theory of Economic Development*, Marshall is cast in the role of Schumpeter's foil, the economist he rests upon in order to differentiate his own approach.

It is simply the case that *TED* is a complexity grounded evolutionary account of economic change precisely because of the role it gives to emergent novelty and its resort to a variation-cum-selection mode of reasoning. Innovations are variations, the introduction of rival goods and ways of producing them, the new combinations of ex-

---

3. Schumpeter (1941).

4. Gerald Shove, writing at much the same time (1942), is the other exception who grasped the deeper evolutionary content of Marshall's thought.

isting resources. They encompass much more than technical innovation in the narrow sense, new forms of organizing business, new forms of organizing the marketing process are just as valid sources of economic variation. If one wanted a generic description of innovation in Schumpeter's work it is surely that every example constitutes a different model of business. Yet innovation is only a necessary potential for transformation, it is not of itself sufficient. We also require the system to respond, to adapt to the possibilities created by innovation and to do so through a competitive process in which resources, formerly used in the production of the old goods and services, are switched to the production of the innovations. The putting of existing resources to new uses is at the heart of Schumpeter's scheme, it is the element which means that innovations displace old methods absolutely as well as relatively. As a practical matter Schumpeter well understood that there are always unemployed resources (labour and raw materials) and unused capacity in any economy and that these idle resources are available for use by an entrepreneur. Moreover, past development provides an explanation of why resources may be idle as a consequence of the disruptive effects of innovation. But to understand the contours of the problem of adaptation Schumpeter insists on an analysis in terms of full employment, a case where, "the new combinations must draw the necessary means of production from some old combinations" (*TED*: 68). In fact it is a question of definition, since development for Schumpeter, means "employing existing resources in a different way, in doing new things with them, irrespective of whether those resources increase or not" (*TED*: 68). It is surely not an accident that his account presumes the system wide full employment of primary inputs, so that the rate and direction of economic evolution is constrained by their availability.

So economic evolution is necessarily a matter of structural change; of different activities growing and declining at widely differing rates, it cannot be captured in any framework that insists on balanced, equi-proportional expansion of all the activities in an economy, as in a regularly expanding circular flow. Uneven development is the necessary corollary to this story of creative destruction; it is a matter of understanding why different activities grow at different rates, of the birth and death of different activities, so that quantitatively and qualitatively the system in view is transformed by economic processes. These phenomena can with care be recorded in the movement of broad economic aggregates but they can never be understood in terms of macroeconomic reasoning. The logic is microeconomic and mesoeconomic that builds from the bottom, the Schumpeterian method is a population method, it is marked by the coexistence of rival ways of acting and the task is to understand why and how those differences matter for the development of the economic system. The variations are not to be treated as noise, as stochastic aberrations, as ephemera that hide from view the essential features of economic life; the variations are the essential features. It is in this emphasis on variation as novelty that a crucial link is made between evolutionary dynamics and the study of complexity (Arthur, 2013).

This is a process story not a state of affairs story; it is a theory of self transformation embedded in a theory of self-organization, a scheme to understand history but not to predict history. There is no point enquiring as to the predictive power of Schumpeter's scheme other than to insist that the past and the future will be different in unspecifiable ways. Rather it is the structural similarity of his theory to the properties of a capitalist economy that matters, for this enables us to understand the threads connecting classes of phenomena despite the great differences that are recorded in particular instantiations. Moreover, structural similarity allows the conduct of "realistic" counterfactual experiments, allows us to conceive of consequences of events that never will be realized, to work through the broad impact of innovations that can only be dimly perceived[5]. Perhaps the central insight is that economic growth is always a product of uneven economic development, that development induces development, that variation is itself induced by the working of the system. Perhaps it is not too bold to claim that Schumpeter's *Theory of Economic Development* is the economics of positive feedback, open, restless systems.

I shall explore this claim by considering briefly three particular aspects of Schumpeter's scheme: namely, the relation between knowledge and enterprise; the process of competition; and, the transience of economic order. The retrospective part of my task over, I can then turn briefly to the principles by which we may analyze the dynamics adaptation to novelty.

Three Schumpeterian elements

Enterprise and knowledge

Schumpeter's treatment of knowledge is a particularly important and distinctive part of his scheme and it is grounded in the contrast between routine action and economic leadership. The broad flavour is as follows. In the circular flow of economic life action is a matter of routine, a habitual response to value data within the context of reliable understandings of cause and effect relationships. This accumulated wisdom has the properties of a capital good (we are invited to equate its features with that of a railway embankment!) the use of which economizes on the need to calcu-

---

5. This is the point at which algorithmic methods such as those implicit in agent based modelling have a great deal to contribute.

late and enables the daily round of decisions to be accomplished in well worn tracks according to custom and experience. To the extent that this is a matter of doing the best one can in the perceived circumstances it is rational but the rationality need only be subconscious not explicit. This does not mean that the data do not change and induce different actions, only that their changes never imply qualitatively new events. So probabilistic risk is fully part of the circular flow but risk implies a complete understanding of the spectrum of possibilities and the likelihood that gains and losses, say due to extreme weather events, will be of a temporary, reversible nature. On average the structure of the system is stationary, sufficiently so that time has hammered economic logic into decision making. Nor does this degree of rationality imply unbounded calculative skills, only that the skills are a sufficient match for the task of the moment. But there is a problem under the surface, the problem of scarcity of means in relation to ends. Indeed, the implication is that reliable knowledge of means-end relationships is scarce; fundamental scarcity of this kind is a problem so why should not attempts be made to solve this problem and broaden the underpinning knowledge for economic action. The very idea of scarcity calls into question any notion that economic knowledge will be stationary. It suggests the most obvious of reasons for the emergence of novelty and it denies that novelty can ever be comprehended in terms of probabilistic risk.

Enter the entrepreneur, whose function is to exercise economic leadership not on the basis of prevailing knowledge but on the basis of conjecture that our reliable knowledge of the world can be rendered different. The entrepreneur thinks beyond experience, operates within the realm of unknowledge as George Shackle expressed it. The act of enterprise is sharply distinguished from routine management (perhaps too sharply from management in general) and sharply distinguished from invention. Enterprise is action that cannot be fully based on what is known, it requires decision in the face of ignorance[6]. In Schumpeter's scheme innovation is consciously and explicitly rational; the entrepreneur must calculate the consequences of his imagined conjectures without the support of past experience and act on the basis of answers that cannot be more than guesses. We need rationality in precisely those circumstances

---

6. Marshall (1919, 1920) provides a far more extensive treatment of managerial tasks and organization than does Schumpeter, and makes innovation one of the tasks that marks a good managerial team. Like Schumpeter he recognizes that there are leaders and followers when it comes to economic action but he also sees innovation as part of the daily routine. This is the continuity theme, the emphasis on the gradual and cumulative as contrasted with the discontinuous and entirely novel. But every innovation that has major transformative effects emerges not *de novo* but in terms of long sequences of gradual improvements as a design space is explored. There is less of a difference between the Marshallian and the Schumpeterian views than might otherwise be imagined. See Metcalfe (2007a & b) for further discussion.

where knowledge is absent, when action cannot be explained in terms of known principles. Rationality comes to the fore precisely when we cannot act as efficient automata, when decision requires not mere calculation but imagination of alternative course of action and their consequences. Just as the human capacity to imagine alternative economic worlds varies across individuals, so does their capacity for calculation. Since calculation is based on rules why should it be thought that these rules are equally within everyone's grasp? Why should entrepreneurs model the effect of their conjectures in the same way? The answer is that they do not and so Schumpeter's appeal to rationality is an explanation of the founding principles of economic variation, an invitation to inquire into the ways that entrepreneurial decisions are made in practice[7].

This capacity for economic leadership is rare, entrepreneurs are a special type, many more find it easier to follow than to venture and it is no surprise to find Schumpeter pointing to the hostility that awaits anyone who seeks to challenge the status quo. Entrepreneurial leadership is subversive action in the face of resistance; it is much more than the exercise of imagination *simpliciter* and it matters that the instituted economic frame is open so that the status quo can be challenged. This is not true of every society but is a peculiarly important feature of the institutions of capitalism that very few activities are rendered sacrosanct from the effects of innovation. The lesson to be drawn is the importance not of rationality itself but that rational thought underpins the diversity of possible courses of action. Rationality in this sense is certainly not the equivalent of Olympian perfect foresight shared in common but rather the highly local, differentiated and fallible understanding of what could be[8].

We begin now to see the link with the wider evolutionary frame. Entrepreneurs are different because they (rationally) believe differently and, while they may base these beliefs on differential knowing of technical possibilities, the fundamental point is that they perceive different economic possibilities and act on the possibilities[9]. This is why invention is not to be equated with innovation, or innovation with matters of physical technique alone. The test for an invention "is does it work?" and this test must be passed before any use of it as an innovation is possible. The test for an innovation

---

7. This is the terrain of the capabilities theory of the firm grounded in the work of Penrose (1959) and Nelson and Winter (1984).

8. While expectations are important in Schumpeter's scheme they are not the uniform expectations of modern macroeconomic discussion. How could they be? No entrepreneur hopes to make a profit by doing what the purported rivals do. As G.B. Richardson (1960) made clear in a different context, a profit opportunity expected by everyone is a profit opportunity available for no one.

9. Ulrich Witt (1998) has rightly insisted on the need for entrepreneurs to mobilize the contribution of others if they are to succeed.

is "is it profitable?" a quite different matter. As Schumpeter went on to express it in *Business Cycles*, most inventions never get off the ground as innovations and, of those that do, ninety percent are unprofitable failures (p. 117). So it is not the supply of inventions that is the rate determining constraint on economic development but rather the supply of innovations, that is to say, the rate at which economic novelty emerges. One is the domain of science and the human built world, the other is the quite different domain of economic action and its social context. Consequently, innovation requires an understanding of more than scientific and technological phenomena. An understanding of how to organise the production process and acquire the requisite inputs, an understanding of what customers will pay for, an understanding of how customers are to be made aware of the innovation, these are vital elements in any entrepreneur's scheme. We may note that deficiencies of knowledge in relation to the prospective customers in particular are perhaps the most frequent sources of failed innovation. To express it rather differently, there is much more to innovation than research and development and much more to research and development than science and technology.

That knowledge and its limitations are central to Schumpeter's vision is perhaps not surprising but what is more surprising is that this connects him not to Walras but to the classical line of thought from Smith and Marshall. By treating innovation in the context of the division of labour, Smith was drawing attention to the highly specialised, uneven nature of human knowing. Even though it may be convenient to call capitalist economies knowledge economies, it is more accurate to say that they are described by shared ignorance rather than by shared knowing. In a modern society, common knowledge is only a small part of the picture required for economic action. We have more in common with other people when we ask what it is that we do not know rather than ask what it is that we do know. All individuals are distinguished by knowing a great deal about a specific narrow sphere of human action and, as a consequence, being reliant on the knowing of others for their standard of life. Being so specialised we can function only in so far as we are connected to the functioning of others who we typically do not know. The economy takes on the property of an open, connected system precisely because human understanding is an open, connected system. The distributed, uneven nature of knowledge cum ignorance is the economic fact that provides the context for entrepreneurial imagination. Thus a world in which many individuals know many different things is a world in which innovative conjecture is likely to arise in many different contexts. Consequently, it is a matter of record that incumbents in a particular market are often surprised when innovations come from quite unanticipated directions and undermine their business model, or when the profitable market for an innovation turns out to be in a quite unanticipated domain. I don't think it at all accidental that Schumpeter emphasized the role of outsiders in the innovation process, it is a phenomena that is commonplace in business history and a deeper reflection of the nature of human ignorance and the barriers this may place to "perceiving the obvious". Schumpeter's scheme further implies that the possibilities for innovation are combinatorial and we might say that we are collectively rich in the economic sense precisely because we have learnt how to profit from our individual ignorance by building connections and interdependence into economic life.

If an economy is to function on the basis of distributed ignorance it is necessary that its individual actors are connected with the purpose of correlating their respective actions and connection requires organization. Whatever their specific form, organizations are humanly instituted devices for harnessing the division of labour in order to capitalise on the distribution of ignorance. Marshall too knew this, when he claimed that knowledge and organization are the most powerful of engines of production. It is only because we have invented the multiple and highly specialised forms of organization required to benefit from our differential knowing that we are able to benefit from the epistemic division of labour. All organizations serve as devices to channel and distribute information in order to coordinate decision making and goal achievement across a wide range of contexts. As Chester Barnard expressed it, "in an important aspect "organization" and "specialization" are synonyms" (Barnard, 1938: 136). That organizations are sets of instructions for coordinating decisions emphasizes their systemic origins in terms of parts and connections functioning within boundaries, and most organizations can fruitfully be thought of as organizations within organizations each with distinctive sub goals so there are matters of connection within and connection between to comprehend. While we typically think of such organizations in terms of firms, hospitals, schools, churches, government agencies and households it is important to recognise that markets are forms of organization too. They veer more towards the spontaneous and undirected forms of association as distinct from the planned and directed forms but they are nonetheless systems for transmitting information as to what is available, what is in demand and on what terms. Markets as specialised systems to correlate behaviors do what firms cannot do and vice versa they are complements not substitutes. Markets too are specific sets of instructions as to how, when and why to trade, rules that are normally generated by the market participants within the wider framework of the governing rules of capitalism. This brings us conveniently to Schumpeter's take on markets and competition as the complement to his treatment of innovation.

The competitive process

The essential point to grasp about *TED* is that the economic effects of innovation flow not from the innovation *per se* but from the adaptive response of the economy to the potential for change opened up by innovation. This process of adapting, of realizing the possibilities contained within any innovation is the domain of markets; they are the organizations that frame the selection side of the evolutionary process. Now Schumpeter's scheme is a market economy scheme and the significance of the market is the particular way in which it channels the process of adaptation. It is a process in which the price system is central, for it is the price system that generates the information not only to correlate choices but to decorrelate choices: innovations, *par excellence* are designed to decorrelate the prevailing economic order, to render different that which is already established. This is what is meant when we say that innovations are invaders of the status quo that they are instabilities rendered manifest. Of course, Schumpeter places his markets within the context of the institution of money and puts the banking and credit system, the money market, at the core of his theory of innovation. The instituting of money and credit gives important flexibility to the economy and through its non neutral effects on enterprise makes innovation the basis for competition. How does competition work?

Well, it is no dull matter of perfect competition, of market structure in an equilibrium circular flow. As in Marshall it is a matter of open competition, driven by differences between rival producers, it is the economic game as sport. Competition is contingent on variation and if innovation is the basis for economic differentiation then it is the competitive process that resolves those variations into economic development. The test for competition becomes not the number of competitors but rather that decorrelation of the status quo is taking place, whether there is an innovator taking business from its established rivals. At its core this is a matter of process of the rate at which the "new" displaces the "old" and these categories, as we have already noted, have no place in equilibrium worlds, it is the process of competition that imparts to economic evolution a velocity and a direction.

In Schumpeter's scheme, this process is deeply connected to the existence of the pure profits that attach to an innovation because of its superior productive characteristics; as such they are a category of realized economic return which is quite inconceivable in the equilibrium circular flow. But profits presuppose prices and the prices in question are the prices, including factor prices, that sustain the old technology, or more generally, the least effective of the productive alternatives that are available, just as Marshall taught. It is because the novel is evaluated economically in terms of the methods it will challenge that it is possible to conceive of profits as a surplus above contracted payments for inputs. Consequently, when the new has entirely displaced the old those profits will have disappeared and the price system will be adjusted in support the characteristics of the new technology. Profit is conditional and transient, it "has the most lamentable similarity with the drying up of a spring" (*TED*: 209). If it is to be sustained as an economic category, it can only be because of further innovation somewhere in the economy. Like all evolutionary processes, competition consumes the fuel that sustains it and, unless yet further innovations occur, competition comes to an end and with it economic development. This is the vision that emerges later in *Capitalism, Socialism and Democracy*, capitalism decays because the conditions for sustained innovation are undermined.

Schumpeter is less forthcoming than he might have been about the precise nature of this evolutionary process. His preference is to rely on imitation, once the innovator has pointed the way the less venturesome are induced to follow by the prospect of the profits in view. This is not unimportant, especially when we take account of the possibility that imitation itself may impose some new innovative twist, but it masks the real issue. This is the need to build productive capacity to produce with the innovation, investment is the core process and this is as true for the innovator as it is for the imitators. The link between prices, profits and investment behavior is the centre piece of Schumpeterian dynamics that is how he explains the uneven nature of economic development.

There is a further connection between Schumpeterian competition and the price system that merits brief discussion. The prevailing price system is not only the basis for the current innovator's profits it is also the value scheme against which future innovations are rationally judged. As the spread of innovation displaces and destroys old methods, it changes the terms on which future innovations will be possible. There is an inevitable historical dependency about such a process, even if production is conditioned by constant returns to scale the system as a whole operates with positive feedback.

How can one sum up the nature of Schumpeter's competitive process? It is that one cannot understand economic change solely in terms of movements in average behaviors. Competition is a matter of deviant behavior; it is the far from average outliers that drive the evolution of the system. Such a system is clearly not a system of equilibrium relations. What then is it?

The transience of the prevailing economic order

The fact that Schumpeter is dealing with an economy that is out of equilibrium does not mean that economic principles have ceased to be relevant. Quite the contrary, Schumpeter's world is not chaotic, it is strongly ordered by market forces but the order that ensues is not to be treated as an equilibrium structure. Equilibrium states are states or sequences of states that have exhausted all reasons to deviate from their jigsaw-like internal consistency, they cannot, by definition, bring into play further change from within, they can only reconfigure via the action of external forces. What Schumpeter is coming to terms with is a system that develops from within, an out of equilibrium system in which every pattern of economic order is transient and the problem is to uncover the rules that transform one order into its successor. Schumpeter gathers his sense of order from the Walrasian scheme in which, at each moment in time, preferences, technologies and the available resources interact to give coherence to economic action. The resultant order is caused, it has the inner logic of demand and supply relationships but, in the presence of innovation, it cannot endure. Investment and innovation provide the twin long period processes of self destruction.

Profits are the sign that the system is out of equilibrium but the issue runs much deeper than that and connects with the theory of economic knowledge discussed above. We have alluded already to the fact of scarcity as an economic problem but it is the dynamic significance of scarcity that underpins Schumpeter's scheme. Scarcity as a problem invites the search for solutions in the expectation that effective solutions will reap entrepreneurial profits. The solutions modify the pattern of scarcity but do not eliminate it; they only suggest new problems on which to work. Thus every solution changes the way in which future problems are posed and solved, not only because it alters relative prices but because it has also altered the prevailing pattern of understanding. Knowledge and ignorance are differently distributed after each innovation. This is why we cannot satisfactorily capture Schumpeter's thought in terms of exercises in comparative statics, enumerating the properties of a post-innovation equilibrium with the situation without the innovation. We are not dealing with a transition between fixed points but rather with a process of transformation that in its *movement* alters the knowledge that underpins the end point and the beginning state. Movement generates information, information revises beliefs and reliable knowledge and it is an irreversible process[10]. So all we ever have is the order of the prevailing moment and the forces of innovation and investment that seek to transform it. Indeed it is this insight that brings Schumpeter and Marshall much closer to one another than might otherwise be expected.

---

10. It is the process sketched in Marshall's (1920) Appendix H.

We are led to a striking contrast. The stability of the prevailing order in terms of the immediate solution to coordination problems of demand and supply is part and parcel of Schumpeter's scheme. As Peter Allen has emphasized, every such order is unstable in the sense that it is open to invasion by novelty in the form of innovation. If it were stable in this second sense, variety could not be generated and economic evolution would be impossible. How then to comprehend the rate and direction of economic change in an open system when we cannot specify points of rest independently of the path of movement? That is the question that Schumpeter poses, the deeper meaning of his insistence that change is taking place from within. This is the essence of his understanding of economic complexity in modern capitalism. It is not that we are dealing with systems of connected parts but that these systems inevitable have specific dynamic properties, properties that generate wealth from knowledge in schemes of self transformation in the presence of self organization.

Adaptation, growth rates and evolutionary change

Complexity, emergent novelty and evolution are naturally dynamic phenomena, they invites to think in terms of rates and directions of change. But novelty matters only in proportion to the impact it has in transforming the status quo and so, if the new is to displace the old, we have to explain why the new activities account for an increasing share of economic activity, that is to say, we have to explain why the growth in their utilization is faster than that of the methods they challenge.

As it turns out, modern evolutionary understanding has developed a series of ways to understand how systems respond to novelty, ways of comprehending how quickly evolution is taking place and in what directions. From the 1930s onwards evolutionary theory developed at a renewed pace as a deepening understanding of genetics was integrated into the structure of Darwin's theory of evolution in the natural world. Now economic evolution has nothing to do with evolutionary biology, it rarely helps to mention Darwin because it is not a Darwinian process. Evolution it is a mode of thinking in its own right and its characteristic feature is the variation cum selection logic that we call population dynamics. Innovations are the root source of the variations in economic behavior which are then selected for or against by specific economic processes. The evolutionary outcomes may or may not be progressive in a wider sense. Economic evolution reflects a different command of human knowing but, as we have seen, solutions merely lead to new problems, and we simply do not know, to take a topical example, whether our reliance on inanimate energy and the knowledge that underpins it will prove to be sustainable for our children's children. As in any open, emergent system the imponderables are too great, we have, as it were, entered into a Faustian bargain with knowledge and we cannot know where this leads. What will happen to the employment generating capacity of capitalism, how will the distribution of income, nationally and internationally develop are just two of those questions that should temper any discussion of progress. But they are not our concern here; rather our aim is to get to the fundamentals of the evolutionary dynamic, namely how variety drives adaptation.

Differential adaptive response takes us to changing structures and the facts of structural change are self evident; no economy has ever developed in the balanced proportional way that makes a macro economic analysis possible. The more we disaggregate the more we find the evidence for the persistent alteration in the relative importance or economic weight, of different goods and methods of production, dif-

Rank	Best Performing	Growth
1	Voice Over Internet Protocol Providers (VoIP)	See Note
2	Search Engines	1655.9%
3	e-Commerce & Online Auctions	468.9%
4	Online Dating & Matchmaking	248.8%
5	Tank & Armored Vehicle Manufacturing	244.7%
6	Petrochemical Manufacturing	221.2%
7	Mining Support	186.7%
8	Wireless Telecommunications Carriers	183.4%
9	Biotechnology	182.1%
10	Warehouse Clubs and Supercenters	146.5%

Worst Performing Industries In The Past Decade (2000-2009)

Rank	Worst Performing	Growth
1	Men's & Boys' Apparel Manufacturing	-89.1%
2	Clothing Accessories Manufacturing	-76.2%
3	Money Market & Other Banking	-73.3%
4	Broad Woven Fabric Mills	-72.7%
5	Women's & Girls' Apparel Manufacturing	-71.4%
6	Apparel Knitting Mills	-70.9%
7	Leather Tanning & Finishing	-70.0%
8	Manufactured Home Dealers	-67.4%
9	Circuit Board & Electronic Component Manufacturing	-63.9%
10	Recordable Media Manufacturing	-63.7%

Table 1 *Fitness Variations in the US Economy, 2000-2009.*
(Source: IBIS World, Inc, California, Santa Monica.)

ferent business units, different firms, different industries, and different economies. The motion is unceasing; it transforms our ways of life almost beyond recognition, so that successive generations live in increasingly different worlds. Even the most cursory understanding of economic and business history makes this plain. But to speak of structural adaptation is to speak in the same breath of differential growth. Moreover, in any modern economy the growth of some entities normally corresponds to the absolute decline of others, growth rates are negative as well as positive and the distributions of growth rates around some relevant population average are frequently quite remarkable. The following table provides an illustration taken for a recent decade in the US economy, they are value data that combine price and quantity movements but they are no less instructive for that. They provide a striking picture of very far from average rates of growth and decline. Of course, the very rapidly changing activities have to be a small part of an economy that expands in the round at single figure rate but it is obvious what changes in structure these data imply. They are Schumpeterian data (or Kuznetsian data) and they are easily replicated for different periods and different data sets, the phenomenon is pervasive it tells us that growth is never found without structural development (Kuznets, 1977).

If the characteristic feature of an evolving economy is a distribution of rates of growth, how is this to be treated by evolutionary theory? It is done by working in terms of populations of economic activities that are different in numerous dimensions but are acted on by common selective forces such that the growth rates of the different entities and the associated activities are different and mutually determined. The Schumpeterian connection is to link the different activities to particular business units producing distinctive goods with distinctive methods of production and to tie investment in capacity in those activities to differences in their profitability within the context of market processes. How this is done is an extremely important aspect of evolutionary economic explanation but, whatever its form, it always requires a prior step to understand the consequences of different growth rates for the structure of the relevant population.

To fix ideas let the businesses in a given population be producing a uniform good, so there is no ambiguity about what we mean by the relative importance of the businesses in terms of their shares in aggregate production. Each of them has in the past innovated in Schumpeterian fashion and so operates a distinctive method for producing the good. At the ruling prices each business has a different level of profitability according to the characteristics of its methods. If all businesses invest the same proportion of their profits in expansion then the firms will have correspondingly different growth rates. It is simply a matter of arithmetic that the proportional rates of change of the output shares are equal to the difference of each growth rate from the industry average growth rate. A business that grows at the average rate maintains a constant share, faster growing business units increase their share and conversely for slower growing units. It is because structural change depends on deviations from average growth that we are led unavoidably to the idea of a replicator dynamic process and to what Marshall meant by economic flux. This is the variation cum selection logic that is embodied in the Fisher/Price principles of evolutionary dynamics.

Fisher/Price dynamics

The starting point is elementary; the growth rate of the population is a weighted average of the growth rates of the individual members, the weights being the shares in population output of each business. How does the average growth rate of a population change over time? Since the average is defined by, $g_s = \sum s_i g_i$, the values, s_i, being the output shares and the values, g_i, being the firm's (exponential) growth rates, it follows that,

$$\frac{dg_s}{dt} \equiv \sum \frac{ds_i}{dt} \cdot g_i + \sum s_i \cdot \frac{dg_i}{dt}$$

CHANGE IN POPULATION AVERAGE ≡ FISHER EFFECT + PRICE EFFECT.

In modern evolutionary theory the first effect is known as the Fisher effect, the consequence of the selection process, and the second effect is the Price effect, the consequence in this case of changes (or innovations) in the individual firm growth rates (Anderson, 2009; Andersen & Holm, 2012; Hodgson & Knudsen, 2010; Frank, 1998, Okasha, 2006; Metcalfe, 2008). Elaborating further, it follows as a matter of definition from the replicator principle that:

$$\frac{ds_i}{dt} \equiv s_i (g_i - g_s)$$

Whence, we can write the Fisher/Price formula as:

$$\frac{dg_s}{dt} \equiv V_s(g_i) + E_s\left(\frac{dg_i}{dt}\right)$$

The first term, the Fisher term, is the variance in the growth rates across the population; the second term, the Price term, is the mathematical expectation of the acceleration or deceleration in those growth rates. This statistical structure runs through all variation cum selection-cum innovation models of economic evolution, whatever may be the underlying explanation of the differences in the growth rates. It tells us that the direction and velocity of change is conditional on the present variety in growth rates contained within the population. Let us take some limiting cases to illustrate.

If, for example, some further underlying process maintains constant the average population growth rate then it would follow directly that:

$$E_s\left(\frac{dg_i}{dt}\right) \equiv -V_s(g_i)$$

Even though the aggregate growth rate is fixed, selection still operates, so the average of the individual growth rates cannot then be constant but must decline at an average rate equal to their variance. This has long been known in heterodox growth theory as the retardation principle, first enunciated by Kuznets and Burns in the 1930s. Consequently, when the aggregate market growth rate is constant, for example, the average growth rate of the population of business cannot be constant[11]. We might also reflect that retardation in rates of growth connects us to logistic and related processes and to the patterns of structural change that are so frequently uncovered in studies of innovation diffusion. It should not be lost on the reader that logistic processes play a significant role in evolutionary thought more generally (Lotka, 1924/1956)

If instead, all the business growth rates are held fixed by some further extraneous argument, the Price term disappears and we have Fisher's fundamental theorem

$$\frac{dg_s}{dt} \equiv V_s(g_i)$$

The average growth rate (what Fisher calls average fitness) increases over time even though the growth rates of all the businesses in the population are constant. So far this is mere tautology; it all follows from the definition of output shares and (exponential) growth rates. In this sense the Fisher/Price principles play the same role in evolutionary theory that Harrod's fundamental identity plays in modern economic growth theory, that is to say, it becomes the basis for deeper explanation.

What is perhaps less fully appreciated is that the Fisher/Price logic applies to the higher moments and co-moments of the population distribution. Thus, for example, with respect to the variance of growth rates it is the case that:

$$\frac{dV_s(g)}{dt} \equiv S_s(g)^3 + 2C_s\left(g_i - g_s, \frac{dg_i}{dt}\right)$$

---

11. Which, depending on the specifics, means that average profitability of the businesses cannot be constant.

The Fisher effect reduces to the third moment about the population mean and the Price effect to the covariance between the rates of change in the growth rates and the first order deviations of the growth rates about the population mean. As we move to the change in higher moments the same logic applies but instead of working in terms of moments it is more instructive to work in terms of the cumulants of the distribution of growth rates, so that the general rule for the nth cumulant, $\kappa_s(g)^n$ can be expressed as:

$$\frac{d\kappa_s(g)^n}{dx} = \kappa_s(g)^{n-1} + n \cdot C_s\left[(g_i - g_s)^{n-1}, \frac{dg_i}{dt}\right]$$

Since the first three cumulants correspond to the first three moments about the mean, we can work in terms of moments for these lower orders of change. But for the higher orders the cumulant formula is far more compact and direct. The rate of change of any cumulant is proportional to the value of its immediate predecessor in the chain, the Fisher effect, while the Price component is equal to the covariance between the changes in the growth rates and the deviations of the growth rates around the when mean raised to the power, $n - 1$.

The Fisher/Price structure permeates all the possible instantiations of the variation cum selection dynamic with degrees of sophistication that are at the analyst's command. The central lesson is that the economic system changes because of the variety that is contained within it, that homogeneity, uniformity are the antithesis of evolution. Its economic content then depends on the manner in which we explain the differences in growth rates in terms of the underlying variation in various selective characteristics of the business units, namely those characteristics that appertain to the generation of profit and the disposal of profit, whether in investment in capacity or investment in further innovation. Variety cascades into variety so the approach is naturally operative at multiple levels but always dependent on an underlying theory of order to generate the prices and costs on which profitability depends. In the wider scheme, the differential profitability of firms is linked not only to their different methods of production but to the different qualities of goods that they produce, to the nature of the goods and factor markets in which they operate and in particular to the inducements to invest that mean that firms with the same profitability may not grow at the same rate nor invest in innovation at the same rate. Ultimately it is to the differences in the innovation record of businesses that the pattern of evolution is traced. The distribution of novelty generation is the vital connection between complexity and economic evolution.

Reprise

I have suggested that the modern evolutionary agenda and of its foundations in Schumpeter's great work contains a very important lesson for the analysis of complex evolving systems, namely, the importance of deviant behavior. Evolution is not a product of average behaviors but a product of the outliers that are distant from the prevailing averages. Evolution means differential growth, the logic of the Fisher/Price principles but differential growth is to be found in the openness of the economic system to invasion by novelty and the way the economic system values and rewards the creation of novelty. Capitalism is an open evolving system because the knowledge on which it is based constitutes an open evolving system, but how creative it can be depends on the openness of its operating rules. It need not always be an open system as Schumpeter realized when he wrote *Capitalism Socialism and Democracy*.

It is of no service to Schumpeter's legacy to suggest that he solved the problem of complexity and economic evolution for he did not. As with Marshall the conceptual tools were not available to him but that does not matter at all. What he did do was point economic reasoning down an evolutionary and complexity path, a path which is only partly trod a hundred years on. This is also the path that Peter Allen's work has richly illuminated.

References

Allen, P M. (1976). "Evolutionary population dynamics and stability," *Proceedings of the National Academy of Sciences*, ISSN 0027-8424, 73(3): 665-668.

Allen, P.M. (2001). "Complex systems approach to learning in adaptive networks," *International Journal of Innovation Management*, ISSN 0219-8770, 5: 149-180.

Allen P.M, and McGlade, J.M. (1987). "Modelling complex human systems: A fisheries example," *European Journal of Operational Research*, ISSN 0377-2217, 30(2): 147-167.

Andersen, E. (2011). *Joseph A Schumpeter's Evolution: A Theory of Social and Economic Evolution*, ISBN 9781403996275.

Andersen, E., and Holm, J.R. (2012). "Extensions of Price's equation for evolutionary economics: The cases of directional, stabilizing and disruptive selection," paper prepared for the Schumpeter Society Conference, Brisbane, July, 2012. Mimeo, Dept of Business and Management, Aalborg University.

Arthur, B. (2013). "Complexity economics: A different framework for economic thought," *SFI Working Paper 2013-04-012*. Santa Fe Institute.

Barnard, C.I. (1938). *The Functions of the Executive*, ISBN 978-0674328037 (2001).

Burns, A.F. (1934). Production Trends in the US Since 1870, Boston, NBER,

Dopfer, K., Foster, J. and Potts, J. (2004). "Micro-meso-macro," *Journal of Evolutionary Economics*, ISSN 0936-9937, 14(3): 263-280.

Dodgson, M. (2011). "Exploring new combinations in innovation and entrepreneurship: Social networks, Schumpeter and the case of Josiah Wedgewood," *Industrial and Corporate Change*, ISSN 0960-6491, 20(4): 1119-1152.

Dopfer, K., and Potts, J. (2008), *The General Theory of Economic Evolution*, ISBN 0415279437.

Frank, S.A. (1998). *Foundations of Social Evolution*, ISBN 0691059349.

Hodgson. G., and Knudsen, T. (2010). *Darwin's Conjecture: The Search for General Principles of Social and Economic Evolution*, ISBN 0226346900.

Kaldor, N. (1985). *Economics Without Equilibrium*, ISBN 087332336X.

Knight, F. (1935). "Statics and dynamics," reprinted in Knight F., 1977, *The Ethics of Competition*, ISBN 1560009551.

Kuznets, S. (1977). "'Two centuries of economic growth: Reflections on US Experience,'" *American Economic Review*, ISSN 0002-8282, 67: 1-14.

Lotka, A.J. (1924). *Elements of Mathematical Biology*, New York, Dover Publications.

Marshall, A. (1919). *Industry and Trade*, London, Macmillan.

Marshall, A. (1920). *Principles of Economics* (8th edition), London, Macmillan.

Metcalfe, J.S. (2007a). "Alfred Marshall and the general theory of evolutionary economics," *History of Economic Ideas*, ISSN 1122-8792, 15(1): 81-110.

Metcalfe, J.S. (2007b). "Alfred Marshall's Mecca: Reconciling the theories of value and development", *Economic Record*, ISSN 0013-0249, 83(Supplement): 1-32.

J.S. Metcalfe, (2008). "Accounting for economic evolution: Fitness and the population method," *Journal of Bioeconomics*, ISSN 1387-6996, 10: 23-50.

Murman, P. (2003). *Knowledge and Competitive Advantage*, ISBN 0521684153 (2006).

Nelson R.R. and Winter, S. (1982). *An Evolutionary Theory of Economic* Change, ISBN 0674272285.

Okashi, S. (2006). *Evolution and the Levels of Selection*, ISBN 0199267979.

Penrose, E. (1959) *The Theory of the Growth of the Firm*, Oxford, Blackwell.

Richardson, G.B. (1960). *Information and Investment*, Oxford, Oxford University Press.

Shove, G. (1942). "The place of Marshall's *principles*, in the development of economic theory," *Economic Journal*, 52: 294-329.

Schumpeter, J.A. (1912). *The Theory of Economic Development*, Oxford, Galaxy Books (1934)

Schumpeter, J.A. (1928). "The Instability of Capitalism," Economic Journal, 38: 361-386.

Schumpeter, J.A. (1939). *Business Cycles: Volumes I&II*, New York, McGraw Hill.

Schumpeter, J.A. (1941). "Alfred Marshall's *Principles:* A Semi-Centennial Appraisal," *American Economic Review*, ISSN 0002-8282, 31: 236-248.

Schumpeter, J.A. (1943). *Capitalism, Socialism and Democracy*, London, George Allen and Unwin.

Schumpeter, J.A., (1947a). "The creative response in economic history," *Journal of Economic History*, 7(2): 149-159.

Schumpeter, J.A. (1947b). "Theoretical problems of economic growth," *Journal of Economic History*, 7: 1-9.

Smith, A. (1776). *The Wealth of Nations*, ISBN 0865970076 (1994).

Witt, U. (1998). "Imagination and leadership: The neglected dimension of an evolutionary theory of the firm," *Journal of Economic Behavior and Organization*, ISSN 0167-2681, 35:161-177.

Stan Metcalfe is Emeritus Professor at Manchester Institute of Innovation Research (MIoIR). He was Stanley Jevons Professor of Political Economy and Cobden Lecturer at the University of Manchester between 1994 and 2008. During his career he has lectured at the Universities of Manchester and Liverpool, and has held visiting positions at Cambridge University, the University of New England, Queensland University and Curtin University. He has been actively involved in the development of science and technology policy in the UK, being a member first of ACARD and subsequently ACOST. He was until 1996 a member of the Monopolies and Mergers Commission.

Chapter Λ

Innovation systems, economic systems, complexity and development policy

Norman Clark

This chapter summarizes how Peter Allen's pioneering work on building computerized models of complex economic systems has helped to generate new approaches to development possibilities for very poor countries. Using the example of Senegal he and colleagues built an interactive model designed to facilitate discourse among diverse relevant professional groups, thereby facilitating much needed cross-disciplinary dialogue. The chapter goes on to describe how such a model is consistent with modern approaches that focus on the use of "innovation systems" in understanding processes associated with the analysis of technology development in very poor economies. The case of a recent British aid programme is used to illustrate the key properties involved.

Introduction

Innovation systems analysis as an aid to development in poor countries is comparatively new though its use is consistent with classical general systems theory which has a much longer pedigree. The science of general systems theory has been explored in some detail by a series of authors starting with von Bertalanffy in the pre-war period, through Lotka, Koestler, Ashby, Emery and others in the post-war period and then adapted by Allen, Checkland, Holling and others in the 80s and 90s to explore the evolutionary properties of natural and social systems (such as urban development and fisheries). The introduction of the "innovation systems" notion was proposed by Freeman in the mid-1980s to help explain differential GDP growth patterns that could not be explained by investment activity across countries and adapted by Hall and others to a specific underdevelopment agenda over the past decade or so.[1]

---

1. A good general reference for the early material is contained in Emery (1970) which contains a number of classical papers. The application to management of innovation in socioeconomic and natural systems is comprehensively covered in Clark and Juma (1992/2013). A summary of use of innovation systems as applied to rural technology development is summarized in Hall *et al.* (2003) and in World

This chapter will take as its starting point Peter Allen's attempts to build evolutionary models to aid in the practical process of development policy. Indeed much of his contribution has centred on the evolutionary behavior of natural and socioeconomic systems. Really for the first time and drawing on the work of his mentor, Prigogine, he has shown how it is possible to use interdisciplinary dynamical models as an aid to understanding the properties of such systems and how this understanding may be used in system intervention. His earlier work on the Senegal model and later work on complex systems theory was further developed in a series of more recent papers[2]. This chapter will try to link much of this thinking to the notion of an "innovation system", which has increased presence in development discourse, particularly in relation to possibilities for poverty reduction in very poor countries. I shall call these the least developed countries (LDCs) typically enjoying incomes per head of less than some $10 per month. While the idea of innovation as a systemic phenomenon is comparatively recent and still not fully understood or accepted by development analysts, nevertheless it is very much on the current agenda. My objective will be to show how the ideas it encapsulates are very close to what Allen has been arguing for some time. In particular it opens the way to a fresh exploration of how economic change takes place and how related public policy interventions might be better managed.

I shall start by summarizing Allen's original Senegal model[3] as an attempt to put some realism behind long term economic development policy. In so doing, however, it revealed what many had come to realize, namely that specifying the behavior of evolutionary systems in the long term is an unduly complex activity. Although Allen's aim was always that the model should be used mainly as a planning tool designed to integrate policy discussion across professional groups its impact has been much weaker than hoped. Section "Innovation systems" goes on to explore how the notion of an innovation system is very much a tool that has its roots in evolutionary complexity albeit that its use is more qualitative. Nevertheless it can perhaps be used productively from an aid policy perspective. The next section "Research into Use Programme" explores these ideas with a specific case, the DFID funded "research into use" project while the final section draws some general conclusions.

Bank (2006). See also Freeman (1991) and other references cited at the end of the text.
2. See for example Allen (2009), Allen and Vargis(2009).
3. The Senegal model itself is described in some detail in IERC (1991) and Clark, Perez-Trejo and Allen (1995).

Economic systems as complex systems

In the development of the Senegal model Allen took as read the proposition that standard economic models could not by their very nature be useful guides for development policy. Instead what was needed was a tool that would focus on long term behavior that would at the same time capture knowledge contributions from all relevant professional disciplines. But to do this he returned implicitly to basic neoclassical economics in the form of the circular flow of income. Here the macro economy is conventionally divided up into two sub-systems: viz (i) the production system (P) which transforms resources (inputs) into commodities (outputs) and (ii) the household system (H) which owns the means of production (natural resources, labour and capital), earns income (Y) from them by selling them to the production system and (finally) uses this income to purchased the commodities (C) produced by that system. Where both the P and H sectors spend all their incomes accruing to them over a specific time period, the macroeconomic system is a closed one in which all resources are conserved. Graphically these relationships may be portrayed in terms of the well known circular flow of income diagram which is often then modified to include specific categories of "systemic openness", viz. foreign trade, government activity, and disturbance arising out of the behavior of savers and investors.

In the first case households are taxed (T) and the state uses these resources to spend resources (G) on behalf of the consumer. In the second case households spend incomes on imports (M) which are of course produced by other (overseas) economic systems, and in order to pay for these our economic system has to export (X) commodities of an equivalent aggregate value. In the third case households save (S) part of their incomes and these savings are used (via the capital market) for the purchase of investment goods (I) by the productive sector. Equilibrium is established where

$$Y + T + M + S = C + G + X + I \qquad (1)$$

but since the economic agents are not identical there is a constant tendency to instability mediated by negative feedback mechanisms (mainly price). The theory of macroeconomics is concerned largely with the determinants of economic behavior on the part of the agents involved and with the conditions under which the macro system will achieve and maintain equilibrium consistent with full utilization of available resources. Where equilibrium does not obtain, macroeconomic analysis (and controversy) focuses on what measures need to be taken to re-achieve it, for example through monetary, fiscal and foreign trade policy. The constituent elements of the H & P sectors (i.e., the individual households and firms) represent the "microstates" of the wider macro system. As with the macro economy these are analyzed in relation to their mutual interactions, i.e., as exchanges of commodities and resources at prices which broadly reflect the forces of supply and demand. The economic study of these individual market relationships is called microeconomics and tells us how the market for any particular commodity (or resource) behaves, what properties it has and how efficiently it does its job. It also tells us something about the behavior of the agents involved - how much, for example, the firms in any specific industry are able to monopolize production and thereby raise prices above marginal costs of production.

What economic analysis does not do, however, is provide any direct insight into how the behavior of microeconomic agents affects the macroeconomic context. In particular there are important questions about how the (macro) economic system evolves. How, for example, does its capacity to produce goods and service increase? Or how does its structure change? How do new industries arise and how does new technology impinge on the economic fabric? The answer to these and many similar questions is, sadly, that the apparatus of economic analysis, at least in its pure (or "mainstream") form, has little to say since most of its attempts to model the behavior of economic systems have tended to be unduly mechanistic. It is here that Allen made a key contribution through the development of a regional dynamical model for Senegal. In contrast he portrays complementary long term behavior as migratory behavior, on the part of households, and investment behavior, on the part of firms—in each case across regions or zones into which the country has been divided. Such behavior is clearly strategic in nature since it is only normally undertaken with a view to payoffs expected to accrue over many years, and may thus be distinguished from the shorter term "optimizing" behavior of conventional economic analysis.

The Senegal Model was developed as a decision tool for policy makers in Senegal. It consisted of sets of interacting equations each of which represents the change in some characteristic variable occurring at a particular time and place. This could be, for example, the change taking place in a zone (or region), of the resident population, in an economic or subsistence activity, in the amount and quality of the soil there, or in the water availability at the location.[4] These changes occur because of the existence of processes, events and mechanisms operating at the different locations within the system. For example, population change in any region occurs because of births, deaths, and in-and-out migration, and similarly water which is used flows into and out of the region, and may also be produced there by wells and springs. Each of these terms reflects the rate of occurrence of different processes and individual events, of which

---

4. For a detailed discussion of this type of modelling procedure, see for example, Allen and Sanglier (1979; 1981).

there may be different types, and this rate of occurrence may itself be influenced both by the values of internal variables (e.g., the size of the population, the area of some crop) and parameters (e.g., soil fertility, the fecundity of the local population), as well as by exogenous parameters such as rainfall, river flow, and migration.

Each equation in the model is made up of terms representing the effects of such mechanisms and processes. Most of the equations fall into two main groups corresponding to the types of agents whose behavior is fundamental to spatial evolution— (i) the households whose migratory behavior determines population movements (ii) the producers whose behavior determines the nature and extent of economic activity in each region and the flow of resources between regions. When fully developed it was expected the model would also include a third group of equations concerning changes in ecological populations, e.g., of water availability, of soil and minerals, and of noxious deposits. The Senegal model and its underlying principles has made some impression on public policy interventions but in development terms has not made make much impact both because of its underlying complexity and because it use cannot be understood within standard macroeconomic modelling.

Innovation systems

An alternative way of looking at public policy for development from a systems standpoint is to go back to analysts like Bob Rosen who contributed significantly to the understanding of the mechanics of complex systems.[5] Rosen was a systems ecologist whose original background lay in the analysis of complex biological systems, but who also tried to show that similar metaphors are applicable to the dynamical behavior of socioeconomic systems as well. He took the view that conventional science works best with physical systems or biological systems that are reasonably stable. Conversely in the case of living systems (ecologies) that are creative, and as a result experience evolution of internal structure, we need an experimental approach that allows us to capture the inevitable indeterminism of their behavior[6]. Such an approach needs above all to avoid constraining the analysis of the system under investigation by presuppositions about its "true nature" and in particular, that of imagining that it can be modelled deterministically.

In formal terms Rosen (1987) put the issue in terms of ecological creativity. Imagine an ecosystem in which there are n species each having either a predator-prey or a

---

5. See also Kline (1985) who was writing about the non-linear properties of innovation systems at roughly the same time.
6. See also Allen (2008).

symbiotic relationship with its fellows, and we wish to construct some kind of model of that system. The first conclusion we shall immediately come to is that to model such an ecosystem in terms of deterministic relationships amongst its constituent species, would not be a useful exercise. Rosen showed that this is because determinate systems in general are merely limit cases of a wider class of complex system whose relationships are informational but not exact. For example, in a predator-prey model the relationship between predator and prey can be pictured as the quantity u_{ij} where:

$$u_{ij}(x_1......x_n) = d/dx_j(dx_i/dt) \qquad (2)$$

x_i, x_j are examples of various species $x1, x2 ....xn$ coexisting in a given ecological space and t = time. The uij represent values giving the rate of change of the production of x_i as a result of a change in the production of x_j over time. If the two species are symbiotic u_{ij} will be positive, but if x_j is a predator on x_i then u_{ij} will be negative. Rosen showed that the equation set (2) is indeterminate—it cannot be solved for specific values of the u_{ij}. However, under specific limiting conditions (2) can approximate to a conventional dynamical system which takes the form:

$$dx_i/dt = f_i(x_1, x_2, x_3, ... x_n) \qquad (3)$$

and is then a determinate system. However, no living system could survive if it behaved according to the relationships outlined in (3), simply because unpredictable and creative behavior would be impossible by definition. All species would be wiped out. The point is that for a system to be alive and to evolve creatively, its behavior must be relatively indeterminate. It follows that to try to model it deterministically is tantamount to turning it from a living system into a dead system, and conversely that we need to develop models which somehow capture the essential creativity of living systems. Now if this is true with natural ecologies it must be even more the case with socioeconomic systems since these are both more complex and subject to much more rapid evolution. In addition whereas the "information" relevant to predator-prey relationships is relatively simple (species x_i very quickly learns whether species x_j is friend or foe and can take appropriate action), that for creative socioeconomic systems is infinitely more complex.[7] That being the case it is likely that the capacity to assimilate, process and use relevant information in a rapidly changing context will be crucial in any development context.

But how is this to be done? In another article Rosen attacked the problem by focusing on the lack of clarity (fuzziness) surrounding many of the concepts associated

---

7. See Clark (2002) for a detailed discussion of this point in relation to underdevelopment.

with development policy. He argued that not only have analysts no clear idea about what constitutes "development", they are often actually incompatible and contradictory. In fact "...even among those who happen to share the same views as to the ends of development, there are similarly incompatible views as to the means by which the ends can be attained"[8]. But, he went on, if "so many distinct and contradictory views ... can be held by (so many) able people ... (then) ... a first step in dealing with such concepts is to try to identify and remove the source of the fuzziness"[9]. And an important source in his view is that different analysts "live intellectually" in a variety of different "analytical worlds" where the underlying concepts used often have different meanings to each analyst. Sometimes, unfortunately, they are not fully understood by the analysts themselves even in terms of their own "world". However, if we begin to separate out these worlds and characterise them unambiguously, then we shall have made an important first step in clearing up much of the confusion.

Rosen's solution was to abstract from the Newtonian mechanics which underlie most models of system behavior (including system dynamics), and in particular to introduce the notion of "anticipation". This is done by imagining a model world in which time trajectories are allowed to move faster than "real time", but where the real system (S) and the model system (M), are coupled in a policy sense through an effecter system (E). He then formalized the analysis by allowing informational feedback from E to M depending upon whether the trajectory of M is held to be "desirable" or not. Notice that since S + M + E is in total an anticipatory system (S*), the M trajectory will always tend to forecast that of S, although the forecast will never be perfect unless M is a perfect model. We know of course that, that can never be the case because S is an evolutionary system in the strict sense that the future is unknown. Now a moment's thought will show that there are therefore an infinite number of possible planning models to choose from, since every analyst will have his/her own "world" in mind in relation to the "real system" that happens to be under investigation. Each analyst, or perhaps it is better to concentrate on each analytical ideology, has his/her (its) own social construction of reality which guides the policy questions asked and the answers that are ideologically acceptable. Rosen's argument was that at least the recognition of this fact should help to clear (some of) the intellectual baggage that obscures the policy process.

We all know of course examples of such ideological confusion. Economists are well known for seeing any particular issue in terms of resource allocation and the price mechanism. What usually differentiates them from a policy standpoint is the position

---

8. See Rosen (1974: 245).
9. Ibid, p. 246.

they take up on market failure in relation to that issue. Those on the right tend to believe that a market solution is the better option, whereas those on the left will tend to favour intervention on the part of some central authority to correct market failure. Their worlds are predetermined in this narrow, ideological sense. Similar differentiation occurs across disciplines. For example on issues of environmental degradation, physical geographers will not see the problem in terms of the efficiency of the price system but will tend rather to concentrate on purely physical processes such as climatic changes and soil and water stress. Political scientists will eschew both in favour of an analysis concentrating upon power structures. Finally it should be noted that such social construction goes well beyond intra- and inter-disciplinary battles, but relates more widely to strongly held views on the part of many powerful and highly motivated pressure groups. As Thompson (1993) has pointed out in graphic detail, the passion with which rival adherents typically cling to their respective worlds can potentially waste many billions of dollars. It is in this profound sense that policy analysis is not a costless exercise.

From my point of view, however, there are two important inferences to be drawn from this discussion:

1. If development policy is to be successful there has to a way of allowing permanent model revision. Not to do so (and therefore to rely on one timeless model) is to open the door once again to confusion amongst the analysts (and even more so, those further down the hierarchy) whose worlds will continue to vary and to evolve even though the model itself does not.
2. Information is a crucial component of the whole exercise since necessary (and continuous) model revision will depend on the accuracy, speed and general efficiency of information flows from S to E to M, and vice versa. And it is for this reason that "information search", sometimes called "policy research", has become a significant factor in modern development analysis.

Allen of course was very well aware of Rosen's argument. Indeed the Senegal model was an attempt to build a facilitating tool that would precisely build bridges across professional disciplines. The model's failure in this respect was due in my view probably to the problem that many disciplines have in relating to evolutionary dynamics. It is hard to relate to such dynamics when your world view is Newtonian. It is here, however, that the notion of an innovation system can prove useful. By thinking of technological change as the result of systemic use of information from many sources it is possible to craft policy interventions in an interdisciplinary and cross-professional way. Also interventions can be used crucially as vehicles for learning by ensuring that they are part of a design that promotes comparative analysis across projects and continuous feedback throughout the life of each project. It is this approach that has guided the case study outlined below, the Research into Use Programme (RIU) funded by the UK's Department for International Development (DFID). The RIU was completed in December 2012 and provides an excellent example of the qualitative use of systems principles to inform the conduct of public policy.[10]

Research into use programme

In 2006 DFID published its third white paper[11], setting out the UK government's policies for eliminating poverty worldwide. The white paper was preceded in 2005 by DFID's Agriculture Policy Paper on Growth and Poverty Reduction, which focused on promoting growth in this all-important sector through the spread of new technologies. This paper also outlined DFID's commitment to enhancing the resilience of farming households to external shocks such as drought or disease, which can plunge already vulnerable households into deeper poverty. It emphasized the need to improve access of poor people to knowledge and technology, through both public- and private-sector institutions. In the same year DFID's Central Research Department (CRD)[12] published its Research Funding Framework (2005-2007) which identified agriculture as one of its priorities. Subsequently, a Strategy for Research on Sustainable Agriculture (SRSA) was prepared. The Research into Use (RIU) programme became the first to be developed under this new strategy. It proposed to contribute to this objective by adopting a pro-poor innovation systems approach to getting research into use and to increase the understanding of how this is done.

The research that would be "put into use" derived from projects funded under DFID's Natural Resources Research Strategy (RNRRS), which consisted of some 1600 projects running from 1995 to 2006. The focus of RNRRS had been to improve the livelihoods of the poor through better management of natural resources. The ten research programmes launched under the strategy were designed to generate new knowledge and promote its uptake and application. They addressed the needs of people living in a range of agro-ecologies including semi-arid areas, high-potential areas, highlands and tropical moist forests; and those at the forest/farm, land/water and rural/urban interfaces. The breadth of projects reflected the multiple routes by which research can have an impact on poverty. The RNRRS saw significant evolution over its life. This in-

---
10. For those interested in looking more closely at the RIU a good source is Clark *et al.* (2013).
11. DFID (July 2006) Eliminating World Poverty: Making Governance Work for the Poor.
12. Now re-named as the Research and Evidence Division (RED).

cluded a shift in focus from generating research and producing scientific publications to emphasizing the impact of research on poverty. The focus also moved from outputs to outcomes and long-term impacts. At the same time, interdisciplinary research, the policy environment and the livelihoods of the poor began to receive greater attention. One of the most influential legacies of the RNRRS was the use of innovation system principles in the development of new partnerships, products, processes, markets, institutions and organizations that are better equipped to put research into use.

In the context of RIU, innovation has meant the use of new ideas, new technologies or new ways (processes) of doing things in a place or by people where they have not been used before. Of course innovations in this sense have been happening for millennia by the actions of rural communities themselves, but the intense pace of global change and the threats from climate change and environmental degradation means the poor in developing countries must rapidly adapt just to cope. Innovation, meaning the use of new knowledge, often involves working with and re-working the existing stock of knowledge. It is often the key to building better and more sustainable livelihoods, because new knowledge is required to deal with the rapidly changing environments that face farmers and other rural people. It often involves local creative imitation and adaptation, rather than the development of something radically new. It is usually achieved through many small improvements (e.g., in production technologies, processing, and institutions) rather than through a few big sweeping changes. And it involves greater ownership of the process by poor people themselves.

The RIU approach was therefore been to shift the focus of attention away from the important tasks involved in the generation of new knowledge to the way in which that knowledge can be put to productive use. An innovation system is usually seen as a network of organizations and individuals involved in generating, modifying, and using new knowledge. The networks might be national, sub-national, regional or international. They comprise not only the users of the knowledge (farmers, consumers, artisans, labourers and traders) and the producers of new knowledge (researchers) but a host of intermediary organizations including extension workers, NGOs, enterprises in the supply chain, credit agencies and government. This systems approach considers not only the totality of the entire research, development and extension spectrum, but also the institutions, systems of production, and social relations in which these activities take place.

In practice the RIU has been focused on south Asia and sub-Saharan Africa. It started by carrying out a series of country assessments to match local demand in selected countries to the supply of RNRRS derived technologies. These assessments were conducted as an interactive activity with local stakeholder and government groups. On this basis a range of project interventions were chosen according to the following broad criteria:

1. Links to earlier RNRRS research networks whose tacit knowledge would aid in relevant development;
2. Likely impact on incomes for poorer sections of populations;
3. Likely impact on improving gender balance on these groups;
4. Likely impact on employment creation;
5. Ability of projects to foster the creation of networks that cut across professional hierarchies;
6. Capacity of projects to attract private sector interests;
7. Capacity of projects to foster local innovative capabilities, and;
8. Capacity of projects to be sustainable after withdrawal of DFID aid.

There followed a phase of establishing national teams to scale out chosen technologies and procedures for enabling policy dialogue with government and other related agencies. Careful attention was also given to proper accounting procedures and other due diligence requirements so that technological "scale-out" could be carried out efficiently and effectively. However, two fundamental requirements for the whole programme relate to the points above derived from Rosen's anticipatory model; that is the need for procedures to enable continuous project revision as projects unfold and the handling of new information and knowledge as it impinges on project development. For key properties of the RIU have been those of action learning and capacity building. These properties were handled by treating all projects also has research and learning projects run by a team of innovation research specialists[13]. This team carried out the following functions:

1. Gathering immediate quantitative and qualitative data on the current state of projects;
2. Conducting regular (monthly) workshops with local teams;
3. Collating regular data on project progress;
4. Assisting in in-country interaction with local and regional policy groupings;
5. Providing regular monitoring reports to RIU management and advisory staff;

---

13. This team was led by Andy Hall, at that time senior research fellow at UNU/MERIT, University of Maastricht, Netherlands, and visiting research professor at the Open University, UK.

6. Presenting periodic review papers at 6-monthly RIU workshops where all projects were expected to present and discuss progress;
7. Playing a key role in any decisions to be made on adaptation/modification of project activity, and;
8. Interacting closely with RIU advisers and management on the preparation of working papers and other outputs.

It was expected that in this way not only would the RIU as a whole learn about how better to put research into use but at the same it would effectively mentor local staff such that by the end of projects they would a good position to ensure sustainability and be more autonomous in the development of related policy and practice. Finally it is worth emphasizing the principles of complex systems thinking that guided the programme as follows:

- In all countries the projects were expected to run as discrete "innovation systems." Such systems have both an "economic" and a "knowledge" system with flows of resources and information taking place among their component nodes and across their respective boundaries. The resource flows comprise finance, materials and labour inputs.
- These knowledge flows would include formal and tacit knowledge associated with the technologies concerned.
- Systems would be evolutionary since new knowledge was constantly entering them and leading to behavior modification. There would be no return to an already established "equilibrium".
- Undoubtedly system would exhibit complexity in that knowledge and resource flows were constantly moving across many stakeholder groups
- This in turn would require careful organization to minimise and manage complexity.
- Systems would be adaptable and resilient while resources flowed across their boundaries.
- Systems would behave holistically. That is each system would behave as a totality and therefore analytically its behavior would not be reducible to that of its component nodes.
- Networking would take place and would be designed to facilitate information interactivity that in turn would improve system efficiency

Some concluding points

In this chapter I have tried to highlight how one of the many contributions Peter Allen has made has indirectly begun to affect development policy and practice in and for very poor countries. While more general use of the Senegal model itself (or one like it) is still to come, I believe its second-best equivalent can be well illustrated by the DFID RIU. What Allen did with the Senegal model was to show how it is possible to understand the evolution of socioeconomic systems in ways that allow different professional groups to talk to each other in ways never before really accomplished and in so doing allow development policy-making to achieve new levels of integrated verisimilitude. In a sense the Senegal model became a crossroads where all disciplinary pathways could meet. Its time, however, has yet to come. In the meantime we should do the best we can with policy analysis procedures that are qualitative but disciplined. The innovation systems approach grounded in complex systems theory is well placed to contribute both in terms of technology development itself but also as a device to ensure its long term sustainability.

References

Allen, P.M. and Sanglier, M. (1979). "Dynamic model of growth in a central place system," *Geographical Analysis*, ISSN 1538-4632, 11: 256-72.

Allen, P.M. and Sanglier, M. (1981). "Urban evolution, self-organization and decision-making," *Environment and Planning A*, ISSN 0016-7363, 21: 167-83.

Allen, P.M. (1984). "Ecology, thermodynamics, and self-organization: Towards a new understanding of complexity" in Ulanowicz and Platt (eds.), Ecosystem Theory for Biological Oceanography, *Canadian Bulletin of Fisheries and Aquatic Sciences*, ISSN 0706-652X, 213: 3-26.

Allen, P.M. (1989) "Modelling innovation and change" in S.E. van der Leeuw and R. Torrance (eds.), *What's New: a Closer look at the Process of Innovation*, ISBN 0044451431.

Allen, P.M. and Phang, H.K. (1993). "Evolution, creativity and intelligence in complex systems," in E Michailov (ed.), *Interdisciplinary Approaches to Complex Systems*, ISBN 3642510329, pp. 12-31.

Checkland, P. (1981). *Systems Thinking, Systems Practice*, ISBN 0471279110.

Clark, N.G. and Juma, C. (2013). *Long Run Economics*, ISBN 1472514467.

Clark, N.G., Perez-Trejo, F. and Allen, P.M. (1995). *Evolutionary Dynamics and Sustainable Development: A Systems Approach*, ISBN 1858982731.

Clark, N.G. (2002). "Innovation systems, institutional change and the new knowledge market: Implication for Third World development," *Economics of Innovation and New Technology*, ISSN 1043-8599, XI(1-2).

Clark, N.G., Frost, A., Maudlin, I. and Ward, A. (2013). *Technology Development Assistance for Agriculture: Putting Research into Use in Low Income Countries*, ISBN 0415826977.

Emery, F.E. (ed.) (1970). *Systems Thinking*, ISBN 0140800719.

Freeman, C. (1991). "Japan: A new national system of innovation," in G. Dosi, C. Freeman, R. Nelson, G. Silverberg and L. Soete (eds.), *Technical Change and Economic Theory*, ISBN 9780861879496, pp. 330-348.

Hall, A., Clark, N.G., Yoganand, B. and Sulaiman, R.V. (2003). *Post-Harvest Innovation in Innovation: Reflections in Partnerships and Learning*, ISBN 0953927482.

Holling, C.S. (1985). "Perceiving and managing the complexity of ecological systems" in *The Science and Praxis of Complexity*, ISBN 9280805606.

IERC (1991). "An integrated strategic planning and policy framework for Senegal," EC Final Report, Article 8 946/89, Cranfield University, UK: IERC, October.

Kline, S.J. (1985). "Innovation is not a linear process," *Research Management*, ISSN 0034-5334, (July/August): 36-45.

RIU (2007). Interim Inception Report July-December 2006, NRIL, Aylesford, Kent, January.

Rosen, R. (1974). "Planning, management, policies and strategies," *International Journal of General Systems*, ISSN 0308-1079, 1: 245-252.

Rosen, R. (1987). "On complex systems," *European Journal of Operational Research*, ISSN 0377-2217, 30: 129-134.

Thompson, M. (1993). "Good science for public policy," *Journal of International Development*, ISSN 1099-1328, 5(6): 669-679.

Von Bertalanffy, L. (1950). "The theory of Open Systems in Physics and Biology," *Science*, ISSN 0036-8075, 111: 23-9.

World Bank (2006). "Enhancing agricultural innovation: How to go beyond the strengthening of research systems," *Economic Sector Work report*, The World Bank: Washington, DC, pp. 149.

Norman Clark is Professor of Innovation Systems and Development at the Open University, UK, and Director of Research and Technical Adviser to the Director at the African Centre for Technology Studies (ACTS), Nairobi. Previously he was Vice Chancellor of Kabarak University, Nakuru, Kenya, and before that Professor of Environmental Studies and Director of the Graduate School of Environmental Studies at the University of Strathclyde where he is now an Emeritus Professor. He is a development economist specialising in science, technology and innovation policy issues with particular relevance to Third World problems, a field in which he has published extensively. He has lived and worked in many countries with particular concentration on Kenya, Nigeria and India. Previously he held academic posts at the Universities of Glasgow and Sussex. While at Sussex he acted as the Founding Director of Graduate Studies at the Science Policy Research Unit (SPRU) where he worked for some 15 years and now holds the post of Honorary Professor. He has also acted as Founding Director of the Technology Planning and Development Unit, University of Ife, Nigeria; Visiting Professor, Institute for Advanced Studies, University of Sao Paulo, Brazil; and Director of the Capacity Development Programme at the African Centre for Technology Studies (ACTS). Professor of Innovation Systems and Development at the Open University, UK. He has also recently completed a period as a senior economic adviser to the DFID-funded Research into Use Programme (RIU).

Chapter M

Fixing the UK's economy

Bill McKelvey

London's growth to become the world's #1 financial centre drove up the value of the Pound (one British Pound equalled ~$2.10 in 2007), with the result that most of the UK's industrial production of goods (and even many services) have been driven out of the UK or into oblivion by the exorbitant value of the Pound. Power-law distributions of cities by rank/size are good indicators of the kinds of economic self-organization developments that produce strong rather than weak economies. Japan, China, India, and the US are hot economies. The UK outside of London is cold and sluggish by comparison; UK's is the most stagnant economy of ~50 countries studied—except for The City. Power laws are also applied to the current discussion about whether a "Tobin Tax" is a useful idea—and more fundamentally how to re-energize the UK's economy outside of London. The foregoing suggest that the best means of rebalancing the UK's economy is for the UK to replace the Pound with the Euro. Finito. It's about time the British woke up to realize how important it is for the UK to join the Euro community.

Introduction

My point of departure is a quote from Peter Allen's Conclusion in his book, *Cities and Regions as Self-organizing Systems: Models of Complexity* (1997: 254):

Despite the advantages of...market based systems, however, they do not in themselves solve all the problems, and indeed are the cause of many new ones. For example, it is not clear how 'objective' knowledge can be made available for decision making.... This promotes the piecemeal nature of investment decisions.... 'Risk avoidance' strategy consists of imitating what others are doing and joining in a speculative boom that is in reality very risky. This leads to the paradox of caution leading to crowd-following behavior, and thence to prodigious failure. The London Docklands scheme with, in particular, the vast office spaces of Canary Wharf is an example of the kind of error that can occur during a boom in office building.... Canary Wharf is not by any means the only such disaster. There have been a multitude of investments like this, sometimes in office space, sometimes in hotel rooms, booms and slumps that have affected many major cities.... Speculation and positive feedback are an inherent problem in a 'free market' economy.... It has to be said that this [Canary Wharf] 'mistake' was not just a minor problem for a few investors. Fundamentally, it sucked an enormous amount of investment capital away from other possible ventures, including real business investments in manufacturing and commerce. It resulted in the mistaken building of the largest building in Europe.... Markets are driven by people, generally wealthy and powerful people, and therefore markets are as blind and as greedy as their participants.

Canary Wharf was the mistaken creation of free-market action within the 'City' of London. Peter notes that it was a mistake by the wealthy, powerful, and greedy. Now just jump up one level in scale: Instead Canary Wharf within the City, now think of the City with the UK. The City is now "swollen beyond its socially useful size" as Lord Turner (2009) puts it; it is a mistake in the UK created by the 'wealthy, powerful, and greedy' in Finance, in London, in the UK. It has been sucking the economic life out of much of the rest of the UK.

In this chapter, I argue that various methods could be used to rebalance the UK's economic structure. Put simply, London's growth to become the world's #1 financial centre (in 2009) drove up the value of the Pound (1 British Pound equalled ~$2.10 in 2007), with the result that most of the UK's industrial production of goods (and even many services) have been driven out of the UK or into oblivion because of the exorbitant value of the Pound. To support my argument, I use power-law graphs of cityscapes in various countries. The power-law distribution of cities by rank/size is a good indicator of the kinds of economic self-organization developments that produce strong rather than weak economies (Krugman, 1996; Batty, 2005; Boisot & McKelvey, 2007, 2013).[1] By this measure, Japan, China, India, and the US are hot economies. The UK—not including London—is cold by comparison; it is the most stagnant economy of ~50 countries I studied—except for the City.[2]

Power-laws (PLs) now constitute the branch of complexity science that deals with phenomena characterized by heterogeneity and multiplicative interdependence, which can produce extreme variance, specifically Pareto distributions and PLs. The latter explain entire classes of phenomena that are difficult or impossible to explain via 'normal' research and Gaussian statistics (which requires finite variance and stable

---

1. I do agree with Batty (2005: 514), however, that specifics of this theory are inadequately developed at this time.
2. This chapter was originally drafted in 2009. Five years later, however, the GDP of the UK is still mostly generated by The City.

means) (Andriani & McKelvey, 2007, 2009, 2010, 2011), such as extreme events (West & Deering, 1995) and the proliferation of micro-niches (Anderson's *Long Tail*, 2006). Despite the fact that PL findings date back to 1913, applications of PLs in social science are sparse. In this chapter, I apply PLs to the current discussion about whether a 'Tobin Tax' is a useful idea—and more fundamentally how to re-energize the UK's economy outside of London.

Cityscapes as self-organizing economies

First, I briefly define what a PL is. Then I present a graph showing PL distributions of cities in various 'hot' economies. This builds from Nobel Laureate Paul Krugman's 1996 book titled: *The Self-organizing Economy*. Then I take a close-up view of cities, specifically the top 100 ranked by population size. My conclusion is that more than the Tobin Tax is needed to fix the UK!

Defining power-law distributions

In Figure 1, I use the stock market-capitalization values of the 30 largest software firms to illustrate the difference between Pareto and power-law distributions.[3] The Data table gives the market-capitalization values; the upper left graph shows a Pareto distribution; the lower left shows a fairly straight power-law slope resulting from the shift to log scales for the *X*- and *Y*-axes. As one can see, Google is 'hotter' than the other firms (it is about to take over 2nd place) and hence is above the line; Amazon is obviously not keeping up. PLs take the form of rank/size expressions such as $F \sim N - \beta$, where F is frequency, N is rank (the variable) and β, the exponent, is constant. In exponential functions, e.g., $p(y) \sim e^{ax}$, the exponent is the variable and e (base of natural logs, i.e., the Euler number) is constant.

Rank vs. Size of firms' market-capitalization values

PLs are ubiquitous. They apply to word usage, papers published, book sales, and web hits (Newman, 2005); and size of villages, traffic jams, the structure of the Internet, firm size, sexual behavior, movie profits, industry market capitalization and price movements on exchanges (Andriani & McKelvey, 2007, 2009); they list ~140 kinds of PLs that range in application from atoms to galaxies, from DNA to species, and from networks to wars; there are many more. McKelvey and Salmador (2011) list 70 more just in Finance. Brock (2000) says PLs are the fundamental feature of the Santa Fe In-

---

3. The data and graphs come from a paper written by Pedro Glaser (2009).

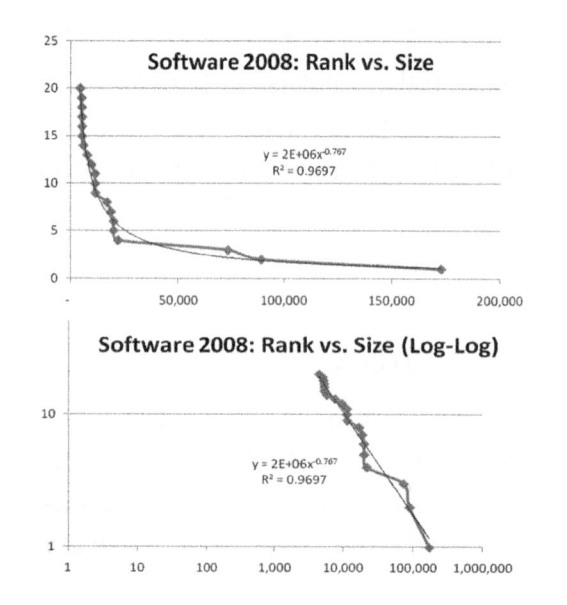

	The Data	
1	MICROSOFT CORP	172,929,939
2	ORACLE CORP	89,465,580
3	GOOGLE INC	73,692,633
4	AMAZON COM INC	21,990,505
5	AUTOMATIC DATA	19,979,960
6	ACCENTURE LTD	19,915,039
7	THOMSON REUTERS	18,774,145
8	YAHOO INC	16,930,147
9	ACTIVISION	11,442,738
10	ADOBE SYSTEMS INC	11,304,053

Figure 1 *Gaussian vs. Pareto distributions.*

stitute's complexity science. Mitchell (2009: 269) just recently says:

> ...power laws are being discovered in such a great number and variety of phenomena that some scientists are calling them "more normal than 'normal'." In the words of mathematician Walter Willinger and his colleagues: "The presence of [power law] distributions in data obtained from complex natural or engineered systems should be considered the norm rather than the exception."

In theory, PL distributions are sloping straight lines. In reality, PLs are not always straight lines. What does it mean if they aren't straight? How to interpret a distribution that is not a straight line, or is flatter or steeper than the so-called '–1 inverse slope'? In the econophysics literature (e.g., Mitzenmacher, 2003; Goldstein *et al.*, 2004; Newman, 2005; Clauset *et al.*, 2009), the usual responses are either to argue that the distribution is not a PL at all but some other kind of distribution (such as lognormal, exponential, stretched exponential, etc.), or truncate the distribution, saying that part (usually one end) of it follows a PL, but the rest fits some other distribution (Goldstein, Morris & Yen, 2004).

Others take the view that straight-line PLs signify adaptively-effective self-organization, i.e., systems successively adjust themselves to better meet changing adaptive constraints (Bak, 1996). Bak labels this 'self-organized criticality' (SOC) in that well-working systems constantly change so as to retain their efficaciously adaptive situation relative to competitors. This idea dates back to Zipf (1949), Preston (1950) and MacArthur (1960). Bak argues that SOC extends across physical, biological, and social systems. In this tradition, PLs show natural processes effectively at work toward efficacious adaptation (McKelvey *et al.*, 2012). Recently we see studies in *Physica A* showing that PLs signal changing language (Dahui *et al.*, 2005, 2006), entrepreneurial as opposed to static industry groups (Ishikawa, 2006), growth vs. static markets (Dahui *et al.*, 2006), and transition economies (Podobnik *et al.*, 2006).

PLs as indicators of self-organizing economies and industries

In 1913 Auerbach discovered that the rank/size plot of cities obeys a PL; Zipf updated this in 1949. Building from them, Krugman (1996), initiates the proposition that PL-distributed cities—as the dominant self-organizing entities in economies—offer a strong indication of well-working economies. He updates their work with 1993 data showing cities more or less on an inverse PL sloping line. These findings are so remarkable that he concludes:

> *We are unused to seeing regularities this exact in economics—it is so exact that I find it spooky.* (p. 40)

Krugman suggests the city PL signature signifies self-organizing economies. I offer a closer look using 2005 data[4] in Figure 2, where I show city PLs for China & India (C/I); Japan & US (J/US); Germany, Turkey, and Italy. What is indeed 'spooky', is that Japan and the US almost totally overlap, yet the US is geographically much larger and without high-speed trains between cities. I show a dashed line through the J/US plots, which overlap so much that the line is almost obscured. The J/US line appears alongside the C/I plots and Germany. While the top 700 or so cities in China and India line up more or less with the J/US line, Germany is pretty much right on it. Turkey is pretty much on the PL line that is a bit flatter in slope. Italy, which is currently in economic and political difficulty (Dinmore, 2008), shows nothing close to a power law distribution.

Krugman (1996) defines self-organization simply as a collectivity of agents creating order from chaos. Holland (1988, 1995, 2002) and other Santa Fe Institute scholars see self-organization giving rise to emergence of structure, interactions, hierarchy, and learning in complex adaptive systems—all serving to enhance continual innovation and adaptation (Arrow *et al.*, 1988; Kauffman, 1993; Bak, 1996; Arthur *et al.*, 1997). McKelvey *et al.* (2012) add additional reasons for viewing PLs as valid indicators of

---

4. Cities are plotted by rank (biggest to smallest) and frequency; one-to-many cities/towns of a specific size. The city-population data for 2005 data come from http://population.mongabay.com/ (accessed March 31, 2007). Rather than use "binning" or "cumulative probability" to make the line look straighter, I use cities systematically below the PL slope as an indication of a country's aggregate economic weakness. These graphs were created by Colin Drayton.

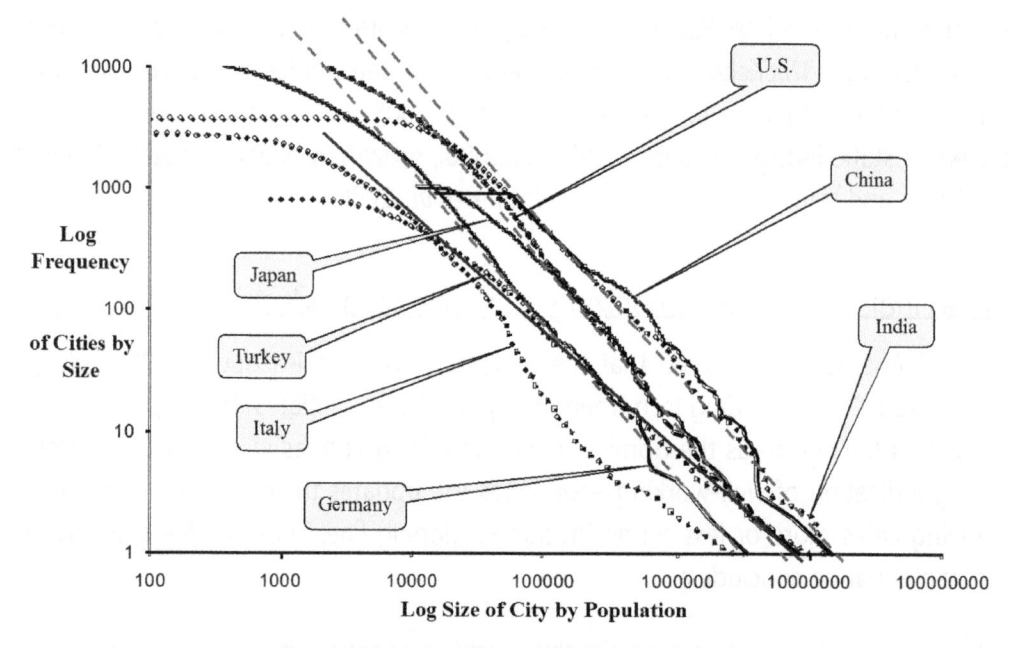

Figure 2 *Some 'hot' economies of 2005.*

efficaciously adaptive self-organization. Adding in the Dahui, Ishikawa, Podobnik *et al.* findings (above), that PLs signify change, we now have a variety of reasons for presuming that efficaciously adaptive self-organization in economies could, indeed, be signified by PLs (McKelvey, 2008; Andriani & McKelvey, 2011).

In his book, Cities and Complexity, Batty shows that while US cities fit a PL, cities in Mexico and the UK do not (2005: 464; based on 1990-1991 data). In Figure 3—using 2005 data on city population sizes—I replicate Batty's finding that UK cities are not PL distributed. For comparison purposes, I also show the India city-plot along with two hot European economies, Ireland and Slovenia. Instead of showing cities all the way down to small towns, I magnify the view by plotting approximately the 100 largest. For interpreting this 'zoom in' look at the UK's cityscape, it is useful to think in terms of the Gini index.

A well-known method (dating back to 1912) for evaluating economies is to use the Gini index measuring income inequality—see Figure 4A.[5] The larger the curved section, the more income inequality in a society. In Figure 4B, the main diagonal is reversed so that it parallels an inverse PL slope. Given this, then, the more that a city rank/frequency sequence falls below the PL 'diagonal', the less well-working a country's economy appears to be.

---

5. The Gini coefficient is a measure of country-level income inequality. A good definition is offered at: http://en.wikipedia.org/wiki/Gini_coefficient. The larger the Gini coefficient, i.e., the larger the area within the curve, the worse income inequality.

In Figure 3, then, and looking at the UK PL cityscape, we see the 'Gini-space' as the large (grey) area below the dots-dash line, which duplicates the role of the principle diagonal for defining one edge of the area measured via the Gini index. Of ~50 countries I have studied,[6] only Mexico (not shown) shows considerable 'Gini-space'—but less than the UK. In short, the UK has some 85+ large cities well below the dots-dash line—which is where the PL slope would be if the UK had the same kind of 'hot' self-organizing economy as India, China, Japan or the US.

Figure 3 also compares the UK with the hot economies of Ireland and Slovenia. It also shows the Malta cityscape PL; let's call this the 'Malta line' (dotted), which is nearly vertical. It is a good indication of what the PL of a dead economy looks like. Outside of London, the next ~85+ UK cities line up pretty well along the 'Malta line'. Not Good!

Is the UK as economically 'broken' as the Gini-space suggests? Outside of London, the UK shows considerable evidence that self-organization towards a 'well-working' economy is lacking. Taking a more detailed look at Figure 3, what do we see?

- UK has the largest Gini-space of any of the ~50 countries I have studied.
- After the 2nd largest UK city, we see a more vertical slope in the UK cityscape; in fact, it is nearly parallel with the dotted line—the 'Malta line'. This suggests that after London, the self-organization of the next 85+ UK cities is essentially frozen.
  - Note the dots-dash line going from London (on the X axis) to the upper end of the plot. This is the PL counterpart to the diagonal from which the Lorenz curve spreads out and which then defines the Gini coefficient. The area between the UK-city plot line and this counterpart line to the Gini diagonal is analogous to the area represented by the Gini index.
  - All of the largest 85+ cities in the UK (except London) show lack of economic self-organization—leaving UK with the lowest level of city-based SOC adaptive-success of any of the ~50 nations studied.

---

6. The '50' includes the 27 EC nations, 10 resource-rich commodity-based economies leaning in the "hot" direction, such as Russia, Saudi Arabia, Nigeria, etc., and other economically hot and cold nations, industrialized or not.

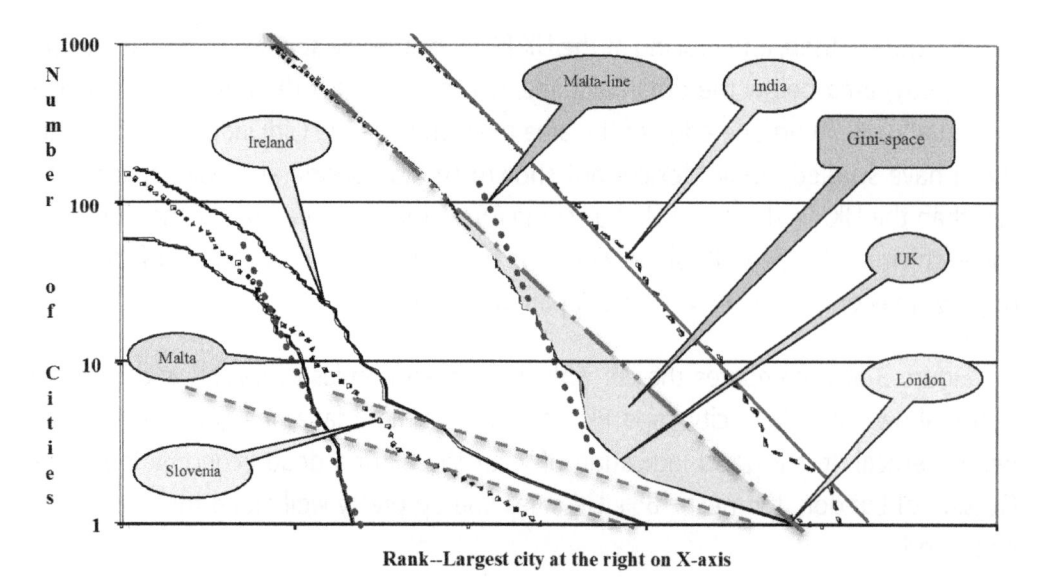

Figure 3 *Larger Cities Shown in Rank/Size PL Distributions*

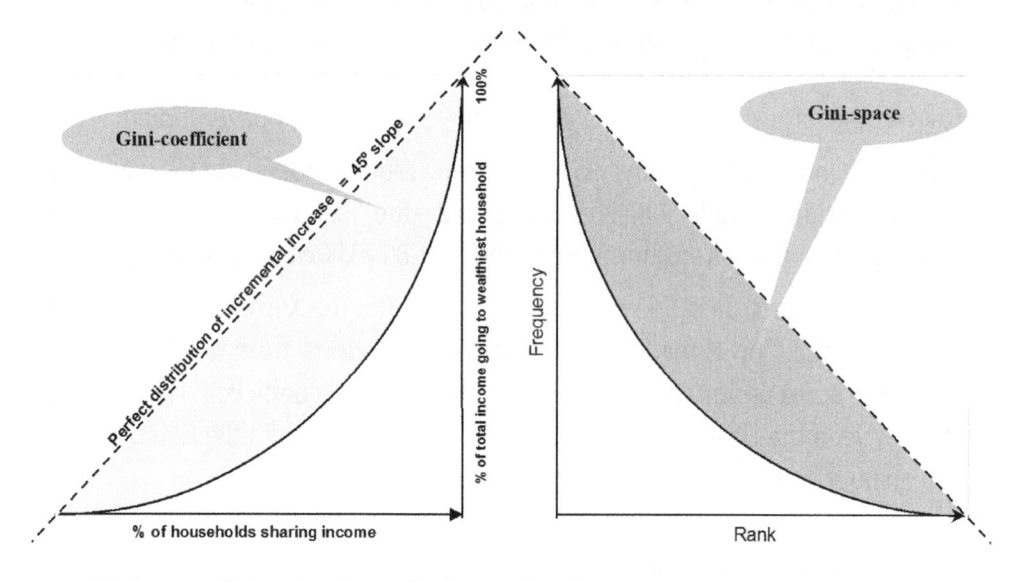

Figure 4 *Gini Index On Left; Power-law 'Deviation' Indicator on Right*

- In Figure 3, I also show a dashed line that is parallel to the line between London and the next largest city—let's take this dashed line as a representation of the London 'hot-slope'; being relatively flat it represents hot economic activity (and it is pretty much opposite to the much more vertical Malta line). Note especially that the slope of the largest 5 cities in Ireland is fairly close to the London hot-slope. Whereas UK has just one hot city, Ireland has five before the Malta-line break in self-organization appears. Slovenia, the hottest economy in Europe (in 2005), also shows the flatter 'hot' PL slope across multiple cities, as opposed to the 'Malta line', which is much more vertical; Slovenia looks much more like Ireland than the UK.

Does the city-based PL indication that the UK is broken with respect to self-organizing show up in other data about the UK economy? Consider the following (city GDP listings by Wikipedia; http://en.wikipedia.org/wiki/List_of_cities_by_GDP):

- London's GDP (at £458bn) is 30% of the UK's GDP (at £1.552tn; 2012 data)
- New York City's GDP (at $1.21tn) is only 9% of the US's GDP (at $13.67tn; 2012 data);
- London's GDP growth rate has been 33% higher than for UK as a whole over the past 15 years;
- The Midlands lost 1.1 million manufacturing jobs between 1995 & 2006 (*London News*, Sept. 6, 2006) and many more since then. Another 1.1 million public-sector jobs are expected to disappear (The Independent, Nov. 3, 2013).

Is the UK as economically 'broken' as the PL equivalent of the Gini coefficient suggests? Outside of London, the UK shows some evidence that self-organization towards a "good" economy is lacking. In theory, a well working economy works for everyone from top to bottom; job opportunities for everyone; equal opportunity for all to improve their life style, etc. "A well working economy must make productive use of its resources—both human and material...[It should be] a stable source of a plentiful livelihood for its members" (Zarsky *et al.*, 1986: 122). Perhaps we are on the verge of being able to use PLs as indicators of the economic viability of disparate regions and cities within national economies.

In general, much data suggests that healthy ecosystems exhibit PL effects, fractal structures, and scale-free dynamics (Andriani & McKelvey, 2007, 2009; Zanini, 2008, 2009). This suggests, in turn, that well-functioning self-organizing processes—i.e., increasing connectivity and SOC under adaptive tension—underpin economic self-organizing success, as Ishikawa (2006) and Podobnik *et al.* (2006) have found elsewhere, and as data I present in this chapter indicate. It is not surprising that we see recent confirmation of the classic studies of cityscape PLs by Auerbach, Zipf, and Krugman. I also confirm with some empirical data, that the UK cityscape is broken, as compared to 'hot' economies. What is good for London appears not good for the rest of the people in the UK.

Lord Turner asks:
"Is the city too large to be socially useful?"

I don't know what evidence Lord Turner was aware of when he asked whether the City is "swollen beyond its socially useful size" (2009). I don't suppose one really needs to see the foregoing PL data indications to know that the UK economy is broken—for sure, Lord Turner had not seen my Figures before he made his remark! About the only things the UK seems to be exporting are intellectuals looking overseas for better jobs—and of course paper profits by investment banks. Recognizing that London's vast size is due, mostly, to its current status as one of the world's #1 financial transaction centers (the other being New York City), Lord Turner quite reasonably wondered, then: Would imposition of a Tobin tax help rebalance the UK's economy?

Calls for a Tobin tax

The Tobin Tax (TT) (Sandbu, 2009) was initially suggested by Nobel Laureate James Tobin (1978) as a small tax on international currency trades and speculation for the purpose of curbing destabilization. Since then, it has been advocated as a means of reducing poverty by raising money rather than simply reducing speculation. Thus, in 2005 in the UK, the Sterling Stamp Duty was proposed. Even though proposed at 200 times lower than what Tobin envisaged, given the currency market's $2trillion/day trading volume (in 2007), huge sums could be raised for other uses.[7]

In the mid-2009 discussion in the Financial Times (FT), the initiating mission of the TT was to curb bonuses. But all of a sudden, "particularly in a quiet August" as Tett (2009) puts it, discussion rages (Corrigan, 2009; Hosking, 2009; Parker, 2009a) about Lord Turner's suggestion that the TT could/should be used to shrink London's wholesale financial sector that has "swollen beyond its socially useful size". But this is not the only rationale for putting the TT in play. Besides the original suggestion of using it as the weapon of choice for curbing outrageous bonuses, it is also proposed for curbing dysfunctional speculation, funding international development, and funding the fight against global warming (Hill, 2009a).

On September 1st an Editorial in the FT (2009) agrees that "parts of the City grew too big". It then zeros in on (1) wanting to curb excess leverage; and (2) making banks "safe-to-fail". The FT also quotes Chancellor Angela Merkel as saying "No bank should be allowed to become so big that it can blackmail governments". And then the FT

---
7. See 'Tobin Tax' at http://en.wikipedia.org/wiki/Tobin_tax (accessed Nov. 3, 2013). For addition information about currency taxes see http://en.wikipedia.org/wiki/Currency_transaction_tax.

prints a letter to the Editor by a Professor Morriss worrying that reducing the size of the City will "make many people in New York Paris, Shanghai, Singapore, Tokyo and elsewhere" happy, and that they and others will jump on board offering further suggestions to reduce various UK industries. The key question, thus, becomes: Is a smaller City good or bad for the UK? The main flaw in Morriss' complaint is that the growth of The City has driven up the value of the British Pound so much that all the other industries has already been reduced. His logic is backwards. In fact, since the Pound has reduced in value relative to the Euro since 2006 the amount of tourism from the Continent has considerably increased! But not industries.

Disagreements

Elliot (2009: 2), quotes Tobin on his two main motives for suggesting the TT: (1) "make exchange rates reflect to a larger degree long-term fundamentals" (we have all recently seen what speculation can do to petrol prices); and (2) "preserve and promote autonomy of national macroeconomic and monetary policies", i.e., produce greater economic stability (Tobin, 1996: xii-xiii). His idea was that it would penalize short-term trading speculations while having little effect on longer-term free-market activities. Tobin soon discovered, however, that his "tax" idea had fallen on deaf ears: "It did not make much of a ripple. In fact, one could say that it sunk like a rock. The community of professional economists simply ignored it" (quoted in Elliott, 2009: 1).[8] François Mitterrand touted the money-raising value of the TT back in 1995. Jacques Chirac, in 2005, proposed a "Tobin tax" to address rampant "liberal globalization" but it was "scorned by Britain" (quoted in Parker, 2009b). Most people against the TT see it is just one more (unnecessary) tax imposed on the rich to help the poor (Elliot, 2009). Economists mostly see it as interfering with free markets (Elliot, 2009). Hill (2009b) suggests, instead, that the TT would drive "useless bankers" getting huge bonuses out of the City. But Buiter (2009) asks: "What problem would a TT...solve?" and "What distortion is...[it] targeted at?"

As noted above, the TT discussion rather quickly focuses on its use as a means of reducing "outrageous" bonuses in the financial sector of the City. This it may or may not do. Besides, the bonuses are better treated as symptoms rather than causes of financial bloat. More broadly, and more importantly, the key issue is the swollen size of The City relative to economic activities in the rest of the UK. The swollen financial growth of London is rooted in the growth of "fixed-income securities, trading, derivatives and hedging...and fund management and share trading" (Hosking, 2009: 1). Short-term economic destabilizations are not really the main issue. Rather, it is the

---

8. The Economist, bastion of economists, gives Lord Turner's idea two sentences of space (2009, 392 (Aug. 29: 66).

weight of the London-based GDP relative to the GDP produced in the rest of the UK (London's GDP is 30% of the UK's). In the US, more specifically, "from the 1950s to the 1980s, the finance sector accounted for 10 percent of all profits earned by US corporations; in the first half of this decade it reached 34 percent" (Friedman, 2009) (NYC's GDP is only 9% of the US's). This disparity still holds in the UK as of 2012.

We should all agree that at some point countries' currencies should not be held hostage to quick-trick traders, and now the financial engineers' computers trading in micro-seconds. But in the case of the TT, there is a much more fundamental use-advantage in the UK. This is illustrated in Figure 4 by the size of the UK's city Gini-space. The large financial magnet that is London has attracted so much foreign money for investment that it has driven up the value of the Pound. The result is that while London makes vast amounts of (paper) money, the rest of UK has lost its industrial market power. Industries have left UK like rats leaving a sinking ship. People leave the UK to go shopping. Retirees left UK to buy nicer houses in Spain (and eat better/healthier food!)—at least until the Great Recession). The problem, folks, is not that swollen City-size has produced outrageous bonuses; the problem is that swollen City-size attracts more and more foreign investment, which drives up the value of the Pound and, then, raises the cost of all other UK products for everyone else around the world—thereby ruining exports.

How to fix the UK?

Why not tax money? What is the best argument favouring the TT? VAT!! In the European Union, most products sold to end users come with a Value Added Tax, which is added sequentially at all stages of production; Norway, Denmark, and Sweden have VATs at 25%. All end users pay a sales tax on all products sold in the US, excepting some food and safety items (States differ). Bottom line: almost all products sold come with a sales tax or VAT added on.

Why is it that sales of money are not taxed when almost everything else is? Given the 'motivation by greed' factor in financial districts—seemingly uncontrolled—why not compare money sales with sales of addictive products like alcohol and cigarettes, which usually come with even higher sales taxes? While the latter are, indeed, good ways for Nations to make money, high taxes are also seen as 'de-motivators' against partaking of bad substances. This fits very well with Tobin's original idea for the TT—reduce the greed-based disease of short-term speculation. Where is it written—by God or anyone else—that all people on Earth should be taxed for selling things but people who sell money via currency exchanges or other financial transactions should remain exempt? If this isn't the fundamental reason for imposing the TT, I don't know what is. With the VAT, governments earn 'value' as a product's value increases. So, OK, why shouldn't governments earn 'value' as monetary transactions take place? But what if money loses value, you ask? Well, in the US, at least, car dealers pay sales taxes when selling used cars, even as they are going down in value. The value of some thing is taxes when it is sold. Period.

So much for arguing in favour of the TT. While it is surely 'fair' to tax money sales like all other sales, this is not the best solution for re-balancing the UK cityscape and underlying economy. Having the UK even more dependent on the Finance industry in London—because of all the money pouring in from a tax on money sales—simply exacerbates the treachery of financial-market bubbles and crashes (though a TT would reduce much of the grossly speculative currency trades). The UK needs a better fix.

UK should switch from Pound to the Euro. Really!

In 2009 Wolf said:

> It would still take until 2031 before the economy [is] as big as it would have been if the 1997–2007 trend had continued. The cumulative loss of output [will] be 160% of 2007 gross domestic product.... Indeed, the fiscal deterioration in the UK has been far bigger than in any other member of the Group of Seven leading high-income countries.... The reason...is that in the UK, the financial sector played a huge role in supporting consumer expenditure, property transactions and corporate profits. No less than a quarter of corporate taxation came from the financial sector alone.

While the cumulative GDP-contraction of the US was 3.2%, it was 5.6% in Germany, 5.9% in Italy, 7.7% in Japan and 4.75% in the UK (by 3rd quarter of 2009). As bad as the financial crash was in the US, the latter's large economy did not suffer as much as smaller economies (and especially economies that depended on exports to the US). The UK's loss in GDP was strictly due to the crash of The City, not loss of exports; with the UK Pound at ~$2.10/Pound in 2007, there were already minimal exports (except tourists and Scotch whiskey) to the US.

GDPs in most other major economies around the world are turning at least mildly positive—but not in the UK. Given that a strong financial City drives up the value of the Pound against all other currencies, what is good for The City is definitely *not good* for the rest of the UK—this is what the Gini-space indicates in Figure 3. The financial crash and the subsequent devaluation of the Pound down to its current "4 pennies more than the Euro" (on Jan. 1st, 2009) has led to tourists flowing *into* the UK from the Continent rather than just cash flowing in from the Mideast Oil States. While the financial sector in London was suffering in 2009, the effects of a lower-valued Pound led to improvements in the rest of the UK. And, the UK would obviously benefit over the longer term if the Pound were to be held to parity with the Euro. Held to Parity you say? In fact it hasn't happened. Here is the value of the Pound relative to the Euro in the past six years:

Date	Pound to Euro
Jan. 1, 2009	1.04
Jan. 1, 2010	1.12
Jan. 1, 2011	1.18
Jan. 1, 2012	1.20
Jan. 1, 2013	1.23
Jan. 1, 2014	1.20

The most recent high was 1.27; occurred on August 5th, 2013.

As you can see from the above table, the Pound has a high of 23% relative to the Euro in the past 6 years. But it is still well below what it was on January 1st, 2006, well before the Crash started.

What better way to accomplish economic balance quickly than for the UK to join the Euro community? Surely you joke, my UK readers will say. Not at all. There is no better argument for the UK to adopt the Euro than this. This simple Euro-move will do more to re-balance the UK economically than any kind of imposition of the TT. And, given the 30 years of history showing nothing but resistance to the Tobin Tax, including more of the same in the recent 2009 discussion, why not get rid of what Lord Turner aptly calls a 'swollen London' much faster and more expeditiously: Just switch the UK to the Euro! The best motive Scotland has for leaving the UK is to join the Euro: The cost of Scotch Whiskey for the entire rest of the world would drop immediately—and profits to Scots would increase!

Immediately, of course, many readers will jump on the current economic weakness of the Euro sector, and especially the weakness of the Mediterranean countries to argue that joining the Euro would be a big mistake. But wait, as the Germans finally learned, Germany's exports are vastly cheaper in the US and China with a weak Euro than with an even stronger Euro. Think what a BMW would cost in the US if Germany went back to the Deutschmark! If the UK were to join the Euro the effect of Mideast oil-money flowing into The City would not drive up the value of the Euro the way it does so with the Pound. It's that simple!

In a World of giant economies, the UK Pound and the now relatively small UK economy—especially if crunched down into one-financial-transaction entity, The City—are increasingly vulnerable to currency and other economic volatilities and imbalances. In a land of giants, it is better to be one. This, rather than any other reason, supports abandoning the Pound in favour of the Euro: The latter is much less sensitive to instabilities than even the Dollar these days.

But wait! If the UK joins the Euro community, won't the money from the Oil States drive up the Euro relative to the Dollar, as opposed to just driving up the value of the UK Pound? Given this possibility, why would the Euro states let the UK join?

Again, the same basic logic applies: First, the Euro community is vastly larger than the UK so the effect of inflowing Oil-State money won't boost the Euro as much as it inflated the Pound. Second, any move that makes London less desirable as a money attractor will send Arab oil money to New York or Hong Kong, thereby inflating the Dollar or Yuan rather than the Euro. By this logic, reducing a swollen City by switching to the Euro could actually benefit the entire Euro community's product and service sectors: As the Dollar gains value relative to the Euro, EU products and services sell better to the US at more competitive prices—and to China since its currency is pegged to the Dollar. By this logic the Euro community should pay the UK to adopt the Euro!!

Conclusion

There are two striking findings from Figure 3:

- First, it is indeed spooky, if not amazing, that something as simple as a power-law Gini-equivalent would show so obviously that much of the UK is economically broken.
- Second, it refocuses the TT discussion toward a better UK-rescue platform: It is not just about London being swollen or not, nor just about speculation or outrageous bonuses, nor upon what the TT revenues should be spent.

How to best improve UK? The Pound is the problem. The best way to shrink a swollen City and actually improve the UK's GDP to replace the Pound with the Euro. From this one strategic decision:

1. The UK economy is much less vulnerable to financial speculations of all kinds;
2. The cost of UK goods and services is reduced with respect to the two other major world economies, US and China (with Brazil and India in the wings);

3. By joining, the UK can help lead the EU toward becoming a true 4th economic power, along beside the US, China, and eventually India. Sitting on the north side of the Strait just watching doesn't compute.

The foregoing add up to the best strategy for "fixing the UK" bar none. Given that the Euro community is currently in a mood for change, the timing couldn't be better. The UK citizens living and working in the non-London parts of the UK deserve better than a future in which The City continues to dominate the economy by producing un-taxed money sales, continues unrestrained speculative buying and selling of money, and controls the dominant portion of total GDP.

But of course, adding a Tobin tax would calm down needless financial speculation. Who benefits from being in a safe haven like The City? Well, needless to say, those who have the most to lose, the Rich of The City. Do they really think they would be better off in Russia, India, Argentina, Zimbabwe or Kazakhstan? Probably not. Well then, shouldn't they pay for the cost of their safety?

References

Allen, P.M. (1997). *Cities and Regions as Self-Organizing Systems: Models of Complexity*, ISBN 90-5699-071-3.

Anderson, C. (2006). *The Long Tail*, ISBN 9781401309664.

Andriani, P. and McKelvey, B. (2007). "Beyond Gaussian averages: Redirecting organization science toward extreme events and power laws," *Journal of International Business Studies*, ISSN 0047-2506, 38: 1212-1230.

Andriani, P. and McKelvey, B. (2009). "From Gaussian to Paretian thinking: Causes and implications of power laws in organizations," *Organization Science*, ISSN 1047-7039, 20: 1053-1071.

Andriani, P. and McKelvey, B. (2010)."Using scale-free theory from complexity science to better management risk," *Risk Management: An International Journal*, ISSN 1460-3799, 12: 54-82.

Andriani, P. and McKelvey, B. (2011). "From skew distributions to power law science," in P. Allen, S. Maguire and B. McKelvey (eds.), *Handbook of Complexity and Management*, ISBN 9781847875693, pp. 254-273.

Arthur, W.B., Durlauf, S.N. and Lane, D.A. (1997). *The Economy as an Evolving Complex System, Vol. II*, ISBN 9780201328233.

Auerbach, F. (1913). "Das gesetz der bevolkerungskoncentration," Petermanns Geographische Mitteilungen, ISSN 0031-6229, 59(1): 74-76.

Bak, P. (1996). *How Nature Works: The Science of Self-organized Criticality*, ISBN 0387947914.

Batty, M. (2005). *Cities and Complexity*, ISBN 0262025833.

Boisot, M. and McKelvey, B. (2007). "Extreme events, power laws, and adaptation: Towards an econophysics of organization," Best Paper Proceedings, Academy of Management Conference, Philadelphia, PA, August.

Brock, W.A. (2000). "Some Santa Fe scenery," in D. Colander (ed.), *The Complexity Vision and the Teaching of Economics*, ISBN 1840642521, pp. 29-49.

Buiter, W. (2009). "Forget Tobin tax: There is a better way to curb finance," *Financial Times*, Aug. 31, http://www.ft.com/cms/s/0/76e13a4e-9725-11de-83c5-00144feabdc0.html.

Clauset, A., Shalizi, C.R. and Newman, M.E.J. (2009). "Power-law distributions in empirical data," *SIAM Review*, ISSN 0036-1445, 51: 661-703.

Corrigan, T. (2009). "Lord Turner's answer to the financial crisis raises more questions," *Telegraph*, Aug. 27, http://www.telegraph.co.uk/finance/comment/tracycorrigan/6101533/Lord-Turners-answer-to-the-financial-crisis-raises-more-questions.html.

Dahui, W., Menghui, L., and Zengru, D. (2005). "True reason for Zipf's law in language," *Physica A*, ISSN 0378-4371, 358: 545-550.

Dahui, W., Li, Z., and Zengru, D. (2006). "Bipartite produce-consumer networks and the size distribution of firms," *Physica A*, ISSN 0378-4371, 363: 359-366.

Dinmore, G. (2008). "In with the old? New party names but few fresh choices for Italy," *Financial Times*, April 11: 7.

Editorial (2009). "Too much of a very good thing," *Financial Times*, Aug. 31, http://www.ft.com/cms/s/0/ed28c2fc-9659-11de-84d1-00144feabdc0.html.

Elliott, L. (2009). "The time is ripe for a Tobin tax," *Guardian*, Aug. 27, http://www.guardian.co.uk/business/2009/aug/27/turner-tobin-tax-economic-policy.

Friedman, B. (2009). "Overmighty finance leavies a tithe on growth," *Financial Times*, Aug. 26, http://www.ft.com/cms/s/0/2de2b29a-9271-11de-b63b-00144feabdc0.html?SID=google.

Glaser, P. (2009). "Fitness and inequality in an increasing returns world: Applying the tools of complexity economics to study the changing distribution of US stock market capitalizations from 1930 to 2008," Working paper, UCLA Anderson School of Management, Los Angeles, CA.

Goldstein, M.L., Morris, S.A., and Yen, G.G. (2004). "Problems with fitting to the power-law distribution," *European Physical Journal B*, ISSN 1434-6028, 41: 255-258.

Hill, A. (2009a). "Lord Turner's pay provocation deserves proper hearing," *Financial Times*, Aug. 26, http://www.ft.com/cms/s/0/3538a990-9272-11de-b63b-00144feabdc0.html.

Hill, A. (2009b). "Fix the system and useless bankers will flee the city," *Financial Times*, Aug. 28, http://www.ft.com/cms/s/0/2d845e3c-9403-11de-9c57-00144feabdc0.html.

Holland, J.H. (1988). "The global economy as an adaptive system," in P.W. Anderson, K. J. Arrow and D. Pines, (eds.), *The Economy as an Evolving Complex System, Vol. 5*, ISBN 9780201156850, pp. 117-124.

Holland, J.H. (1995). *Hidden Order*, ISBN 9780201442304 (1996).

Holland, J.H. (2002). "Complex adaptive systems and spontaneous emergence," in A.Q. Curzio and M. Fortis (eds.), *Complexity and Industrial Clusters*, ISBN 9783790814712, pp. 24-34.

Hosking, P. (2009). "FSA chairman Lord Turner says City too big," *The Times*, Aug. 27, http://business.timesonline.co.uk/tol/business/industry_sectors/banking_and_finance/article6811548.ece.

Iansiti, M. and Levien, R. (2004). "Strategy as ecology," *Harvard Business Review*, ISSN 0017-8012, 82(3): 68-78.

Ishikawa, A. (2006). "Pareto index induced from the scale of companies," *Physica A*, ISSN 0378-4371, 363: 367-376.

Kauffman, S.A. (1993). *The Origins of Order*, ISBN 0195058119.

Krugman, P. (1996). *The Self-Organizing Economy*, ISBN 0262062046.

London News. (2006). "Midlands hit most by manufacturing decline," September, 6th.

MacArthur, R.H. (1960). "On the relative abundance of species," *American Naturalist*, ISSN 0003-0147, 94: 25-36.

McKelvey, B. (2008). "Pareto-based Science: Basic Principles—and Beyond," presented at Organization Science Winter Conference, The Resort at Squaw Creek, CA, Feb. 8.

McKelvey, B., Lichtenstein, B.B. and Andriani, P. (2012). "When organizations and ecosystems interact: Toward a law of requisite fractality in firms," *International Journal of Complexity in Leadership & Management*, ISSN 1759-0256, 2(1-2): 104-136.

Mitchell, M. (2009). *Complexity: A Guided Tour*, ISBN 9780195124415.

Mitzenmacher, M. (2003). "A brief history of generative models for power law and lognormal distributions," *Internet Mathematics*, ISSN 1542-7951, 1(2): 226-251.

Newman, M.E.J. (2005). "Power laws, Pareto distributions and Zipf's law," *Contemporary Physics*, ISSN 0010-7514, 46(5): 323-351.

Oxford Economic Forecasting (2006). "London's place in the UK economy, 2006–07," London, UK.

Parker, G. (2009a). "FSA chief backs City curbs with global tax," *Financial Times*, Aug. 27, http://www.ft.com/cms/s/0/6ac58734-92a1-11de-b63b-00144feabdc0.html.

Parker, G. (2009b). "Treasury frowns on "Tobin" proposal," *Financial Times*, Aug. 27, http://www.ft.com/cms/s/0/04ff9d22-92a1-11de-b63b-00144feabdc0.html.

Podobnik, B., Fu, D., Jagric, T., Grosse, I., and Stanley, H.E. (2006). "Fractionally integrated process for transition economics," Physica A, ISSN 0378-4371, 362: 465-470.

Preston, F.W. (1950). "Gas Laws and wealth laws," Scientific Monthly, 71: 309-311.

Sandbu, M. (2011). "The Tobin tax explained," http://www.ft.com/cms/s/0/6210e49c-9307-11de-b146-00144feabdc0.html.

Tett, G. (2009). "Could 'Tobin tax' reshape financial sector DNA?" *Financial Times*, Aug. 27, http://www.ft.com/cms/s/0/980e9ec8-92f2-11de-b146-00144feabdc0.html.

The Economist (2009). "Britain's public finances: Class warrior," ISSN 0013-0613, (December 12): 12.

Tobin, J. (1978). "A proposal for international monetary reform," *Eastern Economic Journal*, ISSN 0094-5056, 4: 153-159.

Tobin, J. (1996). "Forward," in M. ul Haq, I. Kaul and I. Grunberg (eds.), *The Tobin Tax: Coping with Financial Volatility*, ISBN 019511180X.

Turner, A. (2009). "FSA chairman Lord Turner says City too big," *Prospect Magazine*, ISSN 1359-5024, Aug. 27, http://www.accessinterviews.com/interviews/detail/fsa-chairman-lord-turner-says-city-too-big/16311.

West, B.J and Deering, B. (1995). *The Lure of Modern Science: Fractal Thinking*, ISBN 9789810221973.

Wolf, M. (2009). "How to share the losses: The dismal choice facing Britain," *Financial Times*, (December 16): 13.

Zanini, M. (2008). "Using 'power curves' to assess industry dynamics," *McKinsey Quarterly*, ISSN 0047-5394, (November).

Zanini, M. (2009). "'Power curves': What natural and economic disasters have in common," *McKinsey Quarterly*, ISSN 0047-5394, (June).

Zarsky, L., Bowles, S., and Ells, S. (eds.). (1986). *Economic Report of the People*, ISBN 0896083152.

Zipf, G.K. (1949). *Human Behavior and the Principle of Least Effort*, ISBN 161427312X (2012).

Bill McKelvey—Ph.D. MIT 1967. Professor Emeritus of Strategic Organizing, Complexity Science & Econophysics the UCLA Anderson School of Management. His book, *Organizational Systematics* (1982) remains a definitive treatment of organizational taxonomy and evolutionary theory. He chaired the building committee that produced the $110,000,000 Anderson Complex at UCLA—opened in 1995. In 1997 he became Director of the Center for Rescuing Strategy and Organization Science (SOS). From this Center he initiated activities leading to the founding of UCLA's Inter-Departmental Program, *Human Complex Systems & Computational Social Science*. He has directed over 170 field study teams on 6-month projects concerned with strategic and organizational improvements to client firms. Coedited: *Variations in Organization Science* (with J. Baum, 1999), a special issue of *Emergence* (with S. Maguire, 1999), and a special issue of *J. Information Technology* (with J. Merali, 2006). Edited a special issue of *Int. J. of Complexity and Management* (2013). Coeditor of *SAGE Handbook of Complexity and Management* (2011); Editor of a Routledge Major Works Series: *Complexity Concepts* (2012; 5-volumes, 2447 pgs.). Has ~70 recent articles and chapters applying complexity science to organization science and management.

Chapter N

Complexity and the evolution of market structure

Paul Ormerod

Peter Allen was one of the earliest innovators in the application of complex systems principles to the social sciences. In this contribution, I develop key themes of complexity in the context of two different models of firms and the evolution of competition and market structure. Agents, both firms and consumers, act with purpose and intent, but under conditions not only of imperfect information at any given time, but of an environment which constantly evolves. There is a great deal of contingency in the outcomes of any individual solution of the model, and time and process are important. Simple rules of thumb in general give agents better outcomes than do attempts to follow rational expectations. Agent based modelling of complex systems compels modelers from the outset to confront empirical evidence. Not only does this place the models on a genuinely scientific basis, but it overcomes the standard objection of economists that, once the principle of rational maximization is relaxed, any outcome becomes possible. Peter was a great pioneer of agent based modelling in the social sciences.

Introduction

Peter Allen was one of the pioneers of the application of complex systems to problems in the social sciences. This view of the world stands in marked contrast to the rational agent, rational expectations (RARE) model of conventional economics.

A key feature of the complex systems approach is that the understanding which individual agents have of the world is inevitably imperfect. They cannot be ascribed the cognitive powers of gathering and processing information which exist in conventional economic theory.

Within economics itself in the late 20th/early 21st centuries there has been increasing recognition of this. From the conventional paradigm of the fully rational agent with full information and using a universal behavioral rule of maximization, economics initially relaxed the assumption of full information, creating the concept of bounded rationality. Now, experimental and behavioral economics point to the use of limited information and rules of thumb, each one customized to particular circumstances (for example, Akerlof, 2002; Kahneman, 2003; Smith, 2003).

In a complex system, there is a low (or even zero) ability to predict the state of the system at any given point in the future. There may very well be stable statistical distributions from which describe the range of behaviors of the macroscopic factors, so that we can reasonably estimate the proportion of time which the system spends in any particular state. But we cannot predict consistently at particular points in time with any reasonable accuracy.

Further, complex systems will typically exhibit multiple possible histories. By definition there can only ever be one actual history, but at any point in time the system has the potential to move in a variety of different ways.

Again, complex systems restore the concepts of time and process into modelling. Orthodox economics is essentially concerned with comparing successive equilibriums. A system is in equilibrium, a change of some kind takes place, and economic theory deals with the properties of the new equilibrium following the change. But in reality, systems spend much of their time out of equilibrium, and understanding the time path and the evolution of the system along it is crucial.

Peter was developing models as long ago as the 1980s which exhibited these features. In case anyone should doubt the empirical realism of this approach, here is an account of what life is actually like inside Microsoft, one of the world's largest companies, given in Marlin Eller's book *Barbarians Led by Bill Gates*. Eller was from 1982 to 1995 Microsoft's lead developer for graphics on Windows. Windows now of course dominates the PC operating systems world. But its success was based far more on a series of accidents than on a far-sighted, planned strategy.

Eller's introductory remarks are worth quoting at some length:

> *There was a great disconnect between the view from the inside that my compatriots and I were experiencing down in the trenches, and the outside view... in their quest for causality [outsiders] tend to attribute any success to a Machiavellian brilliance rather than to merely good fortune. They lend the impression that the captains of industry chart strategic courses, steering their tanker carefully and gracefully through the straits. The view from the inside more closely resembles white-water rafting. "Oh my God! Huge rock dead ahead! Everyone to the left! NO, NO, the other left!"*

In this contribution to the *Festschrift*, I discuss two models of how firms operate which have more realistic behavioral foundations than does the standard RARE model. Firms and consumers both react to incentives, but do so on the basis of imperfect information and use of rules of thumb to make decisions rather than so-called 'optimal' decision rules. Time and process are both important features of the models, and there is a great deal of contingency around the outcomes.

I first of all offer some brief remarks on the concept of competition and market structure in the history of economic thought, before setting out the first of the models. The results raise important methodological issues for agent-based models, which I discuss before going on to consider the second model.

Competition and market structure

During the 19th century and the opening decades of the 20th, a great deal of work had been done on the standard paradigm of firm behavior and markets in economics, namely that of so-called perfect competition. The key simplifying assumption in this model is that the number of firms in a particular market is so large that no single firm can influence the price by its decisions on output. The firms are 'price-takers'. This of course immediately raises the question that if no firm can influence price, how is price determined? This may seem an obvious point, but it is one which is so neglected by mainstream economists on an everyday basis that Vernon Smith referred to it specifically in his Nobel prize lecture: 'As a theory the price-taking parable is also a non-starter: who makes price if all agents take price as given?' (Smith, op.cit.).

A substantial amount of work had also been carried out on a market with a single supplier, and a market with a very small number of suppliers, oligopoly. Cournot's analysis in 1838 is a classic example of the latter. The extension to a market with a large number of suppliers each of whom could exercise some influence over price, so-called imperfect competition, took until the 1930s to develop (Chamberlin, 1933; Robinson, 1933).

This analysis completed the set, as it were, in economic theory which classifies markets according to the number of firms. In general, price would be lower under perfect competition, and then gradually rise through imperfect competition, oligopoly and finally monopoly.

However, doubt about the neatness of this paradigm began to emerge within economics. Joan Robinson herself, for example, writing in 1960 stated that 'the number of firms in any particular market is largely a matter of historical accident'. John Sutton, a leading industrial economist and former head of department at LSE, takes up a similar theme four decades later (Sutton, 2000). Writing in the context of market structure, Sutton notes that even relatively mild relaxations of the assumptions of the basic model rapidly lead to indeterminate outcomes characterized by multiple equilibria. He notes that in most practical contexts, the 'search for a true model becomes futile', not least because of the huge amount of precise information which would be needed about the competing firms and their strategies.

I illustrate the above themes with two models of the decisions of firms. First, a market in which the various products on offer are in general identical, but they are differentiated with respect to a single attribute. Consumers have preferences with respect to this attribute. Firms have to consider the offer which they make in order to maximise their market share.

Second, a model of entry into a market in which there is a single monopoly supplier. A change in regulation permits other firms to enter the market. Using reasonable rules of behavior in which both firms and consumers react to incentives, a wide variety of outcomes is possible in terms of the evolution of price, quality, and the number of firms in the market.

For example, there is little or no connection across the individual solutions between the price which eventually obtains and the number of firms which survive in the market. This is much more in keeping with the experiences of deregulated industries than the standard economic view that the fewer the firms, the closer is price to that set by the initial monopolist.

Entry into a simple market

Hotelling (1929) introduced an interesting model in this context. Imagine a crowded beach at the height of summer. Two rival ice-cream sellers are deciding whereabouts to locate on the beach. They know three basic facts. First, the bathers are spread completely evenly across the entire range of the beach. This much may be obvious from simple inspection. Second, each person on the beach will at some point during the day want an ice-cream, although no-one will buy more than one. This is much less obvious, but we might suppose that the sellers are able to deduce this on the basis of previous experience. Third, no matter where the sellers locate on the beach, everyone will still want an ice cream. This is even less obviously true, but suppose for the moment that it is, and that it is known to the prospective vendors.

Where should the rivals choose to set up their stalls? We make a final assumption, that there are no costs to the sellers of re-locating from any location they might choose. In this highly simplified model, the firms (the ice cream sellers) are assumed to have accurate, detailed knowledge of consumer preferences, and their products are absolutely identical except for the place at which they are located. In the jargon of economics, their products are homogeneous except with respect to one attribute. This latter assumption is actually very restrictive when one thinks of the proliferation of ways in which even staple products such as milk are differentiated in practice: whole fat, semi-skimmed, skimmed, flavoured, available in bottles, cartons and plastic containers of different sizes.

Under these assumptions, Hotelling demonstrated that there is only one equilibrium location. Both firms locate exactly half way along the beach. At this point, each firm obtains half the total market, and neither can increase its market share by re-locating. For descriptive purposes, the consumers are distributed evenly in the interval [0, 100], and this point is 50. There is an unstable equilibrium at (25, 75), where each firm also obtains a market share of one-half. It is obvious that this is more convenient for consumers, since no-one would have to travel a distance of more than 25, compared to a maximum of 50 if the firms locate at (50,50). However, a firm can always increase its share by moving towards 50.

This very simple model becomes much more complicated purely by extending the number of potential entrants to more than two firms. The theoretical work on this is summarized by Huck *et al.* (2002): "only the two and four agent cases yield unique pure and symmetric equilibrium configurations that give identical payoffs to each agent". For six or more firms, the equilibria cease to be unique.

In the case of four firms, the unique equilibrium involves two located at 25 and two at 75. However, as Huck *et al.* (2002) note, this is 'not entirely intuitive and is also conflicting with casual empirical evidence'. They point out that there are no cities without shops in the central location, nor are there democracies with no political parties in the centre ground. They go on to carry out experiments with economics and business school students, and find that a W-shape distribution tends to emerge, with agents clustered around not just the two equilibrium points but also around 50. So although a stable equilibrium exists, in practice it appears to be hard to discover. The situation is even more complicated in the case of three and five firms, where the equilibrium involves entrants using a randomized strategy. And it is "a well established fact that experimental subjects have difficulties in randomizing".

So far, the assumption has been made that the firm can re-locate without cost. This is clearly unrealistic in most contexts, even in the current one where many simplifying assumptions are being made. Hotelling used for illustrative purposes a physical location decision. But the phrase 'location' can obviously be used more generally to indicate the positioning of an attribute of a product or offer which differentiates it from its competitors. Once the firm has selected the attribute and entered the market with it, changing the position of the attribute may not be easy. Apart from any direct costs involved, consumer perceptions of the attribute of the product may be difficult to alter.

As it happens, by making the model somewhat more realistic and assuming that the initial location which is chosen cannot be altered at all, it becomes easier to derive a practical rule of thumb for would be entrants. With two firms, for example, we can make a random draw for each from a uniform distribution on [0,100] as to where they locate, and observe the market shares which result. We can repeat this exercise many times, and plot market share against location. A rather well-defined inverted 'U' shape is observed, which is well fitted by a regression of market share on location and location squared. With the resulting regression coefficients, it is easy to show that in general market share will be maximized by locating at 50.

This result continues to hold as the number of firms is increased, but not as strongly. Figure 1 plots for illustration the result of 200 separate solutions of the model in which each time 5 firms locate at a fixed position at random

The inverted U-shape of the plot is apparent, but with a great deal of noise around it. The choice of 50 again turns out to be the best. But as the number of entrants gradually increases, the result becomes even less well determined, and by the time we reach 10 entrants it has disappeared altogether.

Even retaining the strong assumptions about consumer behavior and firms' knowledge of it, the results can readily be made much less clear. We could, for example, allow for entrants to fail if they do not obtain a specified number of customers, and their sales are then re-allocated to the nearest survivors. We can give some firms knowledge of the above result, and let others enter at random. We can give all firms this knowledge, but then the problem becomes the well-known one of the order of degree of reasoning used by agents when forming expectations on the expectations of others. So, for example, if a firm believes that the others will not try to anticipate what any other agent does but will locate at 50, then it is sensible to locate at 49 or 51.

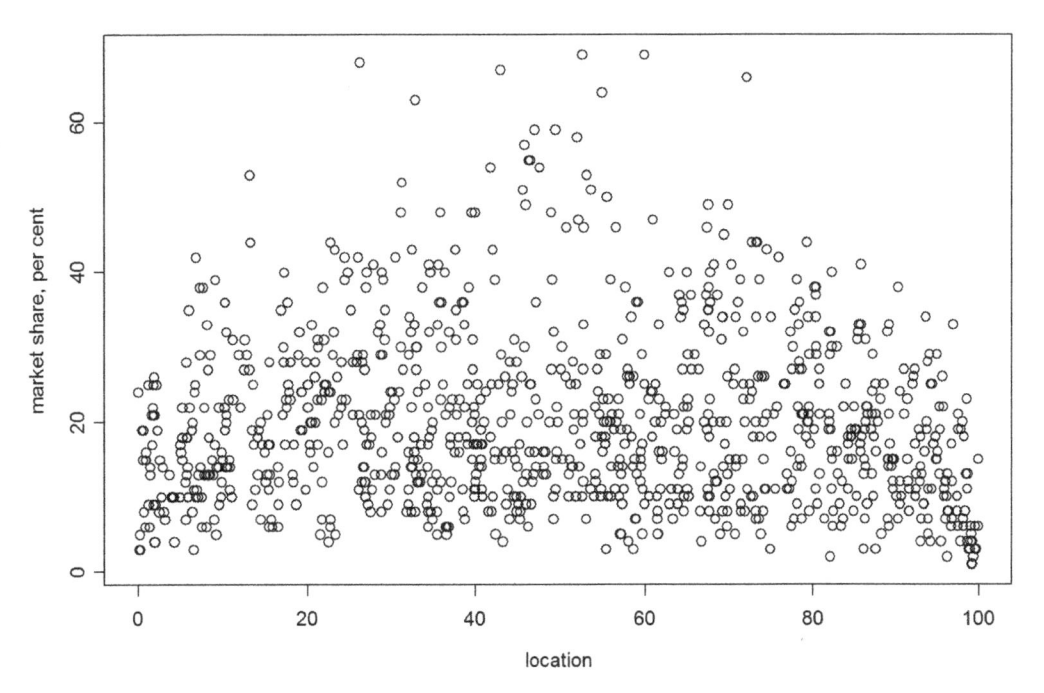

Figure 1 *5 firms entering a market in the Hotelling model, with only one feature differentiating the rival offers. Consumer evenly spread between (0, 100). 200 solutions of the model, no re-location allowed.*

But what to do if the firm believes that others will look one step ahead, as it were, and none of them will locate at 50?

Even if under any particular set of assumptions about the model, a Nash equilibrium could be proved to exist, knowledge of this might be of little or no value in a practical context. For example, in the Beauty Contest game, the Nash equilibrium solution is to choose zero. But played against humans, such a strategy is guaranteed to lose (or more precisely not win), except possibly in the latter stages of an iterated game played with the same agents throughout (Duffy & Nagel, 1997), because humans appear to use a low (but indeterminate) order of reasoning.

Methodological reflections

The Hotelling model can readily be made even more complicated. The above discussion, for example, is all on the assumption that consumer demand is inelastic with respect to the attribute of the product—everyone will buy one no matter where firms locate. But if this ceases to be true, quite different results can be obtained, all of which rely on firms having knowledge of the shape of the demand curve (for example, d'Aspremont *et al.*, 1979).

An implication of the above is that the level of knowledge required, whether about consumers or their potential competitors, to make optimal decisions, even if they can be shown to exist, is so great that in practice a firm cannot possibly hope to obtain it.

A standard response to such arguments in economics is to invoke the 'as if' clause. In this context, it is held that successful firms, although they may not consciously do this, act 'as if' they had the information on which to take an (approximately) optimal decision. However, simple inspection of Figure 1 illustrates the problem with this argument. If other firms are choosing their locations at random, on average 50 is the best location to choose. But selecting this can, on occasions, lead to the poor outcome of a market share of less than 10 per cent in a market containing 5 firms. History is only played once, and even the fittest agent can become extinct in any particular play of the evolutionary game (for example, Newman, 1997; Solé & Manrubia, 1996).

So there seem to be good grounds for the type of behavioral rules which are usually used in complex systems models, such as Peter Allen's model of fishing fleets. Namely, simple rules of thumb which avoid obvious loss.

However, the above discussion can also be seen in a rather nihilistic context. It appears that a very large variety of results can be obtained, often with what may be small changes in the assumptions. In order to be able to form a reasonable view on what assumptions to make in any given context, a very considerable amount of knowledge seems to be required.

An important methodological approach in agent based modelling of complex systems offers a potential way out of this problem. These models have as their foundation a set of behavioral rules for individual agents. The rules themselves should be capable of justification using evidence from outside the model. But the rules, and the way in which agents react to each other's decisions, should also be capable of accounting for the emergent properties of the system. In other words, the micro level rules for individual agents should give rise to key macroscopic empirical features of the system being analyzed.

A way of selecting an appropriate set of agent rules, which avoids the problem of having to acquire huge amounts of information about the system, is to use *micro* level rules which a) seem plausible judged by evidence outside the model, and; b) are able to generate selected key *macro*scopic features of the system. I discuss these issues at greater length in two papers coauthored with Bridget Rosewell, a pioneer of complex systems analysis in her role as Chief Economic Advisor to the Mayor of London and the Greater London Assembly (Ormerod & Rosewell, 2004, 2008).

The fact that macro features which are compatible with the empirical evidence emerge from a set of rules on agent behavior does not mean that these are necessarily the only rules which are capable of doing so. But this approach enables us to narrow down dramatically in any given context the potential set of rules to consider, without having to acquire huge amounts of detailed information about the system.

A criticism encountered from mainstream economists of this approach is that there may be a very large number of such rules, all of which can generate the relevant macro properties. In principle this may be true. But good agent based models are by no means easy to construct, and my standard reply to anyone making this argument is to ask them to produce an alternative set of micro rules from which the empirically observed macro behavior emerges. The conventional postulate of the rational, representative agent which is used in real business cycle and dynamic stochastic general equilibrium models, for example, really struggles to generate the most basic features of the time series on, say, American GDP growth, namely its properties in the time and frequency domains, the auto-correlation function and the power spectrum. So to a large extent, this type of criticism of agent based models by the economic mainstream is very much a case of the pot calling the kettle black.

In the next section, I give a simple illustration of this methodology with a model developed to account for the evolution of market structure following deregulation of an industry.

The evolution of market structure and competition

Consider the evolution of a market in which a single product is produced, which can be differentiated both on price and quality. The specific focus is upon the consequences of new entrants into a market in which, initially, there is a single monopoly supplier. The model is an agent-based one of firms and consumers, each following particular rules of behavior for pricing and purchasing. The model extends beyond the comparative statics of conventional theory. In this model, the market evolves over time, and solutions to the model describe the market structure which evolves and emerges from the process of competition.

The model was initially developed for British Telecom and the specific application was landline telephones. In this market, and others across the West, deregulation of the market by legislation undoubtedly led to improvements in the overall offer available to consumers in terms of both price and quality. Yet in general, the initial monopoly supplier retained a large market share, sufficiently large that, under most standard criteria, it would still be deemed to hold a near-monopoly position.

The model was developed to account for this empirically observed phenomenon. The behavioral rules for both firms and consumers are rooted in the principle that they both react to incentives. However, they do not necessarily do so in a completely rational way.

Agents operate with imperfect information under bounded rationality in this model. Yet, in the limit, if the model is allowed to run for a sufficiently long time with a very large number of potential entrants, a result from standard economic theory is recovered. Eventually, under these conditions, both p and q will in general converge towards zero. So from initial conditions with (p,q) at $(1,1)$, with unlimited entry and over unlimited time, the market will converge on the perfectly competitive solution $(0,0)$.

However, a key feature of the model is that it is not intended to describe the outcomes of two equilibria, the first with the monopolist and the second a perfectly competitive outcome. Rather, it is recognized that economic processes are rooted in time. It is not at all useful to say that, once a monopoly has been opened up to competition, *eventually*, at some unspecified point in the future, a different equilibrium will prevail in which (p,q) is no longer at $(1,1)$ but is arbitrarily close to $(0,0)$. We are interested in how the market evolves over a specified and realistic time-scale—up to 10 years, say. And we are interested in the relationship between (p,q) and the market structure which evolves over such a time scale.

The market is populated by n consumers. We assume for simplicity that they each consume an identical amount of the product in any given period. The amount spent per period by each consumer, and hence total sales of the product, may change over time, but our interest is on, amongst other things, the market shares of the producers rather than on the total size of the market. So the amount spent by each consumer is the same in any *given* period.

Initially, the market contains a single monopoly supplier, selling the product at a price of p_{mon} (using the subscript '*mon*' to indicate the incumbent firm's monopoly price), and with a quality q_{mon}. The model evolves on a step-by-step basis, in which each step is a period of time.

We specify a process by which other firms enter the market, both in terms of frequency and in terms of the total numbers entering each period. We specify as an input to the model the maximum number of new entrants into the market. With the incumbent monopolist, this makes a maximum number of k firms, where $k << n$. It is perfectly reasonable to assume that the maximum number of new entrants, $(k-1)$, will in general be relatively small, for two reasons. First, the entry of new firms generally reduces price and improves quality, so that the opportunity for profitable entry of additional firms is reduced. Second, capital stock and skills are by no means malleable in the real world, and even very large companies rarely undertake ventures which are well outside their established spheres of activity.

In the first step of the model, a potential entrant is drawn at random and enters the market. Both this firm and all subsequent new entrants come into the market with a (p,q) drawn at random from a uniform distribution on $[p_{min}, p_{mon}]$ and $[q_{min}, q_{mon}]$. The price p_{min} is the lowest possible price at which, after the process of technological innovation is complete, the product can be offered and a normal rate of profit obtained by the most efficient supplier. The quality q_{min} is the best quality at which the product can be offered, again subject to a normal rate of profit being obtained. Note that, for simplicity, quality is measured inversely, so that the *lower* the quality measure in the model, the *better* the quality is. Note also that quality is also expressed on a single dimension. This does not mean that the product necessarily has only one feature which measures its quality. There could well be several features, which are concentrated into a single measure of overall quality.

In the second step of the model, each of the remaining $(k-2)$ firms enters the market with probability π, where $\pi = (p_{av} + q_{av})/2$, where p_{av} and q_{av} are the average market price and quality which obtain at the relevant time. We specify below how firms capture consumers from competitors. The rationale for this entry process is straightforward. The lower is the value of p, in other words the lower is the market price and the higher is market quality, the less likely it is that a new entrant will be able to make a sufficiently attractive profit to justify the costs and risks of entry into a new market.

In subsequent steps of the model, all potential entrants who have not previously entered consider the prevailing value of p, and decide probabilistically on entry in just the same way.

All *n* consumers are connected on a network to the initial monopoly supplier. This could be in the case of telecommunications quite literally a physical network, but the use of the word 'network' in this physical sense is too limiting. 'Network' in this context means that consumers on the network of firm f_i are both aware of the offer from firm f_i and are willing to consider buying from it.

Each new entrant obtains potential access to a network of consumers. Each new entrant obtains potential access to a proportion of the total number of consumers drawn at random from a uniform distribution on $[v_{min}, v_{max}]$, where $v_{min,max} \in [0, 1]$. Once the group of customers to which a firm has potential access has been chosen, it is set up immediately. A further simplification is that the group then remains fixed during all subsequent steps in that particular solution of the model.

There are three obvious reasons why new firms in the market do not have potential access (in general) to all consumers, which can obtain either singly or in combination. First, the regulator could impose restrictions so that, for example and purely by way of illustration from the telephone market, a new entrant could be permitted to offer international calls but not domestic ones. Second, the marketing strategy of the firm may be such that not all consumers are aware that the firm is making an offer in the market. In reality, marketing strategies vary widely in effectiveness, and this is reflected in our model. Third, the firm itself may deliberately target only a small percentage of consumers. In the context of British land line phone calls, for example, several firms now specialize in offering cheap calls to India, say, or to the United States.

In each period, each consumer reviews the price and quality of each of the firms to which he or she is connected. The consumer at any point in time is only permitted to buy from a single supplier. This is not always completely realistic, but is a reasonable assumption to make in this initial specification of the model. The consumer is allocated from the outset a weight w_i, which expresses his or her preference between price and quality (w_i is chosen from a uniform distribution on $[0,1]$). The consumer calculates for each of the firms on his or her network $w_i^* p + (1-w_i)^* q$. For the kth firm, we describe $w_i^* p_k + (1-w_i)^* q_k$ as being the overall value to the ith consumer of this offer—v_{ik} for short.

The consumer switches all of his or her business to the firm offering the lowest v_{ik} of the *k* firms on his/her network (which may not correspond the lowest $w_i^* p + (1-w_i)^* q$ then on offer to other consumers, because the particular consumer concerned may not be aware of such offers), subject to the following condition. At the outset, each consumer is allocated a 'switching propensity', s_i, which is drawn at random within $[0,$

1]. If the customer identifies a v_{in}, which is lower than that of his or her existing supplier, v_{im}, he or she will then switch to firm n from firm m, with probability s_i.

There are several reasons for introducing this probabilistic element into the choice. Although the product offers of the firms are very similar, they are not perfect substitutes, for two reasons. First, the lowest (p,q) supplier may specialize in an offer which is not very important to a given consumer. Someone who makes only local phone calls will not be interested in a firm which provides only cheap international calls. Second, even within the same segment of the market, such as local calls, the product is not completely homogenous in that consumers may have doubts about the reliability of a previously unknown supplier.

There are two other possible reasons why consumers will not in general switch to the lowest price producer. First, there may be costs involved in switching. To take an obvious example, if changing suppliers involved having to change telephone number—staying with the telecomm example—for most people the savings on price would have to be considerable to offset the inconvenience involved. Second, consumers may simply exhibit inertia and stay with their existing supplier, perhaps because the savings involved are small.

At the start of the next period, each firm already in the market is given the opportunity to reduce its (p,q) offer. Firms are not certain about the distribution of preference across consumers regarding price and quality, and so assign equal weight to price and quality in each of the (p,q) offers which they observe. They aspire to move to the (p,q) of the firm for which $\omega^* p_i + (1 - \omega)^* q_i$ is minimized, where ω is the average of the w_i across all consumers.

However, firms differ in their ability to adapt their organization in order to deliver the desired (p,q) offer. We can think, for example, of firms as differing in their level of X-efficiency. The ability of the firm to do achieve the desired (p,q) depends on the firm's flexibility level φ_i. At the outset, each firm is allocated a flexibility level, φ_i, which is drawn at random from a uniform distribution on $[\varphi_{min}, \varphi_{max}]$, where $\varphi_{min, max} \in [0, 1]$. In each period, each firm switches to the lowest $\omega^* p_i + (1 - \omega)^* q_i$ with probability φ_i.

Consumers then review their choice of suppliers given the revised set of (p,q) from existing suppliers, and given the (p,q) offered by new entrants (if any) in that period.

Once a firm has entered the market, it is able in principle to acquire customers both in the period in which it enters, and in each subsequent period. We do not specify an explicit cost function for firms, but assume there is a minimum level of sales, x, which any firm needs to be able to continue to exist. We specify x in terms of market share, with x being a parameter of the model. If a firm fails to secure x per cent of the market for two successive periods, it is deemed to then exit the market.

The properties of any individual solution of the model are contingent on a number of factors, and no two solutions are the same. For example, an important element is the combination of the (p,q) with which the first competitor enters the market and the proportion of consumers, v , to which it gains access. If the former is sufficiently low and the latter sufficiently high, the probability of any additional entrant deciding to join becomes low, because the average (p,q) which consumers face is low, and so the prospective ability of a further entrant to make a profit is low.

The results below illustrate the properties of the model when it is populated by 1000 consumers and a total of 20 firms can in principle enter the market. 1000 separate solutions of the model are obtained. The model is solved for 40 steps, the rationale for this being that, in the context of a utility such as telecoms consumers are usually billed quarterly, so that it is reasonable to regard each step as a quarter, 40 steps making a total of 10 years.

Figure 2 plots the histogram of average price over the 1000 separate solutions of the model.

The single most frequently observed outcomes for the market price is in the range 0.00-0.05. In other words, price does fall to a level close to the minimum which is feasible. Occasionally, however, the price remains relatively high. In terms of the distribution of outcomes across a number of solutions of the model, the quality of the product evolves similarly to price, although the outcome of the two may obviously differ in any given scenario.

The mean level of market price after 40 periods is 0.147, with a minimum of 0.0001 and a maximum of 0.581. The inter-quartile range is between 0.069 and 0.225.

Figure 3 plots the histogram of the outcomes, again after 40 periods, across the 1000 solutions of the model of the market share of the initial monopolist.

Quite frequently, the incumbent monopolist retains a very high market share. Despite this, as Figure 3 shows, the market price usually falls very sharply. The average market share of the monopolist after 40 periods is 46.2 per cent, with a minimum of 3.5 and a maximum of 95.5 per cent. The inter-quartile range is wide, between 32.4 and 60.0 per cent.

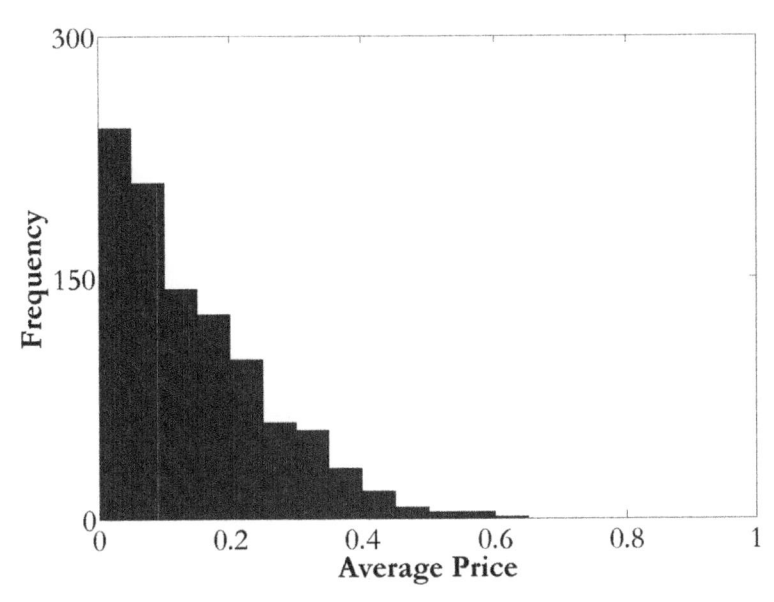

Figure 2 *Histogram of average price outcomes in 1,000 solutions of the model.*

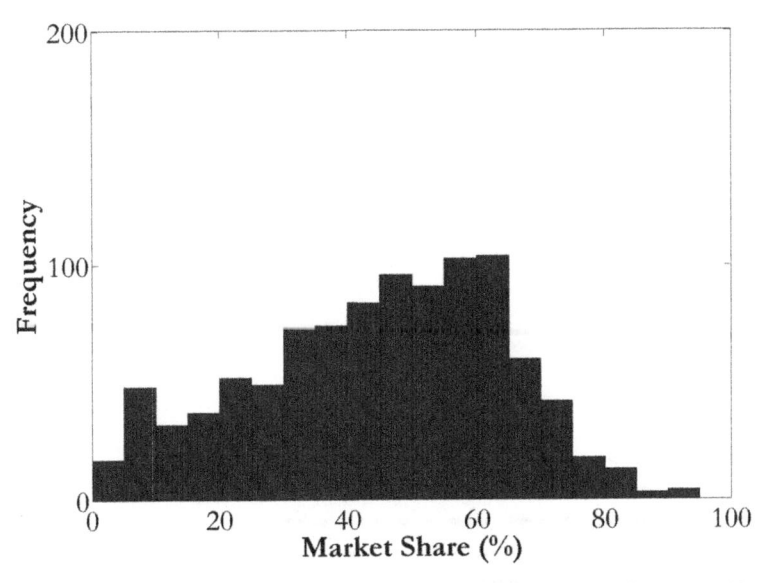

Figure 3 *Histogram of market share of the initial monopolist in 1,000 solutions of the model after 40 periods.*

So the model captures the key stylized fact of deregulated markets, namely that the offer to consumers improves markedly in terms of price and quality, but the market share of the initial monopolist remains high.

This is confirmed in Figure 4, which plots the eventual market price and the market share of the initial monopolist. The simple correlation between the two variables is -0.06.

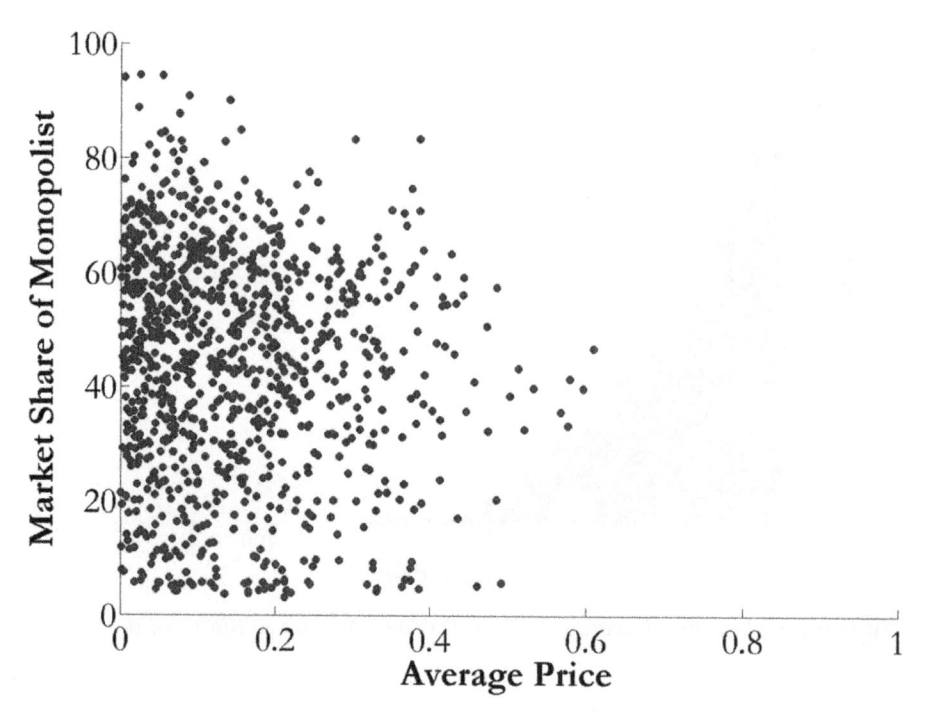

Figure 4 Scatter plot of 1,000 solutions of the model between the eventual market share of the initial monopolist and the average price after 40 periods.

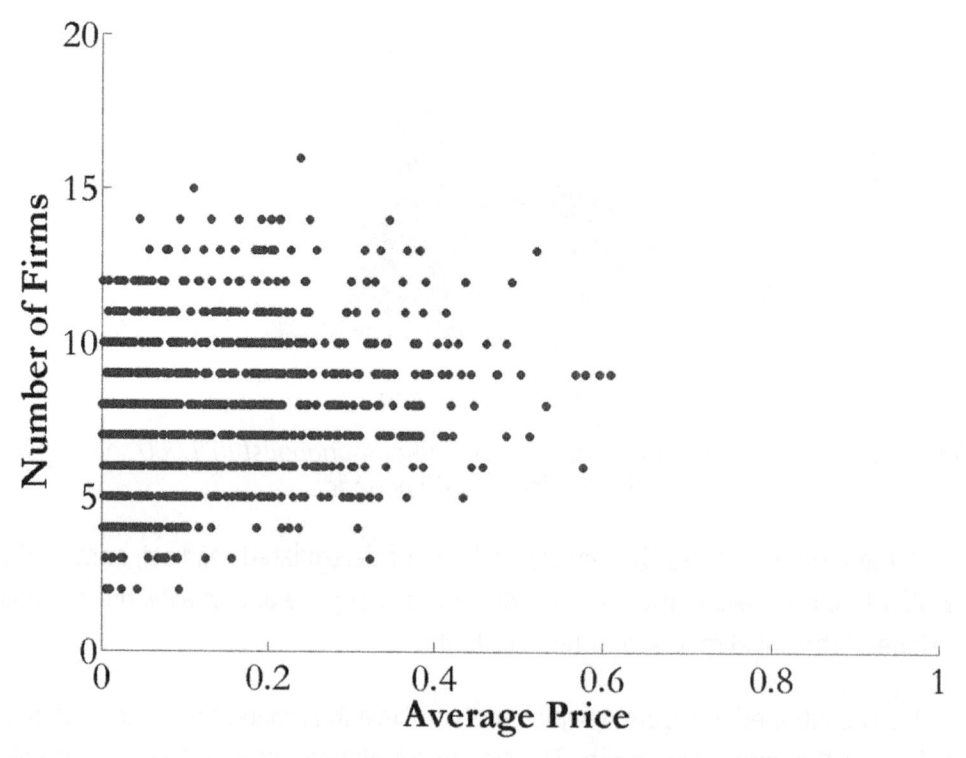

Figure 5 Scatter plot of 1,000 solutions of the model between the number of firms in the market and the average price after 40 periods

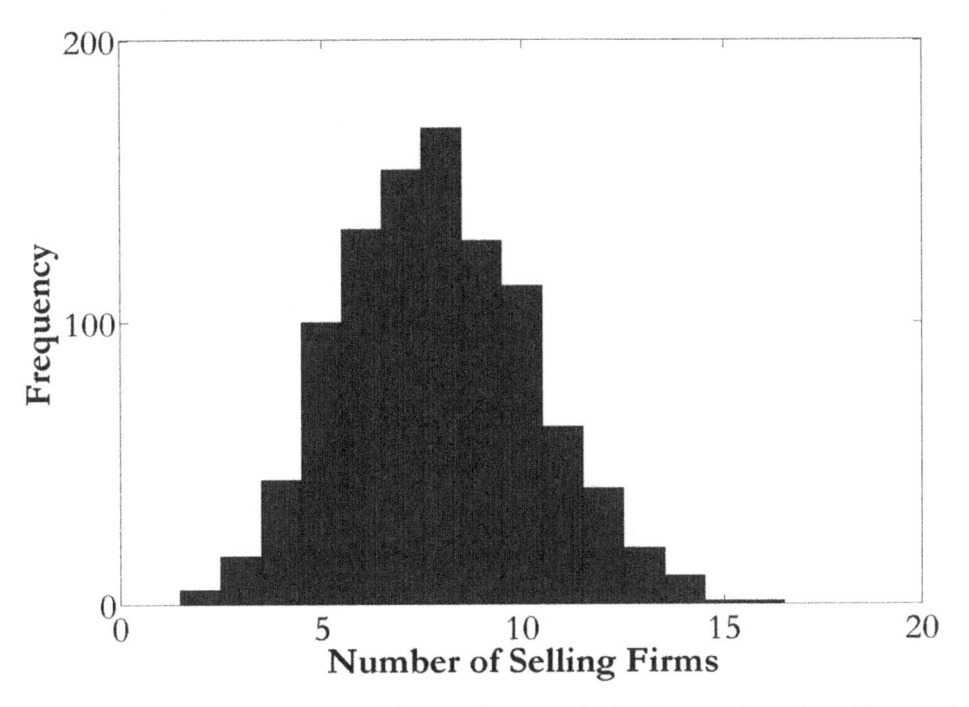

Figure 6 *Histogram of number of firms who remain in the market after 40 periods, 1000 solutions*

Figure 5 shows that, more generally, there is little connection between the market price and the eventual number of firms surviving in the market.

The model also captures two more general properties of evolving markets in which firms both enter and exit over time, as Figures 5 and 6 illustrate. The model was not intended to capture these stylized facts, but nevertheless does so.

A wide range of outcomes is possible for the number of firms who remain in the market after 40 periods, as Figure 6 shows.

The mean number of firms is 7.9, so that on average just over 12 out of the 20 firms exit the market. This seems compatible with the outcomes which are observed in practice (see, for example, Carroll & Hannan, 2000).

Figure 7 sets out the distribution of the average market share of the 8 largest selling firms after 40 periods. (Of course, in a number of the solutions of the model, when fewer than 8 have non-zero sales, the value for some of the 'largest' 8 in these particular solutions is zero).

The figure shows the distribution regardless of the identity of the firm. In the majority of solutions of the model, the largest firm is the initial monopolist, but this is not

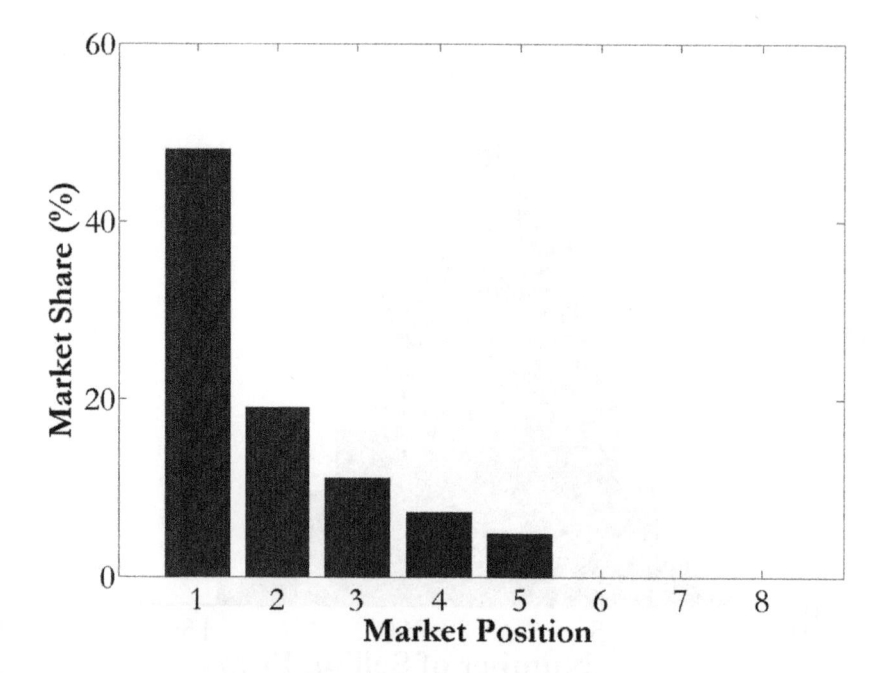

Figure 7 *Distribution of market share of the 8 largest firms in 1,000 solutions of the model after 40 periods*

always the case.

A good approximation to the size distribution of the largest 8 firms after 40 periods is provided by a power law. Axtel (2001) shows that this a general characteristic of the distribution of firm sizes in the United States. A log-log least squares fit of average market share on the rank of the firm by market share (largest has rank equal to 1, etc.) gives an R^2 of 0.982 and an estimated exponent of -1.67 with a standard error of 0.02. An exponential also gives a good approximation to the distribution of firm size, but the power law is better.

Conclusion

Peter Allen was one of the earliest innovators in the application of complex systems principles to the social sciences. His fishing fleet model captures some general features of the behavior of real-life firms. They act with purpose and intent, but under conditions not only of imperfect information at any given time, but of an environment which constantly evolves. There is a great deal of contingency in the outcomes of any individual solution of the model, and time and process are important. Simple rules of thumb in general give agents better outcomes than do attempts to follow rational expectations.

In this contribution, I develop these themes in the context of two different models of firms and the evolution of competition and market structure.

An important development in complex systems analysis over the past 20 years has been the realization that agent based modelling of such systems requires modelers from the outset to confront empirical evidence. This is not only in terms of having plausible rules of behavior for the micro-level agents in the system, but to ensure that the macroscopic properties of the system which emerge are consistent with aspects of reality. Apart from the fundamental principle of scientific validation which this addresses, it enables agent based models to escape from the criticism of mainstream economists that, once the assumption of rational agents following rational expectations is relaxed, any outcome becomes possible. This is far from being the case. Requiring models to be compared with empirical evidence limits dramatically the range of models which can be considered as plausible in any particular context.

I illustrate this with a model of the evolution of competition and market structure of an industry where de-regulation permits entry into a market controlled by a monopoly supplier. Two key features of the real world are that, first, that the price/quality offer to consumers usually improved markedly, yet at the same time the initial monopolist typically retains a large market share. Second, the evolution of price/quality in the market bears little or no relationship to the number of firms active in the market at any point in time. The model is able to replicate these empirical features. In addition, two further stylized facts about firms emerge from the model: the majority of new entrants into a market fail, and the size/rank distribution of firms is heavily right-skewed, indeed approximated by a power law relationship.

References

Akerlof, G.A. (2002). "Behavioral macroeconomics and the macroeconomics of behavior," *American Economic Review*, ISSN 0002-8282, 92: 411-433

Axtell, R.L. (2001). "Zipf distribution of US firm sizes," *Science*, ISSN 0036-8075, 293: 1818-1820

Carroll, G.R. and Hannan, M.T. (2000). *The Demography of Corporations and Industries*, ISBN 9780691120157.

Chamberlin, E.H. (1933). *The Theory of Monopolistic Competition*, Harvard University Press, Cambridge MA.

d'Aspremont, C., Gabszewicz, J.J. and Thisse, J.-F. (1979). "On Hotelling's stability in competition," *Econometrica*, ISSN 0012-9682, 47: 1145-1140.

Duffy, J. and Nagel, R. (1997). "On the robustness of behavior in experimental 'beauty contest' games," *Economic Journal*, ISSN 0013-0133, 107: 1684-1700

Eller, M. and Edstrom, J. (1998). *Barbarians Led by Bill Gates*, ISBN 0805057544.

Huck, S., Müller, G. and Vriend, N. (2002). "The East End, the West End and King's Cross: on clustering in the four player Hotelling game," *Economic Inquiry*, ISSN 0095-2583, 40: 231-240.

Hotelling, H. (1929). "Stability in competition," *Economic Journal*, ISSN 0013-0133, 39: 41-57.

Kahneman, D. (2003). "Maps of bounded rationality, psychology for behavioral economics," *American Economic Review*, ISSN 0002-8282, 93: 1449-1475.

Leibenstein, H. (1966). "Allocative efficiency versus X-efficiency," *American Economic Review*, ISSN 0002-8282, 56: 392-415.

Newman, M.E.J. (1997). "A model of mass extinction," *Journal of Theoretical Biology*, ISSN 0022-5193, 189: 235-252

Ormerod, P. (1994). *Death of Economics*, ISBN 0571171257.

Ormerod, P. and Rosewell, B. (2004). "On the methodology of assessing agent-based models in the social sciences," in J.S. Metcalfe and J. Foster (eds.), *Evolution and Economic Complexity*, ISBN 1847203388 (2007)

Ormerod, P. and Rosewell, B. (2009). "Validation and verification of agent-based models in the social sciences," in F. Squazzoni (ed.), *Epistemological Aspects of Computer Simulation in the Social Sciences*, ISBN 364201108X.

Robinson, J. (1933). *The Economics of Imperfect Competition*, Palgrave, London

Robinson, J. (1960). *Exercises in Economic Analysis*, ISBN 0333069579.

Smith, V.L. (2003). "Constructivist and ecological rationality in economics," *American Economic Review*, 93: 465-508.

Solé, R.V. and Manrubia, S.C. (1996). "Extinction and self-organized criticality in a model of large-scale evolution," *Physical Review E*, ISSN 1539-3755, 54: R42-R45.

Sutton, J. (2000). *Marshall's Tendencies: What Can Economists Know?* ISBN 9780262692793.

Paul Ormerod is an economist and author of best-selling books *The Death of Economics* (1994), *Butterfly Economics* (1998), *Why Most Things Fail: Evolution, Extinction and Economics* (2005) and *Positive Linking: How Networks Can Revolutionise the World* (2012). He was elected as a Fellow of the British Academy of Social Sciences in 2006, and in 2009 was awarded a DSc honoris causa by the University of Durham for "the distinction of your contribution to the discipline of economics". He is currently a Visiting Professor in the Centre for Decision Making Uncertainty, University College London. He publishes in a wide range of journals such as *Nature, Proceedings of the National Academy of Sciences, Proceedings of the Royal Society (B), Evolution and Human Behavior, Advances in Complex Systems, Physica A, Plos ONE, Applied Economics, Economics* e-Journal, *Economic Affairs* and *Journal of Economic Interaction and Coordination*. He was one of the founders of the Henley Centre for Forecasting in the mid-1980s, which the management sold to WPP Group, a FTSE 100 company, in the mid-1990s. He is a partner of Volterra Partners LLP, London, and is currently involved in several hi-tech start-ups analyzing Big Data with innovative techniques.

Index

A

abstraction   27, 29, 31, 33, 35, 37, 39, 41, 43, 45
adaptation   11-12, 19-21, 44, 151, 179-80, 184, 225, 233, 236-7, 256, 258, 267, 279
adaptive systems, complex   133, 150, 153, 179, 181, 267, 280
agents   17, 29-30, 41, 46, 51, 81, 122, 124-5, 133, 135, 143, 180, 199, 205, 228, 249-51, 267, 283, 285, 287-92, 300-1
Allen   1, 6-7, 11-24, 27-9, 34, 39-40, 42-4, 51, 53, 75-6, 107-8, 111, 124-6, 132-6, 139, 147-8, 150, 202-3, 216, 220, 223-4, 236, 242, 247-8, 250-1, 254, 259, 263, 278, 283
analysts   33, 251, 253-4
archaeology   150, 191, 203, 216-18, 220
architecture   80, 104, 173, 175, 201
assumptions   17-19, 27, 134, 136, 283, 286-90, 301
attractor basins   83, 86, 91, 99, 103
attractors   83, 85-7, 89-90, 97-8

B

behavior   17, 20, 40, 42, 44, 89, 93-4, 111-13, 115, 120, 139, 183, 189, 191, 194, 203, 209, 248-52, 258, 284, 286, 291, 300-2
  alternative   124-5
  migratory   19, 250-1
benefits   37, 79, 124-5, 162-3, 183-4, 217, 232, 276-8
bifurcations   51, 53, 75-6, 83, 101
Boolean network model   106
Boolean networks   80-1, 83-6, 89-94, 97, 103-5, 107
  complex   104
  random   93, 95
buildings   18, 24, 35-6, 41, 53, 67-9, 136, 139, 151-2, 256, 264, 267
businesses   70, 117, 170, 188, 227, 233, 238-42, 279-80, 294
  small   36, 172-3

C

capitalism   24, 128, 223, 225-6, 230, 232, 234, 237, 242-3
  modern   223, 225, 236
CBD (central business district)   57
chaos   21-2, 24, 76, 103, 133, 152-3, 180, 217, 224, 267
character-states (CSs)   136, 138-9, 141-4, 147-9
cities   39, 49-58, 60-5, 67-8, 72-7, 150, 194, 197, 205, 216, 263-5, 267-73, 275-81, 287
  size distribution of   52, 54
city systems   49-55, 57-9, 61, 63, 65, 67, 69, 71, 73, 75, 77
clusters   67, 69, 83, 161, 169, 181, 183
coevolution   11, 21-3, 25, 40, 114, 126, 150, 169, 178-81, 184-5, 218
cognitive maps   113, 123, 125
commodities   181, 192-3, 204-6, 215, 249-50
competition   21, 40, 46, 53, 67, 139, 164, 169, 176, 179-80, 188-9, 193, 200-1, 203, 208, 226, 232-4, 243, 285, 291-2, 301
  evolution of   283, 301

competitive process   69, 223, 227, 233-4
complex adaptive systems (CAS)   133, 150, 152-3, 181, 267, 280
complex exchange dynamics   191, 193, 195, 197, 199, 201, 203, 205, 207, 209, 211, 213, 215, 217, 219
complex networks   88-9, 92-3, 95, 104, 106, 201, 203, 214
complex societies   114, 217
complex systems   3-4, 6-8, 12-15, 17, 19, 21, 25, 27-8, 31-3, 35, 40, 46, 79-81, 83, 85, 87, 89, 91-5, 97, 99, 101, 103, 105-7, 124, 131, 139, 249, 251-2, 259, 283-4
  evolutionary   126, 132
Complex Systems Research Centre   11-12, 17, 25
complex systems thinking   131-2, 150, 258
complexity   1, 3-8, 11-12, 16-17, 22-5, 27-8, 30, 33, 43-4, 46, 76-7, 105-7, 125-6, 150-1, 185, 187-8, 214, 216-18, 223-9, 231, 233, 235-7, 239, 241-3, 247, 249, 251, 257-61, 278-81, 283
  evolutionary   23, 248
  transition function   105
complexity analysis   179, 225
complexity approach   159, 161, 163, 165, 167, 169, 171, 173, 175, 177, 179, 181, 183, 185, 187, 189
complexity literature   29
complexity perspective   5, 159, 202, 215, 225
complexity science   7, 11-13, 15-25, 44-5, 51, 107, 264, 278
complexity theory   5, 8, 45, 52, 128
complexity thinking   79, 107
components   6, 16-17, 92, 94, 134-5, 163-4, 174-5, 208, 254
computers   2, 5, 28, 39, 104, 107, 165, 172, 185
concepts   21, 29, 32, 35, 40, 49, 61, 86, 92, 94, 128, 164-5, 176, 184, 187, 224, 252-3, 283-5
connectivity   33, 37, 39, 81, 91, 172, 203
consensus   15, 29-32, 143, 145, 148
consumers   18, 162-3, 180, 249, 256, 283, 285-7, 289-97, 301
context   3, 5-6, 11, 19, 24, 27-8, 30-4, 39, 43, 45, 51, 54-5, 86-7, 99, 101, 111, 132, 136, 149, 184, 199, 228, 230-3, 238, 283, 286, 290, 294, 296, 301
  complex   31, 33
  local   40, 44
  multiple   33-4
  unique   37
control   113, 150, 166, 189, 193-5, 199, 205, 208, 214, 217
  resource priority   140, 144-5
cooperation   179-80, 203
core sizes, dynamic   96-7, 103
customers   117, 133, 140, 161, 175, 179, 181, 184, 231, 288, 294-5

D

decision-making   19, 23, 44, 132-3, 135-6, 143, 149-50, 259
decision-support tools   131-3, 149
decisions   6-7, 36, 38, 41-2, 123, 133, 151, 172, 203, 215-16, 229-30, 232, 258, 263, 285-6, 290
decomposability   94-5
degree   41, 49, 54, 104, 136-8, 144-7, 208, 213, 229, 241, 288

descriptions   3, 15-17, 28, 113, 136, 140-1, 160, 175, 201-2, 226
  qualitative   14-15, 17
design   41, 45, 77, 80, 103-4, 156, 162-4, 169, 171, 173, 178, 180, 184, 220, 255
development policy   18, 247-9, 251, 253-5, 257, 259, 261
distributed networks   203, 213, 215
distribution networks, complex structure of   169
distributions   17, 49, 52, 54-5, 57-61, 65, 67, 69-70, 73, 193, 232, 238, 241, 266, 280, 295-6, 299-300
  power-law   264-5, 279
  uniform   288, 293-5
diversity   6-7, 11, 13, 17, 21-2, 43, 125-6, 132-6, 145, 147-50, 159, 164-6, 169, 175, 178, 230
dominant design   162-4, 174, 180, 184-5
dynamic core   89-93, 95, 97-8, 102-3
dynamic robustness   95, 97, 99-100, 102-3
dynamic structure, emergent   89
dynamical robustness   85-6, 94, 100-1
  network's   101, 103
dynamics   4, 7, 12, 15, 17, 51-4, 68-70, 74-5, 81, 84, 87, 89, 94, 101, 133, 191-2, 194, 198, 203-4, 206, 223, 254
  temporal   50-1

E

economic complexity   45, 223, 236, 302
economic development   11, 15, 51, 188, 223, 225-6, 228, 231, 233-4, 243
economic evolution   223, 225-7, 229, 231, 233, 235-7, 239, 241-3
economic growth   199, 228, 243-4
economic systems   18, 22-3, 72, 202, 224-5, 227, 242, 247, 249-51, 253, 255, 257, 259, 261
economics   35, 73, 76, 119, 126, 153-4, 187-8, 223-4, 228, 242-4, 258, 267, 279, 283, 285-7, 290, 302
economies   51, 72, 184, 194, 204, 217, 224, 227, 231-5, 237-8, 265, 267-8, 271, 275, 278-9
economists   223, 226, 253, 273, 281, 283, 302
ECS model   134-6, 138-9, 142, 149
emergence   1, 11, 14, 21-2, 24, 32-3, 44, 89-90, 105, 107-8, 114, 126, 139, 150, 163, 166, 174-5, 180, 185, 193, 199, 214-15, 223-5, 267, 281
employees   134, 141, 144, 148, 155, 170
emulation, competitive   193, 201, 203, 214-15
entrepreneurial firms   168-9
entrepreneurs   165, 172, 197, 227, 229-30
environments   5, 7, 11-12, 18-21, 23, 27, 39-40, 44-6, 50, 85, 99, 101, 111, 113, 115-16, 118, 123-4, 134, 177, 179, 192, 216, 259, 283, 300
epistemology   12, 16, 21-2
equilibrium   50-1, 124, 226, 233, 235, 243, 249, 258, 284, 288, 292
evolution   15, 18-23, 45, 49, 107, 111, 113, 115, 117, 119, 121, 123, 125-7, 132, 134-5, 137, 150-2, 177, 179, 181, 187-8, 192, 213-14, 217, 220, 226, 234, 236, 241-3, 259
  industrial   162, 165, 182, 186
  organizational   22, 131-2
evolution of complex exchange dynamics   191, 193, 195, 197, 199, 201, 203, 205, 207,

209, 211, 213, 215, 217, 219
evolutionary economics   21-2, 133, 150, 243
evolutionary theory   236, 238, 240, 281
evolving operations, management of   131, 133, 135, 137, 139, 141, 143, 145, 147, 149, 151, 153, 155
exchange systems   192, 194, 203, 214
expectations, rational   283, 300-1
experience   2, 5-6, 19-20, 33, 35, 40, 42-3, 67, 115, 123-4, 139, 153, 159-60, 172, 182-4, 229, 286
experiments   7, 17, 20, 37, 95, 136, 164, 192, 215, 217, 287

F

failures   114, 116, 120, 146, 148-9, 177, 184, 186
feedback loops, structural   87, 90, 97-8
firms   13, 20, 53, 70-3, 131, 133, 136, 149, 160-1, 163, 167-71, 173, 175, 178, 180-3, 185, 187, 232, 238, 241, 249-50, 265, 280, 283, 285-96, 298-301
  models of   283, 301
fishing fleet model   290, 300
flexibility   44, 100, 103, 137, 141, 180
flow   72, 79-80, 84, 87, 90, 94, 139, 141, 200, 205, 216, 251
frequency   49, 53-4, 58, 208, 265, 267, 293
frozen nodes   87, 89-90, 94, 97, 101, 104
futures, sustainable   27, 43

G

group   20, 39, 53, 111, 143, 145, 155, 160, 177, 195, 199, 205-6, 213, 251, 257, 275, 294
growth   76, 131, 139, 142, 162, 171, 207-8, 236-40, 243, 255, 259, 267, 273, 279
growth rates   65, 236, 238-41

H

high buildings   53, 67-9
history   3, 6, 14, 24, 128, 178, 183, 188, 224-5, 228, 244, 276, 280, 284, 290
hubs   53, 58, 69-70

I

identity   14, 22, 105, 120, 122, 161, 179, 184, 187, 299
ignorance   13, 19, 21, 42, 44, 125-6, 229, 231-2, 235
income   44, 49, 53, 55-7, 59-61, 65-6, 237, 248-9, 257
individuals   12, 19, 112, 116-19, 121, 123-4, 131, 135, 194, 225, 230-1, 256
  transitioning   119, 125
industrial activity   159, 181
  organization of   178
industrial development   160-2, 165, 168-9, 171-2, 174, 176-7
industrial systems   180, 185
industries
  evolution of   14, 159, 161, 163, 165, 167, 169, 171, 173, 175, 177, 179, 181, 183, 185, 187, 189
  semiconductor   170-1

industry evolution   161-2, 186
information   1, 6, 79-80, 84, 87, 90, 94, 99, 101-2, 107, 111, 113, 115, 117, 119, 121,
    123-5, 127, 140, 144, 179, 187, 233, 235, 243, 252, 254, 258, 286, 290-1
  imperfect   283, 285, 292, 300
information conserving loops   87, 90-1, 97, 99, 102-3
information flows   79, 84, 87, 90, 106-7, 254
innovation   6, 13, 22, 38, 44, 53, 76, 125, 151, 159-60, 167, 169, 171-2, 178-80, 186-9,
    201, 207, 223, 225-36, 239, 241, 243, 248, 256, 259-60
Innovation systems   18, 247-9, 251, 253-9, 261
inputs   43, 80-2, 86, 91, 101, 149, 185, 234, 249, 293
instrumentation   161, 165, 167-9, 184-5
integrator   11-13, 15, 17, 19, 21, 23, 25, 182
interactions   3-4, 6, 12-14, 17, 28, 31-3, 43, 50, 76, 80-1, 84-5, 90, 94, 113-14, 134-5,
    138-9, 161, 167, 179, 182, 184, 186, 191-2, 194, 198, 200-1, 206, 216, 267
  nonlinear   92, 205, 214
interpretation   4, 6-7, 14, 16, 106, 176, 192
interviews   14, 127, 136-8

K

knowledge   1-7, 11, 13, 16, 19-22, 33, 40-4, 60, 67, 79, 107, 134, 159, 166, 182-3, 185,
    188, 223, 228-32, 235-7, 242, 255-8, 287-90
  economic   229, 235
  network of   3
  recombinations of   159-60
  tacit   42, 166, 257-8
  transfer of   42
knowledge management   1, 4, 6-7, 44

L

landscape   24, 41, 184, 195, 199-201, 203, 205, 209, 211, 213-14
  competitive   198-9, 215
  fitness   152, 184
language   14-15, 17, 35, 92, 104, 113, 126, 224-5, 279
laws   22, 33, 45, 53, 55, 80, 106, 114, 187, 280
learning   5, 7, 12, 16, 19-22, 28-9, 33, 37-8, 44, 126, 133, 150, 163, 172, 177, 180, 183,
    186, 242, 255, 260, 267
  experiential   134-6
levels   18, 20, 28, 33, 38, 43, 89, 105-6, 113-14, 134, 136-7, 141, 160, 164, 178-80, 194,
    202, 206, 208, 211, 214, 238, 264, 290, 295-6
Levels of abstraction   27, 29, 31, 33, 35, 37, 39, 41, 43, 45
limitations   80, 105, 144-5, 165, 185, 231
links   28, 49, 69, 73, 136, 181, 223, 228, 230, 234, 238, 248, 257
locations   18, 33, 37, 53, 212, 250, 287-8, 290
logic   27, 227, 241-2, 273, 277
loops   84, 87, 89, 97, 99

M

management 6-7, 23-5, 45, 132-3, 145, 147, 149-53, 156, 189, 220, 229, 242, 258, 260, 278-9, 281, 303
managers 7, 38, 131-2, 136, 138, 142-5, 147-8, 172
manufacturing 36, 136, 139, 151, 153, 166, 264, 280
maps 32-3, 73-4, 112, 116, 118, 191, 196, 217
market 13, 19, 21, 50, 117, 162-3, 168-9, 171, 173-5, 177, 185, 194, 197, 206, 232-3, 250, 256, 263-4, 285-6, 288-96, 298-9, 301
  diverse 159, 161, 163, 165, 167, 169, 171, 173, 175, 177, 179, 181, 183, 185, 187, 189
market share 286-8, 290, 292, 296-8, 300
market structure 188, 233, 283, 285-6, 291-2, 301
  evolution of 283, 285, 287, 289, 291, 293, 295, 297, 299, 301, 303
members 112-13, 116, 118-19, 122, 124, 183-4, 186, 219, 244, 271, 275
membership 119-20, 122, 183, 198
micro-diversity 20, 40, 43, 134-5, 143
microcomputers 167, 171-5, 184, 186
microprocessors 152, 171-2, 183, 185
modeling 17, 19, 25, 80, 92, 208
modelling 25, 32, 131-3, 135, 137, 139, 141, 143, 145, 147, 149-51, 153, 155, 217, 220, 242, 284
  computational 133
models 17-19, 32-3, 40, 43-4, 51, 58, 86, 92, 106, 111, 113, 131-2, 134-5, 139, 142-3, 187, 191, 194, 200, 203-4, 206, 208-9, 212-15, 248, 250-2, 254, 283, 285-6, 288-94, 296-302
  business 25, 227, 231
  complex 33, 191
  core-periphery 200
  evolutionary 192, 248
  multi-scalar 27, 43
  self-organizational 134-5, 143
  simulation 51, 132, 139
Modern societies 114-19, 124-5, 231
Modernity 116, 118-20, 123-6
modes 85-7, 169, 226, 236
  operational 85, 87, 89
modularization 87, 89, 94-5, 104
  process of 88-9, 93-4, 103

N

narratives 11, 33, 43-4
nature 1-2, 8, 13, 16, 23-4, 35, 50-1, 53, 55, 76, 80, 94, 108, 111-12, 128, 135, 143, 153, 164-5, 185, 191-2, 204, 214, 223-4, 226, 232, 234, 241, 250-1, 302
Network Computer (NC) 161, 176-8, 184-6
network dynamics 85, 90, 203
networks 3-4, 7, 12, 33, 36, 53, 69, 80-7, 89-95, 97-8, 100-1, 103-6, 176, 179, 184, 196, 200-1, 203, 214, 256-7, 265, 294, 302
  reduced 95
  regulatory 81, 85-6
  spatial 69, 76

transaction 208
  unreduced 95, 98
new knowledge 168, 185, 255-6, 258
new technologies 18, 50, 132, 134, 159, 162, 167, 178, 234, 255-6, 259
niches 113, 119, 125, 165, 168-9, 178
nodes 80-4, 86-7, 89-91, 101, 104-5, 203, 214
noise 20, 32, 58, 67, 111, 113, 115, 117, 119, 121, 123-5, 127, 135, 227, 288
non-conserving information loops 87, 89-90, 93-5, 98-9, 103-6
non-conserving loops 89-90, 99, 101, 106
novelty 12, 223-4, 226, 228-9, 236, 242
  emergent 223-4, 226, 236

O

objects 13, 49, 52-3, 55-7, 60-1, 73-4, 193, 206
operation, modes of 87, 89, 101
organizations 16, 18, 21-4, 29, 41, 44, 80, 94, 99, 108, 117, 126, 131, 134-5, 139-40, 142-3, 145, 149-50, 160-1, 179, 181, 187, 201, 203, 229, 232-3, 244, 256, 258, 278-80
outcomes 12, 19, 132, 134-5, 168, 180, 199, 256, 283, 285-6, 292, 296, 299-301

P

participants 136-8, 163, 179-80, 264
patterns 49-53, 148, 162, 166, 216, 225-6, 235, 240-1
PC (Personal Computer) 161, 173-5, 177-8, 185-6, 284
performance 44, 132, 135, 137-9, 142-4, 147, 149, 153, 163, 166, 173, 175
performance criteria 137-8, 143
perspectives 3, 6, 13, 25, 30, 40, 54, 58, 75, 90, 93, 113, 133, 202-4, 224
PLs (Power-laws) 264-8, 271
policies 11, 18-19, 23, 41, 44, 46, 111, 131-2, 134, 136-9, 141, 148-9, 154-5, 174, 189, 258, 260
population 49, 52, 55-61, 65-7, 74-5, 112-13, 125, 161, 199, 204-5, 207, 209, 211, 213, 219, 238-41, 251, 257
population dynamics 150, 204-6, 209-10, 213, 236
power 54-5, 57, 59, 117, 192-3, 195, 198-200, 208, 213-17, 225, 241
power laws 50, 52-5, 57, 67, 69-70, 75, 263, 266, 278-80, 300
PPI (peer polity interaction) 200-1, 214-15
pre-market economy 191-3, 195, 197, 199, 201, 203, 205, 207, 209, 211, 213, 215, 217, 219
preferences 18, 101, 103, 148, 161, 164, 234-5, 286, 294-5
processes 4-5, 7, 13, 15, 17, 29-30, 32, 39, 42-3, 87, 90, 92-3, 95, 111, 113, 123-4, 127-8, 163, 165, 176, 178-9, 181-3, 185-6, 199, 201, 233-5, 250-2, 256, 283-5, 293
  decision-making 131-3, 147
  evolutionary 13, 111, 135, 164, 178, 233-4
product life cycle (PLC) 163-5, 185
production 35, 140, 150-1, 153-4, 160, 163-4, 168-70, 173, 181, 195, 199, 204, 207-8, 213-14, 219, 227, 232, 234, 237-8, 241, 249-50, 252, 256, 274
production processes 137, 141, 163-6, 231
products 21, 28, 36, 72, 99, 141, 155, 160-4, 167, 171, 173, 175, 178, 181, 185, 228, 242, 256, 274, 286-9, 292-3, 295-6

profits   70, 118, 163, 230, 232-5, 238, 241, 272, 274, 276, 293, 296
projects   1, 8, 25, 35, 37-8, 41, 145, 156, 255, 257-8
properties   35-6, 39, 53, 80, 194, 203, 217, 228, 231, 235-6, 248, 250, 257, 284, 291, 296
  emergent   11-12, 15, 101, 191, 194, 223, 290

Q

quality   20, 35, 137, 140-1, 155, 162-3, 170, 241, 250, 286, 291-7, 301

R

rank clock   49, 54, 62-3, 66-8, 71-6
rank clock visualizer   49, 73-5
rank size   54-5, 59, 65, 71, 73
rank size distributions   49, 55-6, 60-2, 65, 70, 72
ranks   49, 52, 54, 58, 60-3, 65, 67-8, 70, 73-5, 138, 237, 263-5, 267, 270, 300
rationality   24, 118, 229-30, 284
reduction   6, 19, 34-6, 79, 155, 163, 180
regions   30, 39, 51, 76-7, 150, 176, 194, 199, 201, 216, 250-1, 263, 278
regularities   49-50, 52-3, 59, 72, 267
relationships   5, 14, 17-18, 27, 39, 41, 46, 55-6, 58-9, 79, 86-7, 89-90, 95, 97-8, 100, 102-6, 108, 115-16, 118-19, 121, 151, 184, 192, 200, 207, 249, 252, 292, 301
relativism   2, 7
religion   111, 115, 117, 119-20, 123, 126
representations   4, 11, 15, 17-19, 43, 92, 135, 206, 217, 270
resources   6, 12, 24, 99, 101, 112, 116, 119, 123, 139-40, 167, 178, 180, 182, 227, 249-51, 258, 271
robustness   79, 81, 83, 85, 87, 89, 91, 93, 95, 97-9, 101, 103, 105, 107, 302
rules   16-18, 20, 81, 84, 91, 112, 155, 226, 230, 232, 235, 290-1
  behavioral   20, 290, 292
  micro level   290

S

scale   29, 32, 36-7, 40, 50-2, 58, 162, 164, 170, 209, 234, 257, 264, 280
scaling   49, 51, 53, 55, 57, 59, 61, 63, 65, 67, 69-71, 73, 75-7, 107, 216
scaling laws   54-6, 76
science   7, 22-3, 27, 46, 76, 107-8, 111, 114, 119, 126-8, 150, 165-6, 168, 187, 216, 231, 244, 247, 260, 278, 301
self-organization   12, 22, 44, 99, 107, 150, 179-80, 182-3, 185, 187, 202, 216-17, 228, 259, 267, 269-71
  adaptive   268
self-organization process   179, 183
self-organizing economies   265, 267, 269, 280
semiconductors   161, 166, 169, 171, 175, 183, 185
separatrix   83, 98, 101
services   161-2, 177, 179, 227, 242, 250, 263-4, 277
simple models, very   106, 287
simulations   23, 131-2, 142-5, 147-50, 209, 213, 217
size, firm   54, 58, 67, 70, 72, 265, 300-1
skills, cross-cutting   27, 29, 31, 33, 35, 37, 39, 41, 43, 45

social sciences   186, 217, 265, 283, 300, 302
social systems   19, 40, 45, 111-14, 116, 120, 135, 150, 191, 203, 217, 247
societies   11, 22, 46, 112-14, 116, 118, 121, 124, 193, 195, 198, 206, 208, 214, 230, 268
software   72, 108, 140, 169, 172-4, 178, 185
solutions   31-2, 163, 235-6, 288-9, 291, 296-300
state space   81, 83, 85-6, 89-91, 95, 97-8, 103, 105
  qualitative structure of   90-1
  size of   97
state space attractors   82-5, 89, 99-101
state space connectivity   83
status   2, 5, 118, 192-3, 195, 198, 272
strategies   11, 20, 28, 40, 83, 99, 101, 105, 120, 133, 150, 171-2, 180, 183-4, 188, 255, 260, 286, 289
structural attractors   21, 40, 87, 89, 103
structural loops   84, 87, 90, 98
suppliers   39, 43, 137, 140, 155, 161, 173, 179, 285, 295
supply   112, 161-3, 231, 236, 250, 256
sustainability   22, 28-9, 38, 43, 46, 258
systems, operating   173-4

T

tail, long   54-5, 265, 278
technologies   13-14, 38-9, 46, 73, 104, 114, 126, 128, 138, 144, 155, 159-60, 162, 166, 168, 171, 177-8, 180, 182, 184, 186-8, 207, 219, 223, 231, 235, 255, 257-8, 260
terrorism   111, 113, 115, 117, 119-21, 123-8
terrorists   111, 119, 121, 123, 125-7
theory   11, 45, 50-1, 94, 134, 150, 178, 185, 228, 233, 235, 243-4, 249, 260, 264, 266, 271, 285
time   14, 20-1, 28-30, 33, 35-9, 49-50, 52-4, 59-62, 64-5, 67-70, 73-5, 81, 83, 134, 140-1, 162, 175-8, 180-1, 184-5, 209, 226, 239-40, 248-9, 251-2, 263-4, 279-80, 283-5, 288, 291-2, 299-301
time periods   15, 52, 54, 60-5, 73, 175, 249, 292
trajectories   51, 61-5, 72, 74, 185, 213, 253
transition   61, 103, 118, 120-5, 164, 167, 180, 184, 193-4, 235
TT (Tobin Tax)   263, 265, 272-6, 278-9, 281
types   15, 33, 39, 56, 80, 87, 89-90, 93-4, 107, 135, 170, 181, 183, 185-6, 193-4, 199, 215, 250-1, 290-1

U

UK   23, 39, 73, 76, 167, 174, 184, 220, 244, 257, 260-1, 263-5, 268-78, 280
understanding   1, 3-7, 24, 27, 29, 32, 36, 45-6, 86, 89-90, 92-3, 97, 104, 107, 111, 127, 132-3, 145, 148, 152, 214-15, 225, 227, 230-1, 236, 248, 251, 255, 259, 283-4
  qualitative   17, 33-4, 43
urban systems   50-1, 53, 60, 62, 65, 134, 150
users   18, 73-4, 159-62, 164, 173, 175, 181, 184, 186, 256, 274

V

variations   13, 37, 60, 67, 73, 75, 113, 124, 159, 163-4, 173, 225-8, 233, 236, 239, 241
volatility   60-1, 63-5, 69, 72

W

wealth   116-17, 188, 195, 198, 206, 211, 213-14, 236
WWW (World Wide Web)   80, 92, 104, 216, 279